Lecture Notes in Computer Science 2251

Edited by G. Goos, J. Hartmanis, and J. van Leeuwen

Springer
Berlin
Heidelberg
New York
Barcelona
Hong Kong
London
Milan
Paris
Tokyo

uan Y. Tang Victor Wickerhauser
ong C. Yuen Chun-hung Li (Eds.)

Vavelet Analysis
ınd Its Applications

:cond International Conference, WAA 2001
ong Kong, China, December 18-20, 2001
:oceedings

Springer

Volume Editors

Yuan Y. Tang
Pong C. Yuen
Chun-hung Li
Hong Kong Baptist University
Department of Computer Science
Kowloon Tong, Hong Kong E-mail:{yytang/pcyuen/chli}@comp.khbu.edu.hk

Victor Wickerhauser
Washington University, Department of Mathematics
Campus Box 1146, Cupples I
St. Louis, Missouri 63130, USA
E-mail: victor@math.wustl.edu

Cataloging-in-Publication Data applied for

Die Deutsche Bibliothek - CIP-Einheitsaufnahme

Wavelet analysis and its applications : second international conference ; proceedings / WAA 2001, Hong Kong, China, December 18 - 20, 2001. Yuan Y. Tang ... (ed.). - Berlin ; Heidelberg ; New York ; Barcelona ; Hong Kong ; London ; Milan ; Paris ; Tokyo : Springer, 2001
(Lecture notes in computer science ; Vol. 2251)
ISBN 3-540-43034-2

CR Subject Classification (1998): E.4, H.5, I.4, C.3, I.5

ISSN 0302-9743
ISBN 3-540-43034-2 Springer-Verlag Berlin Heidelberg New York

Springer-Verlag Berlin Heidelberg New York
a member of BertelsmannSpringer Science+Business Media GmbH

http://www.springer.de

Typesetting: Camera-ready by author, data conversion by DA-TeX Gerd Blumenstein
Printed on acid-free paper SPIN 10845973 06/3142 5 4 3 2 1 0

Preface

The first international conference on wavelet analysis and its applications was held in China in 1999. Following the success of the first conference, the second international conference (ICWAA 2001) was held in Hong Kong in December 2001. The objective of this conference is to provide a forum for researchers working on both wavelet theory and its applications. By idea-sharing and discussions on the state of the art in wavelet theory and applications, ICWAA 2001 is aimed to stimulate the future development, explore novel applications, and exchange ideas for developing robust solutions.

By August 2001, we had received 67 full papers submitted from all over the world. To ensure the quality of the conference and proceedings, each paper was reviewed by three reviewers. After a thorough review process, the program committee selected 24 regular papers for oral presentation and 27 short papers for poster presentation. In addition to these 24 oral presentations, there were 3 invited talks delivered by distinguished researchers, namely Prof. John Daugman from Cambridge University, UK, Prof. Bruno Torresani from Inria, France, and Prof. Victor Wickerhauser, from Washington University, USA. We must add that the program committee and the reviewers did an excellent job within a tight schedule.

We wish to thank all the authors for submitting their work to ICWAA 2001 and all the participants, whether you came as a presenter or an attendee. We hope that there was ample time for discussion and opportunity to make new acquaintances. Finally, we hope that you experienced an interesting and exciting conference and enjoyed your stay in Hong Kong.

October 2001

Yuan Y. Tang, Victor Wickerhauser
Pong C. Yuen, C. H. Li

Organization

The Second International Conference on Wavelet Analysis and Applications is organized by the Department of Computer Science, Hong Kong Baptist Univeristy and IEEE Hong Kong Section Computer Chapter.

Organizing Committee

Congress Chair:	Ernest C. M. Lam
General Chairs:	John Daugman Ernest C. M. Lam
Program Chairs:	Yuan Y. Tang Victor Wickerhauser P. C. Yuen
Organizing Chair:	Kelvin C. K. Wong
Local Arrangement Chair:	William K. W. Cheung
Registration & Finance Chair:	K. C. Tsui
Publications Chairs:	C. H. Li M. W. Mak
Workshop Chair:	Samuel P. M. Choi
Publicity Chair:	C. S. Huang

Sponsors

Hong Kong Baptist University
Croucher Foundation
IEEE Hong Kong Section Computer Chapter

Program Committee

Metin Akay	Dartmouth College
Akram Aldroubi	Vanderbilt University
Claudia Angelini	Istituto per Applicazioni della Matematica
Algirdas Bastys	Vilnius University
T. D. Bui	Concordia University
Elvir Causevic	Everest Biomedical Instrument Company
Mariantonia Cotronei	Universita' di Messina
Hans L. Cycon	Fachhochschule fur Technik und Wirtschaft Berlin
Dao-Qing Dai	Zhongshan University
Wolfgang Dahmen	Technische Hochschule Aachen
Donggao Deng	Zhongshan University
T. N. T. Goodman	University of Dundee
D. Hardin	Vanderbilt University
Daren Huang	Zhongshan University
Wen-Liang Hwang	Institute of Information Science
Rong-Qing Jia	University of Alberta
P. Jorgensen	University of Iowa
K. S. Lau	Chinese University of Hong Kong
Seng-Luan Lee	National University of Singapore
Jian-Ping Li	Logistical Engineering University
Wei Lin	Zhongshan University
Guixing Luan	Shenyang Inst. of Computing Technology
Hong Ma	Sichuan University
Peter Oswald	Bell Laboratories, Lucent Technologies
Lizhong Peng	Peking University
Valrie Perrier	Domaine Universitaire
S. D. Riemenschneider	West Virgina University
Zuowei Shen	National University of Singapore
Guoxiang Song	XiDian University
Georges Stamon	University Rene Descartes
Chew-Lim Tan	National University of Singapore
Michael Unser	Batiment de Microtechnique
Jianzhong Wang	Sam Houston State University
Yueshen Xu	University of North Dakota
Lihua Yang	Zhongshan University
Rongmao Zhang	Shenyang Inst. of Computing Technology
Xingwei Zhou	Nankai University

Table of Contents

Keynote Presentations

Image Compression and Coding

Video Coding and Processing

Theory

Image Processing

Signal Processing

Systems and Applications

Personal Identification in Real-Time by Wavelet Analysis of Iris Patterns

John Daugman, OBE

The Computer Laboratory, University of Cambridge, UK

Abstract. The central issue in pattern recognition is the relation between within-class variability and between-class variability. These are determined by the various degrees-of-freedom spanned by the patterns themselves, and by the selectivity of the chosen feature encoders. An interesting application of 2D wavelets in computer vision is the automatic recognition of personal identity by encoding and matching the complex patterns visible at a distance in each eye's iris. Because the iris is a protected, internal, organ whose random texture is highly unique and stable over life, it can serve as a kind of living password or passport that one need not remember but is always in one's possession. I will describe wavelet demodulation methods that I have developed for this problem over the past 10 years, and which are now installed in all existing commercial systems for iris recognition. The principle that underlies iris recognition is the failure of a test of statistical independence performed on the phase angle sequences of iris patterns. Quadrature 2D Gabor wavelets spanning 3 octaves in scale enable the complex-valued assignment of local phasor coordinates to iris patterns. The combinatorial complexity of these phase sequences spans about 244 independent degrees-of-freedom, and generates binomial distributions for the Hamming Distances (a similarity metric) between different irises. In six public independent field trials conducted so far using these algorithms, involving several millions of iris comparisons, there has never been a single false match recorded. The time required to locate and to encode an iris into quantized wavelet phase sequences is 1 second. Then database searches are performed at a rate of 100,000 irises/second. Data will be presented in this talk from 2.3 million IrisCode comparisons. This wavelet application could be used in a wide range of settings in which persons' identities must be established or confirmed by large scale database search, without relying upon cards, keys, documents, secrets, passwords or PINs.

Y. Y. Tang et al. (Eds.): WAA 2001, LNCS 2251, p. 1, 2001.

Hybrid Representations of Audiophonic Signals

Bruno Torresani

LATP, CMI, Université de Provence, France

Abstract. A new approach for modeling audio signal will be presented, in view of efficient encoding. The method is based upon hybrid models featuring transient, tonal and stochastic components in the signal. The three components are estimated and encoded independently using a strategy very much in the spirit of transform coding. The signal models involve nonlinear expansions on local trigonometric bases, and binary trees of wavelet coefficients. Unlike several existing approaches, the method does not rely on any prior segmentation of the signal. The talk is based on joint works with L. Daudet and S. Molla.

Y. Y. Tang et al. (Eds.): WAA 2001, LNCS 2251, p. 2, 2001.

Singularity Detection from Autocovariance via Wavelet Packets

M. Victor Wickerhauser

Department of Mathematics, Washington University, USA

Abstract. We use the eigenvalues of a version of the autocovariance matrix to recognize directions at which the Fourier transform of a function is slowly decreasing, which provides us with a technique to detect singularities in images. In very high dimensions, we show how the wavelet packet best-basis algorithm can be used to compute these eigenvalues approximately, at relatively low computational complexity.

Y. Y. Tang et al. (Eds.): WAA 2001, LNCS 2251, p. 3, 2001.

Empirical Evaluation of Boundary Policies for Wavelet-Based Image Coding

Claudia Schremmer

Praktische Informatik IV
Universität Mannheim, 68131 Mannheim, Germany
schremmer@informatik.uni-mannheim.de

Abstract. The wavelet transform has become the most interesting new algorithm for still image compression. Yet there are many parameters within a wavelet analysis and synthesis which govern the quality of a decoded image. In this paper, we discuss different image boundary policies and their implications for the decoded image. A pool of gray–scale images has been wavelet–transformed at different settings of the wavelet filter bank and quantization threshold and with three possible boundary policies.
Our empirical evaluation is based on three benchmarks: a first judgment regards the perceived quality of the decoded image. The compression rate is a second crucial factor. Finally, the best parameter settings with regard to these two factors is weighted with the cost of implementation. Contrary to the JPEG2000 standard, where mirror padding is implemented, our investigation proposes circular convolution as the boundary treatment.

Keywords: Wavelet Analysis, Boundary Policies, Empirical Evaluation

1 Introduction

Due to its outstanding performance in compression, the wavelet transform is the focus of new image coding techniques such as the JPEG2000 standard [8,4]. JPEG2000 proposes a reversible (Daub 5/3–tap) and an irreversible (Daub 9/7–tap) wavelet filter bank. However, since we were interested in how filter length affects the quality of image coding, we investigated the orthogonal and separable wavelet filters developed by Daubechies [2]. These belong to the group of wavelets used most often in image coding applications. They specify a number n_0 of vanishing moments: if a wavelet has n_0 vanishing moments, then the approximation order of the wavelet transform is also n_0.

Implementations of the wavelet transform on still images entail other aspects as well: speed, decomposition depth, and boundary treatment policies. Long filters require more computing time than short ones. Furthermore, the (dyadic) wavelet transform incorporates the aspect of iteration: the low–pass filter defines an approximation of the original signal that contains only half as many coefficients. This approximation successively builds the input for the next approximation. For compression purposes, coefficients in the time–scale domain

Y. Y. Tang et al. (Eds.): WAA 2001, LNCS 2251, pp. 4–15, 2001.

are discarded and the synthesis quality improves with the number of iterations on the approximation. Finally, the wavelet transform is mathematically defined only *within* a signal; image applications thus need to solve the boundary problem. Depending on the boundary policy selected, the number of iterations in a wavelet transform might vary with the filter length. Moreover, the longer the filter length, the more important the boundary policy becomes.

In this work, we investigate the effects of three different boundary policies in combination with different wavelet filter banks on a number of gray–scale images. A first determining factor is the visual perception of a decoded image. As we will see, although the quality varies strongly with the selected image, for a given image it remains relatively unconcerned about the parameter settings. A second crucial factor is therefore the expected compression rate. Finally, the cost of implementation weights these two benchmarks. Our empirical evaluation leads us to recommend *circular convolution* as the boundary treatment, contrary to JPEG2000 which proposes padding.

The article is organized as follows. In Section 2, we cite related work on wavelet filter evaluation. Section 3 reviews the wavelet transform and details the aspects that are important for our survey. In Section 4, we present the technical evaluation of the wavelet transform and detail our results. The article ends in Section 5 with an outlook on future work.

2 Related Work

Villasenor's group researches wavelet filters for image compression. In [10], the focus is on biorthogonal filters, and the evaluation is based on the information preserved in the reference signal, while [3] focuses on a mathematically optimal quantizer step size. In [1], the evaluation is based on lossless as well as on subjective lossy compression performance, complexity and memory usage. An interpretation of *why* the observations are made is nevertheless lacking. Strutz has thoroughly researched the dyadic wavelet transform in [9]: the design and construction of different wavelet filters is investigated, as are good Huffman and arithmetic encoding strategies. An investigation of boundary policies, however, is lacking.

3 The Wavelet Transform

A wavelet is an (ideally) compact function, i.e., outside a certain interval it vanishes. Implementations are based on the fast wavelet transform, where a given wavelet (i.e., *mother wavelet*) is shifted and dilated so as to provide a base in the function space. That is, a one–dimensional function is transformed into a two–dimensional space, where it is approximated by coefficients that depend on *time* (determined by the translation parameter) and on *scale*, i.e., frequency (determined by the dilation parameter). The localization of a wavelet in time spread (σ_t) and frequency spread (σ_ω) has the property $\sigma_t\sigma_\omega$ = const. However, the resolution in time and frequency depends on the frequency. This is the so–called *zoom* phenomenon of the wavelet transform: it offers high temporal localization for high frequencies while offering good frequency resolution for low frequencies.

3.1 Wavelet Transform and Filter Banks

By introducing multiresolution, Mallat [7] made an important contribution to the application of wavelet theory to multimedia: the transition from mathematical theory to filters. Multiresolution analysis is implemented via high–pass, respectively, band–pass filters (i.e., wavelets) and low–pass filters (i.e., scaling functions): The detail coefficients (resulting from the high–pass, respectively, band–pass filtering) of every iteration step are kept apart, and the iteration starts again with the remaining approximation coefficients (from application of the low–pass filter). This multiresolution theory is 'per se' defined only for one–dimensional wavelets on one–dimensional signals. As still images are two–dimensional discrete signals and two–dimensional wavelet filter design remains an active field of research [5][6], current implementations are restricted to *separable* filters. The successive convolution of filter and signal in both dimensions opens two potential iterations:

- *standard*: *all* approximations, even in mixed terms, are iterated, and
- *non–standard*: only the *purely* low–pass filtered parts of every approximation enter the iteration.

In this work, we concentrate on the non–standard decomposition.

3.2 Image Boundary

A digital filter is applied to a signal by *convolution*. Convolution, however, is defined only *within* a signal. In order to result in a reversible wavelet transform, *each* signal coefficient must enter into `filter_length/2` calculations of convolution (here, the subsampling process by factor 2 is already incorporated). Consequently, every filter longer than two entries, i.e., every filter except *Haar*, requires a solution for the boundary. Furthermore, images are signals of a relatively short length (in rows and columns), thus the boundary treatment is even more important than e.g. in audio coding. Two common boundary policies are *padding* and *circular convolution*.

Padding Policies. With padding, the coefficients of the signal on either border are padded with `filter_length-2` coefficients. Consequently, each signal coefficient enters into `filter_length/2` calculations of convolution, and the transform is reversible. Many padding policies exist; they all have in common that each iteration step physically increases the storage space in the wavelet domain. In [11], a theoretical solution for the required storage space (depending on the signal, the filter bank and the iteration level) is presented. Nevertheless, its implementation remains sophisticated.

Circular Convolution. The idea of circular convolution is to 'wrap' the end of a signal to its beginning or vice versa. In so doing, circular convolution is the only boundary treatment to maintain the number of coefficients for a wavelet transform, thus simplifying storage management[1]. A minor drawback is that the time information contained in the time–scale domain of the wavelet–transformed coefficients 'blurs': the coefficients in the time–scale domain that are next to the right border (respectively, left border) also affect signal coefficients that are located on the left (respectively, right).

The selected boundary policy has an important impact on the iteration behavior of the wavelet transform. It does not affect the iteration behavior of padding policies. However, with circular convolution, the decomposition depth varies with the filter length: the longer the filter, the fewer the number of decomposition iterations possible. For example, for an image of 256×256 pixels, the Daub–2 filter bank with 4 coefficients allows a decomposition depth of 7, while the Daub–20 filter bank with 40 coefficients has reached signal length after only 3 decomposition levels.

Thus, the evaluation presented in Tables 1 to 4 is based on a decomposition depth of level 8 for the two padding policies, while the decomposition depth for circular convolution varies from 7 to 3, according to the selected filter length.

4 Empirical Evaluation

4.1 Set-Up

Our empirical evaluation sought the best parameter settings for the choice of the wavelet filter bank and for the image boundary policy to be implemented. The performance was evaluated according to the criteria:

1. visual quality,
2. compression rate, and
3. complexity of implementation.

The quality was rated based on the *peak signal–to–noise ratio* (PSNR)[2]. The compression rate was simulated by a simple quantization threshold: the higher the threshold, the more coefficients in the time–scale domain are discarded, the higher is the compression rate. More precisely, the threshold was carried out only on the parts of the image that have been high–pass filtered (respectively, band–pass filtered) at least once. That is, the approximation of the image was excluded from the thresholding due to its importance for the image synthesis.

---

[1] Storage space, however, expands indirectly: an image can be stored with `integers`, while the coefficients in the time–scale domain require `floats`.

[2] When $\text{org}(x, y)$ depicts the pixel value of the original image at position (x, y), and $\text{dec}(x, y)$ denotes the pixel value of the decoded image at position (x, y), then $\text{PSNR [dB]} = 10 \cdot \log \left(\frac{\sum_{xy} 255^2}{\sum_{xy} (\text{org}(x,y) - \text{dec}(x,y))^2} \right)$.

Our evaluation was set up on the six gray–scale images of size 256×256 pixels demonstrated in Figure 1. These test images have been chosen in order to comply with different features:

- contain many small details: *Mandrill*, *Goldhill*,
- contain large uniform areas: *Brain*, *Lena*, *Camera*, *House*,
- be relatively symmetric at the left–right and top–bottom boundaries: *Mandrill*, *Brain*,
- be very asymmetric with regard to these boundaries: *Lena*, *Goldhill*, *House*,
- have sharp transitions between regions: *Brain*, *Lena*, *Camera*, *House*, and
- contain large areas of texture: *Mandrill*, *Lena*, *Goldhill*, *House*.

4.2 Results

Image-Dependent Analysis. The detailed evaluation results for the six test images are presented in Tables 1 and 2. Some interesting observations made from these two tables and their explanations are as follows:

- For a given image and a given quantization threshold, the PSNR remains astonishingly constant for different filter banks and different boundary policies.
- At high thresholds, *Mandrill* and *Goldhill* yield the worst quality. This is due to the large amount of details in both images.
- *House* produces the overall best quality at a given threshold. This is due to its large uniform areas.
- Due to their symmetry, *Mandrill* and *Brain* show good quality results with padding policies.
- The percentage of discarded information at a given threshold is far higher for *Brain* than for *Mandrill*. This is due to the uniform black background of *Brain*, which produces small coefficients in the time–scale domain, compared to the many small details in *Mandrill* which produce large coefficients and thus do not fall below the threshold.
- With regard to the heuristic for compression, and for a given image and boundary policy, Table 2 reveals that
  - the compression ratio for zero padding *increases* with increasing filter length,
  - the compression ratio for mirror padding *decreases* with increasing filter length, and
  - the compression ratio for circular convolution varies, but most often stays *almost constant*.

  The explanation is as follows. Padding an image with zeros, i.e., black pixel values, most often produces a sharp contrast to the original image, thus the sharp transition between the signal and the padding coefficients results in large coefficients in the fine scales, while the coarse scales remain unaffected. This observation, however, is put into a different perspective for longer filters: With longer filters, the constant *run* of zeros at the boundary does not show strong variations, and the detail coefficients in the time–scale domain thus remain small. Hence, a given threshold cuts off fewer coefficients when the filter is longer. With mirror padding, the padded coefficients for shorter filters represent a good heuristic for the signal adjacent to the boundary. Increasing filter length and accordingly, longer padded areas, however, introduces too much 'false' detail information into the signal, resulting in many large detail coefficients that 'survive' the threshold.

Image-Independent Analysis. The above examples reveal that most phenomena are signal–dependent. As a signal–dependent determination of best–suited parameters remains academic, our further reflections are made on the *average* image quality and the *average* amount of discarded information as presented in Tables 3 and 4 and the corresponding Figures 2 and 3.

Figure 2 visualizes the coding quality of the images, averaged over the six test images. The four plots represent the quantization thresholds $\lambda = 10, 20, 45$ and 85. In each graphic, the visual quality (quantified via PSNR) is plotted against the filter length of the Daubechies wavelet filters. The three boundary policies: *zero padding*, *mirror padding* and *circular convolution* are regarded separately. The plots obviously reveal that the quality decreases with an increasing threshold. More important are the following statements:

- Within a given threshold, and for a given boundary policy, the PSNR remains almost constant. This means that the quality of the coding process depends hardly or not at all on the selected wavelet filter bank.
- Within a given threshold, mirror padding produces the best results, followed by circular convolution. Zero padding performs worst.
- The gap between the performance of the boundary policies increases with an increasing threshold.

Nevertheless, the differences observed above with 0.28 dB maximum gap (at the threshold $\lambda = 85$ and the filter length of 40 coefficients) are so marginal that they do not actually influence visual perception.

As the visual perception is neither influenced by the choice of filter nor by the boundary policy, the coding performance has been studied as a second benchmark. The following observations are made in Figure 3. With a short filter length (4 to 10 coefficients), the compression ratio is almost identical for the different boundary policies. This is not astonishing as short filters involve only little boundary treatment, and the relative importance of the boundary coefficients with regard to the signal coefficients is negligible. More important for our investigation is that:

- The compression heuristic for each of the three boundary policies is inversely proportional to their quality performance. In other words, mirror padding discards the least number of coefficients at a given quantization threshold, while zero padding discards the most.

- With an increasing threshold, the gap between the compression ratios of the three policies narrows.

In the overall evaluation, we have seen that mirror padding performs best with regard to quality, while it performs worst with regard to compression. Inversely, zero padding performs best with regard to compression and worst with regard to quality. Circular convolution holds the midway in both aspects. On the other hand, the gap in compression is by far superior to the differences in quality. Calling to mind the coding complexity of the padding approaches, compared to the easy implementation of circular convolution (see Section 3.2), we strongly recommend to implement circular convolution as the boundary policy in image coding.

5 Conclusion

We have discussed and evaluated the strengths and weaknesses of different boundary policies in relation to various orthogonal wavelet filter banks. Contrary to the JPEG2000 coding standard, where mirror padding is suggested for boundary treatment, we have proven that circular convolution is superior in the overall combination of quality performance, compression performance and ease of implementation.

In future work, we will improve our heuristic on the compression rate and rely on the calculation of a signal's entropy such as it is presented in [12] and [9].

References

1. Michael D. Adams and Faouzi Kossentini. Performance Evaluation of Reversible Integer–to–Integer Wavelet Transforms for Image Compression. In *Proc. IEEE Data Compression Conference*, page 514 ff., Snowbird, Utah, March 1999. 5
2. Ingrid Daubechies. *Ten Lectures on Wavelets*, volume 61. SIAM. Society for Industrial and Applied Mathematics, Philadelphia, PA, 1992. 4
3. Javier Garcia-Frias, Dan Benyamin, and John D. Villasenor. Rate Distortion Optimal Parameter Choice in a Wavelet Image Communication System. In *Proc. IEEE International Conference on Image Processing*, pages 25–28, Santa Barbara, CA, October 1997. 5
4. ITU. *JPEG2000 Image Coding System. Final Committee Draft Version 1.0 – FCD15444-1*. International Telecommunication Union, March 2000. 4
5. Jelena Kovačević and Wim Sweldens. Wavelet Families of Increasing Order in Arbitrary Dimensions. *IEEE Trans. on Image Processing*, 9(3):480–496, March 2000. 6
6. Jelena Kovačević and Martin Vetterli. Nonseparable Two– and Three–Dimensional Wavelets. *IEEE Trans. on Signal Processing*, 43(5):1269–1273, May 1995. 6
7. Stéphane Mallat. *A Wavelet Tour of Signal Processing*. Academic Press, San Diego, CA, 1998. 6
8. Athanassios N. Skodras, Charilaos A. Christopoulos, and Touradj Ebrahimi. JPEG2000: The Upcoming Still Image Compression Standard. In *11th Portuguese Conference on Pattern Recognition*, pages 359–366, Porto, Portugal, May 2000. 4

9. Tilo Strutz. *Untersuchungen zur skalierbaren Kompression von Bildsequenzen bei niedrigen Bitraten unter Verwendung der dyadischen Wavelet–Transformation.* PhD thesis, Universität Rostock, Germany, May 1997. 5, 10
10. John D. Villasenor, Benjamin Belzer, and Judy Liao. Wavelet Filter Evaluation for Image Compression. *IEEE Trans. on Image Processing*, 2:1053–1060, August 1995. 5
11. Mladen Victor Wickerhauser. *Adapted Wavelet Analysis from Theory to Software.* A. K. Peters Ltd., Natick, MA, 1998. 6
12. Mathias Wien and Claudia Meyer. Adaptive Block Transform for Hybrid Video Coding. In *Proc. SPIE Visual Communications and Image Processing*, pages 153–162, San Jose, CA, January 2001. 10

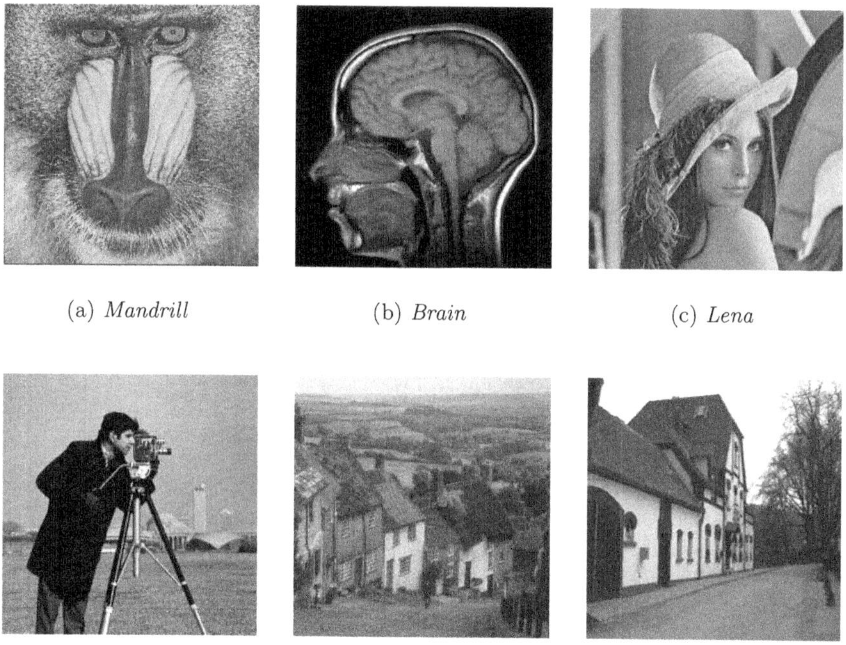

(a) *Mandrill* (b) *Brain* (c) *Lena*

(d) *Camera* (e) *Goldhill* (f) *House*

Fig. 1. Test images for the evaluation

Table 1. Detailed results of the quality evaluation with the PSNR on the six test images. The mean values over the images are given in Table 3

Quality of visual perception — PSNR [dB]									
Wavelet	zero padding	mirror padding	circular convol.	zero padding	mirror padding	circular convol.	zero padding	mirror padding	circular convol.
	Mandrill			*Brain*			*Lena*		
	Threshold: 10 — Excellent overall quality								
Daub-2	18.012	17.996	18.238	18.141	18.151	18.197	16.392	16.288	16.380
Daub-3	18.157	18.187	18.221	18.429	18.434	18.433	16.391	16.402	16.350
Daub-4	18.169	18.208	17.963	18.353	18.340	18.248	16.294	16.355	16.260
Daub-5	18.173	18.167	18.186	18.279	18.280	18.259	16.543	16.561	16.527
Daub-10	17.977	17.959	18.009	18.291	18.300	18.479	16.249	16.278	16.214
Daub-15	17.938	17.934	18.022	18.553	18.543	18.523	16.267	16.304	16.288
Daub-20	17.721	17.831	18.026	18.375	18.357	18.466	16.252	16.470	16.238
	Threshold: 20 — Good overall quality								
Daub-2	14.298	14.350	14.403	16.610	16.611	16.577	14.775	14.765	14.730
Daub-3	14.414	14.469	14.424	16.743	16.755	16.721	14.758	14.817	14.687
Daub-4	14.231	14.239	14.276	16.637	16.628	16.734	14.862	14.918	14.735
Daub-5	14.257	14.216	14.269	16.747	16.751	16.854	14.739	14.946	14.815
Daub-10	14.268	14.274	14.360	16.801	16.803	16.878	14.624	14.840	14.699
Daub-15	14.246	14.258	14.300	16.822	16.810	16.852	14.395	14.631	14.477
Daub-20	14.046	14.065	14.227	16.953	16.980	16.769	14.252	14.597	14.353
	Threshold: 45 — Medium overall quality								
Daub-2	10.905	10.885	10.910	14.815	14.816	14.747	13.010	13.052	12.832
Daub-3	10.988	10.970	10.948	15.187	15.150	15.052	12.766	13.138	12.903
Daub-4	10.845	10.839	10.885	15.014	15.029	15.056	12.820	13.132	12.818
Daub-5	10.918	10.969	10.949	15.036	15.031	14.999	12.913	13.301	12.983
Daub-10	10.907	10.929	10.913	14.989	15.013	15.212	12.447	13.066	12.795
Daub-15	10.845	10.819	10.815	15.093	15.133	15.064	12.577	12.954	12.686
Daub-20	10.784	10.872	10.843	14.975	14.934	14.882	12.299	12.877	12.640
	Threshold: 85 — Poor overall quality								
Daub-2	9.095	9.121	9.135	13.615	13.621	13.783	11.587	11.902	11.577
Daub-3	9.206	9.184	9.124	13.787	13.784	13.759	11.437	11.793	11.516
Daub-4	9.160	9.152	9.168	13.792	13.815	13.808	11.539	11.806	11.636
Daub-5	9.171	9.208	9.203	13.837	13.850	13.705	11.692	11.790	11.872
Daub-10	9.207	9.193	9.206	13.870	13.922	14.042	11.128	11.430	11.555
Daub-15	9.083	9.161	9.126	13.731	13.795	13.917	11.128	11.610	11.475
Daub-20	9.071	9.142	9.204	13.852	13.800	13.974	11.142	11.694	11.597
	Camera			*Goldhill*			*House*		
	Threshold: 10 — Excellent overall quality								
Daub-2	17.334	17.346	17.371	16.324	16.266	16.412	19.575	19.563	19.608
Daub-3	17.532	17.560	17.625	16.322	16.296	16.358	19.640	19.630	19.621
Daub-4	17.529	17.591	17.577	16.241	16.212	16.342	19.560	19.558	19.584
Daub-5	17.489	17.448	17.389	16.214	16.193	16.154	19.613	19.555	19.566
Daub-10	17.539	17.541	17.383	16.307	16.223	16.317	19.482	19.388	19.732
Daub-15	17.747	17.530	17.523	16.012	16.067	16.033	19.653	19.671	19.726
Daub-20	17.474	17.527	17.484	16.322	16.245	16.319	19.550	19.495	19.524
	Threshold: 20 — Good overall quality								
Daub-2	14.387	14.365	14.396	13.937	13.940	13.898	17.446	17.480	17.471
Daub-3	14.473	14.452	14.426	13.872	13.892	13.858	17.525	17.594	17.612
Daub-4	14.438	14.438	14.430	13.828	13.836	13.753	17.468	17.647	17.351
Daub-5	14.460	14.505	14.427	13.743	13.743	13.711	17.454	17.458	17.465
Daub-10	14.468	14.400	14.409	13.762	13.785	13.798	17.592	17.635	17.689
Daub-15	14.408	14.406	14.414	13.687	13.730	13.697	17.260	17.276	17.266
Daub-20	14.384	14.370	14.362	13.700	13.782	13.731	17.476	17.449	17.240
	Threshold: 45 — Medium overall quality								
Daub-2	12.213	12.242	12.131	12.033	12.034	11.876	15.365	15.437	15.155
Daub-3	12.032	12.122	12.188	11.961	12.006	11.889	14.957	15.476	15.118
Daub-4	12.150	12.178	12.145	11.855	11.891	11.925	14.906	15.080	15.180
Daub-5	12.077	12.133	12.120	11.848	11.844	11.801	15.159	15.382	15.244
Daub-10	12.061	12.197	12.093	11.760	11.917	11.726	14.776	15.246	14.872
Daub-15	12.074	12.059	12.176	11.725	11.855	11.753	14.810	15.090	14.969
Daub-20	11.798	11.975	12.048	11.763	11.803	11.703	14.420	15.033	14.609
	Threshold: 85 — Poor overall quality								
Daub-2	11.035	11.161	11.041	10.791	10.805	10.844	13.530	13.804	13.703
Daub-3	11.092	11.176	11.080	10.943	10.916	10.754	13.488	13.726	13.627
Daub-4	10.943	11.152	11.046	10.861	10.904	10.740	13.524	13.613	13.510
Daub-5	11.018	11.148	11.129	10.826	10.935	10.738	13.114	13.903	13.111
Daub-10	10.815	11.064	10.987	10.824	10.972	10.771	13.158	13.695	13.434
Daub-15	10.779	11.005	10.982	10.737	10.838	10.607	13.073	13.357	13.123
Daub-20	10.688	11.031	11.090	10.709	10.819	10.766	13.173	13.257	13.678

Table 2. Heuristic for the compression rate of the coding parameters of Table 1: The higher the percentage of discarded information in the time–scale domain is, the higher is the compression ratio. The mean values over the images are given in Table 4

Discarded information in the time–scale domain — Percentage [%]									
Wavelet	zero padding	mirror padding	circular convol.	zero padding	mirror padding	circular convol.	zero padding	mirror padding	circular convol.
	Mandrill			*Brain*			*Lena*		
	Threshold: $\lambda = 10$ — Excellent overall quality								
Daub-2	42	41	41	83	83	83	78	79	79
Daub-3	43	42	42	84	84	84	78	80	80
Daub-4	44	42	41	85	84	84	78	79	79
Daub-5	45	41	41	85	84	84	79	79	80
Daub-10	53	38	41	87	82	84	79	74	78
Daub-15	59	35	40	88	78	82	82	69	77
Daub-20	65	32	40	89	74	83	83	64	77
	Threshold: $\lambda = 20$ — Good overall quality								
Daub-2	63	63	63	91	91	91	87	89	88
Daub-3	64	63	64	92	91	91	87	89	89
Daub-4	65	63	63	92	91	91	87	88	89
Daub-5	66	62	63	92	91	91	87	90	89
Daub-10	70	58	63	93	89	91	88	83	88
Daub-15	74	56	62	93	86	91	89	79	88
Daub-20	78	51	63	94	82	91	90	74	88
	Threshold: $\lambda = 45$ — Medium overall quality								
Daub-2	86	86	87	96	96	96	94	95	95
Daub-3	86	86	87	96	96	96	94	95	95
Daub-4	87	86	87	96	96	96	94	95	96
Daub-5	87	85	87	96	96	96	95	94	96
Daub-10	88	82	87	97	94	96	94	91	96
Daub-15	90	79	87	97	91	96	95	88	96
Daub-20	92	74	87	97	89	96	96	83	96
	Threshold: $\lambda = 85$ — Poor overall quality								
Daub-2	96	96	97	98	98	98	97	98	98
Daub-3	96	96	97	98	98	98	97	98	98
Daub-4	96	96	97	98	98	98	97	97	98
Daub-5	96	95	97	98	98	98	98	97	98
Daub-10	97	93	97	98	97	98	97	94	98
Daub-15	97	91	97	98	95	98	98	92	98
Daub-20	97	86	98	98	93	99	98	88	99
	Camera			*Goldhill*			*House*		
	Threshold: $\lambda = 10$ — Excellent overall quality								
Daub-2	78	80	79	70	71	70	79	80	80
Daub-3	77	79	78	70	71	71	79	80	80
Daub-4	77	79	78	71	71	70	79	80	79
Daub-5	77	78	78	71	71	70	79	79	79
Daub-10	77	74	76	73	67	69	80	72	78
Daub-15	80	71	75	77	63	68	82	66	77
Daub-20	81	66	74	79	58	68	83	59	76
	Threshold: $\lambda = 20$ — Good overall quality								
Daub-2	86	88	88	85	87	86	87	88	88
Daub-3	86	88	88	85	87	86	87	88	88
Daub-4	86	88	88	86	86	86	87	88	87
Daub-5	86	87	88	86	86	86	87	87	88
Daub-10	86	85	87	86	83	86	87	81	87
Daub-15	88	82	86	89	79	86	89	75	87
Daub-20	88	78	86	89	73	86	89	69	87
	Threshold: $\lambda = 45$ — Medium overall quality								
Daub-2	93	95	95	94	96	95	93	95	94
Daub-3	93	95	95	95	96	95	94	95	95
Daub-4	94	95	95	95	95	95	94	94	95
Daub-5	94	94	95	95	95	96	94	94	95
Daub-10	93	93	95	95	92	96	94	89	95
Daub-15	94	91	95	95	89	96	95	84	94
Daub-20	95	88	95	96	85	96	95	78	95
	Threshold: $\lambda = 85$ — Poor overall quality								
Daub-2	97	98	98	97	98	98	97	98	98
Daub-3	97	98	98	98	98	98	97	97	97
Daub-4	97	98	98	98	98	98	97	97	98
Daub-5	97	97	98	98	98	99	97	97	98
Daub-10	97	96	98	98	96	99	97	93	98
Daub-15	97	95	98	98	93	99	97	89	98
Daub-20	98	93	98	98	90	99	98	84	99

Table 3. Average quality of the six test images. Figure 2 gives a more 'readable' plot of these digits

Average image quality — PSNR [dB]						
Wavelet	zero padding	mirror padding	circular convol.	zero padding	mirror padding	circular convol.
	Threshold $\lambda = 10$			Threshold $\lambda = 20$		
Daub–2	17.630	17.602	17.701	15.242	15.252	15.246
Daub–3	17.745	17.752	17.768	15.298	15.330	15.288
Daub–4	17.691	17.711	17.662	15.244	15.284	15.213
Daub–5	17.719	17.701	17.680	15.233	15.270	15.257
Daub–10	17.641	17.615	17.689	15.253	15.290	15.306
Daub–15	17.695	17.675	17.686	15.136	15.185	15.168
Daub–20	17.616	17.654	17.676	15.135	15.207	15.114
	Threshold $\lambda = 45$			Threshold $\lambda = 85$		
Daub–2	13.057	13.078	12.942	11.609	11.736	11.681
Daub–3	12.982	13.144	13.016	11.659	11.763	11.643
Daub–4	12.932	13.025	13.002	11.637	11.740	11.651
Daub–5	12.992	13.110	13.016	11.610	11.806	11.626
Daub–10	12.823	13.061	12.935	11.500	11.713	11.666
Daub–15	12.854	12.985	12.911	11.422	11.628	11.538
Daub–20	12.673	12.916	12.788	11.439	11.624	11.718

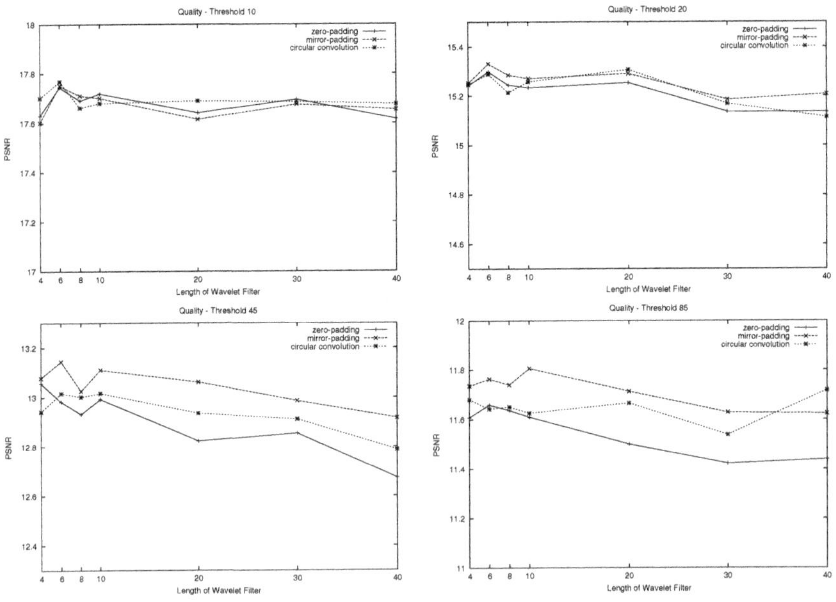

Fig. 2. Visual quality of the test images at the quantization thresholds $\lambda = 10, 20, 45$ and 85. The values correspond to Table 3

Table 4. Average bitrate heuristic of the six test images. Figure 3 gives a more 'readable' plot of these digits

Average discarded information — Percentage [%]						
Wavelet	zero padding	mirror padding	circular convol.	zero padding	mirror padding	circular convol.
	Threshold $\lambda = 10$			Threshold $\lambda = 20$		
Daub–2	72.0	72.3	72.0	83.2	84.3	84.0
Daub–3	71.8	72.7	72.5	83.5	84.3	84.3
Daub–4	72.3	72.5	71.8	83.8	84.0	84.0
Daub–5	72.7	72.0	72.0	84.0	83.8	84.2
Daub–10	74.8	67.8	71.0	85.0	79.8	83.7
Daub–15	78.0	63.7	69.8	87.0	76.2	83.3
Daub–20	80.0	58.8	69.7	88.0	71.2	83.5
	Threshold $\lambda = 45$			Threshold $\lambda = 85$		
Daub–2	92.7	93.8	93.7	97.0	97.7	97.8
Daub–3	93.0	93.8	93.8	97.2	97.5	97.7
Daub–4	93.3	93.5	94.0	97.2	97.3	97.8
Daub–5	93.5	93.0	94.2	97.3	97.0	98.0
Daub–10	93.5	90.2	94.2	97.3	94.8	98.0
Daub–15	94.3	87.0	94.0	97.5	92.5	98.0
Daub–20	95.2	82.8	94.2	97.8	89.0	98.7

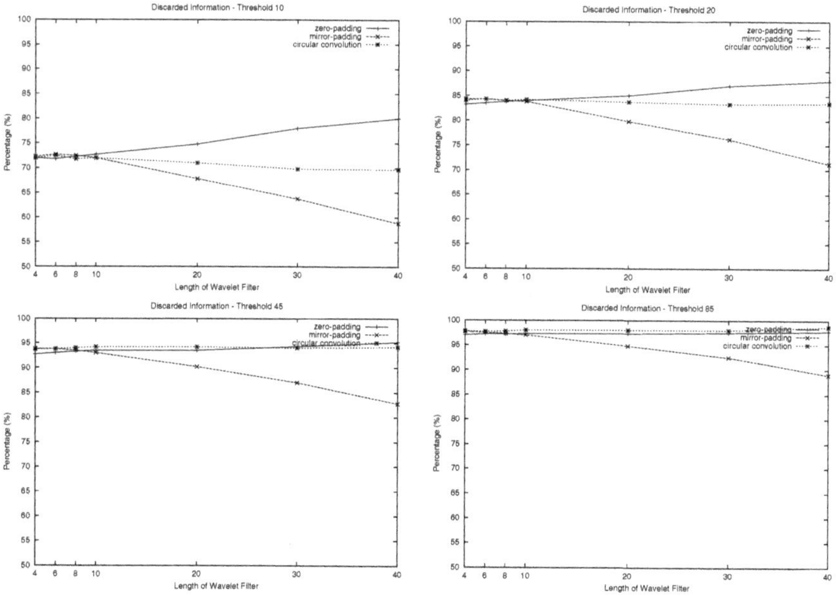

Fig. 3. Average bitrate heuristic of the test images at the quantization thresholds $\lambda = 10, 20, 45$ and 85. The values correspond to Table 4

Image-Feature Based Second Generation Watermarking in Wavelet Domain

Song Guoxiang and Wang Weiwei

School of Science, Xidian University
Xi'an, 710071, P.R.China

Abstract. An image-feature based second generation watermarking scheme is proposed in this paper. A host image is firstly transformed into wavelet coefficients and features are extracted from the lowest approximation. Then a watermark sequence is inserted in all high frequency coefficients corresponding to the extracted featured approximation coefficients. Original host image is not needed in watermarking detection, but the featured approximation coefficients position is necessary for robust detection. The correlation between the embedded watermark and all high frequency coefficients of a possibly corrupted watermarked image corresponding to the approximate coefficients at the same position as the original featured approximation coefficients is calculated and compared to a predefined threshold to see if the watermark is present. Experimental results show the watermark is very robust to common image processing, lossy compression in particular.

Keywords: image feature, digital watermarking, wavelet transform

1 Introduction

Lately, multimedia and computer networking have known rapid development and expansion. This created an increasing need for systems that protect the copyright ownership for digital images. Digital watermarking is the embedding of a mark into digital content that can later be, unambiguously, detected to allow assertions about the ownership or provenience of the data. This makes watermarking an emerging technique to prevent digital piracy. To be effective, a watermark must be imperceptible within its host, discrete to prevent unauthorized removal, easily extracted by the owner, and robust to incidental and intentional distortions.

Most of the recent work in watermarking can be grouped into two categories: spatial domain methods and frequency domain methods. Kutter et al. [1] refered both the spatial-domain and the transform domain techniques as first generation watermarking schemes and introduced the concept of second generation watermarking schemes which, unlike the first generation watermarking schemes, employ the notion of the data features. For images, features can be edges, corners, textured areas or parts in the image with specific characteristics. Features suitable for watermarking should have three basic properties: First, invariance

Y. Y. Tang et al. (Eds.): WAA 2001, LNCS 2251, pp. 16–21, 2001.

to noise (lossy compression, additive, multiplicative noise, ect.) Second, covariance to geometrical transformations (rotation, translation, sub-sampling, change of aspect ratio, etc.) The last, localization (cropping the data should not alter remaining feature points).

In this paper, we deal with the wavelet domain image watermarking method with the notion of second generation watermarking scheme. Previous wavelet domain watermarking schemes [2,3,4,5,6,7,8] added a watermark to a selected set of DWT coefficients in chosen subbands. The methods proposed in [2,3,6,8] requires the original image for detection, while the methods in [4,5,7] does not. However, the method [4] needs the embedded position and the corresponding subband label as well as two threshold value. For the method [5], if the watermarked image is tampered, the number of the coefficients that are greater than the larger threshold may not be equal to the size of the embeded watermark, thus there existed a problem for detection in calculating the correlation between the embedded watermark and the coefficients of a possibly modified watermarked image, whose absolute magnitude is above the larger threshold. The method [7] embedded watermarks into all HL and LH coefficients at levels 2 to 4, resulted in poor quality.

Based on the concept of second generation watermarking scheme, we propose a wavelet domain watermarking method which embeds watermarks into all high frequency coefficients corresponding to the featured lowest approximation coefficients. First, the host image is transformed using DWT and features are extracted from the lowest approximation using the method in [9]. Then the watermark is embedded into all subband coefficients corresponding to the featured lowest approximate coefficients. Finally, the modified coefficients is inversely transformed to form the watermarked image. In the watermark detection, the original image is not needed, but for more robust detection, the featured lowest approximate coefficients position of the original image is required, which can be encrypted using private key encryption and stored in the image header. The correlation between the embedded watermark and all high frequency coefficients of a possibly corrupted watermarked image corresponding to the lowest approximate coefficients at the same position as the original featured approximation coefficients is calculated and compared to a predefined threshold to see whether the watermark is present or not. Experimental results show that the watermark is very robust to common image processing, lossy compression in particular. Even when the watermarked image is compressed by JPEG with a quality factor of one percent, the watermark is still present.

2 The Proposed Method

The original image is firstly decomposed using DWT with 8 taps Daubechies orthogonal filter [10] until the scale N to obtain multiresolution LH_n, HL_n, HH_n $(n = 1, 2, \cdots, N)$ and the lowest resolution approximation LL_N.There exists a tree structure between the coefficients [11] as shown in Fig.1(for $N = 3$). The tree relation can be defined as follows:

$$tree(LL_N(x,y)) = tree(HL_N(x,y)) \cup tree(LH_N(x,y)) \cup tree(HH_N(x,y)) \quad (1)$$

$$tree(HL_n(x,y)) = tree(HL_{n-1}(2x-1,2y-1)) \cup tree(HL_{n-1}(2x,2y-1)) \cup tree(HL_{n-1}(2x-1,2y)) \cup tree(HL_{n-1}(2x,2y)) \quad (2)$$

where $n = N, N-1, \cdots, 2$. For $tree(LH_n(x,y))$, $tree(HH_n(x,y))(n = N, N-1, \cdots, 2)$, the definition is similar to (2).

$$tree(HL_1(x,y)) = HL_1(x,y)$$

$$tree(LH_1(x,y)) = LH_1(x,y)$$

$$tree(HH_1(x,y)) = HH_1(x,y)$$

For the experiments reported in this paper, N is taken as $N = 4$.

2.1 Feature Extraction

We use the method in [9] to extract features of the image. The difference is that we extract features from the lowest approximation components LL_N of the DWT of the image, rather than from the original image. Since the size of LL_N is $1/(4^N)$ times that of the original image, the time needed for extracting features is largely reduced.The feature extraction scheme is based on a decomposition of the image using Mexican-Hat wavelets. In two dimensions, the response of the Mexican-Hat mother avelet is defined as:

$$\psi(x,y) = (2 - (x^2 + y^2))e^{-(x^2+y^2)/2} \quad (3)$$

The isotropic nature of the Mexican-Hat filter is well suited for detecting point-features. Here we briefly describe the feature-detection procedure as follows: Firstly, define the feature-detection function, $P_{ij}(\cdot,\cdot)$ as:

$$P_{ij}(k,l) = |M_i(k,l) - \gamma M_j(k,l)| \quad (4)$$

where $M_i(k,l)$ and $M_j(k,l)$ represent the responses of Mexican-Hat wavelets at the image location (k,l) for scales i and j respectively. For an image A, the wavelet response $M_i(k,l)$ is given by:

$$M_i(k,l) = < (2^{-i}\psi(2^{-i}(k,l))), A > \quad (5)$$

where $< \cdot,\cdot >$ denotes the convolution of its operands. We only consider wavelets on a dyadic scale. Thus, the normalizing constant is given by $\gamma = 2^{-(i-j)}$. The operator $|\cdot|$ returns the absolute value of its parameter. Here we take $i = 2$ and $j = 4$ as in [9]. Secondly, determine points of local maxima of $P_{ij}(\cdot,\cdot)$. These maxima correspond to the set of potential feature-points. A circular neighborhood with a radius of 5 points is used to determine the local maxima. Finally, accept a point of local maxima of $P_{ij}(\cdot,\cdot)$ as a feature-point if the variance of the image-pixels in the neighborhood of the point is higher than a threshold. Here a 7×7 neighborhood around the point is used for computing the local variance. A candidate point is accepted as a feature-point if the corresponding local variance is larger than a threshold, which we take as 20.

2.2 Watermark Inserting

The original image I is firstly decomposed using DWT with 8 taps Daubechies orthogonal filter until the scale $N = 4$ to obtain multiresolution LH_n, HL_n, HH_n $(n = 1, 2, \cdots, 4)$ and the lowest resolution approximation LL_4. Then feature-points are extracted from LL_4 using the method in 2.1. If $LL_4(x, y)$ is a feature-point, then some watermark bits $\boldsymbol{x} \in X$ are added to all the children notes of $tree(LL_4(x, y))$. X stands for a set of watermark $\boldsymbol{x}$ and the elements x_l of $\boldsymbol{x}$ are given by the random noise sequence whose probability law has a normal distribution of zero mean and unit variance. Since for every tree$(LL_4(x, y)))$, there are 255 children in all, except for the root, the size of the watermark $\boldsymbol{x}$, denoted by M, is given by $M = 255\times$ the number of feature-points in LL_4). The specific embedding method is as follows: For every feature-point $LL_4(x, y)$, for every $W_l \in tree(LL_4(x, y))$ and $W_l \neq LL_4(x, y)$

$$W_l' \leftarrow W_l + \alpha|W_l|x_l \tag{6}$$

where w_l and W_l' denotes respectively the DWT coefficient of the original and watermarked image,α is a modulating factor, here we take $\alpha = 0.2$. Finally, inversely transform the modified multiresolution subbands to obtain the watermarked image I'.

2.3 Watermark Detection

The original image is not required in the watermark detection, but for more robust detection, the feature-points position of the original image is indeed necessary. Firstly, A possibly corrupted watermarked image $\tilde{I}$ is decomposed as I in 2.2. Then for every feature-point $\tilde{LL}_4(x, y)$, all coefficients $\tilde{W}_l \in tree(\tilde{LL}_4(x, y))$ and $\tilde{W}_l \neq \tilde{LL}_4(x, y)$ are taken out, where $\tilde{LL}_4$ and $\tilde{W}_l$ respectively represents the lowest resolution approximation and high frequency coefficients of $\tilde{I}$.We calculate the correlation z between $\tilde{W}_l$ and all candidates $\boldsymbol{y} \in X$ of the embedded watermark $\boldsymbol{x}$ as:

$$z = 1/M \sum_{l=1}^{M} \tilde{W}_l y_l \tag{7}$$

By comparing the correlation with a predefined threshold S_z, which is given in [7] to determine whether a given watermark is present or not. In theory, the threshold S_z is taken as

$$S_z = \frac{\alpha}{2M} \sum_{l=1}^{M} |W_l| \tag{8}$$

In practice, the watermarked image would be attacked incidentally or intentionally, so for robust detection, the threshold is taken as

$$S_x = r\frac{\alpha}{2M} \sum_{l=1}^{M} |\tilde{W}_l|, 0 < r \leq 1 \tag{9}$$

3 Experimental Results

In order to confirm that the proposed watermarking scheme is effective, we performed some numerical experiments with some gray-scale standard images. Here we describe experimental results for the standard image "lenna"(512×512 pixels, 8 bits/pixel) shown in Fig.2(a). Fig.2(b) shows the watermarked image with parameters $\alpha = 0.2$, $N = 4$ and $M = 4080$. Next, we tested the robustness of the watermark against some common image processing operations on the watermarked image Fig.2(b). Fig.3 is the result of JPEG compression with quality factor of 1. The image after 11×11 mean filtering is shown in Fig.4. The image after adding white Gaussian noise of power 40db is shown in Fig.5. Fig.6 is the clipped image with only 25% center data left. Fig.7 shows the result of rotation counter clockwise by 10 degrees. The response of the watermark detector and the corresponding threshold for the untampered and attacked watermarked image are given in Tab.1. The threshold is calculated using the equation (10), where $r = 2/3$. As shown in Tab.1, though image degradation is very heavy, the watermark is still easily recovered and the detector response is also well above the threshold. Numerical experiments with the other standard images have also demonstrated similar results.

4 Conclusions

An image-feature based wavelet domain second generation watermarking scheme is proposed in this paper. Experiments show that the watermark is very robust to common image processing, lossy compression and smoothing in particular. Even for the JPEG compressed version of the watermarked image with quality factor of 1%, the feature-points remain salient. Furthermore, we will investigate watermarking method that resistant to geometric attacks.

References

1. M. Kutter, S. K. Bhattacharjee, and T. Ebrahimi, "Towards second generation watermarking scheme," Proc. IEEE ICIP'99, Vol.1,1999 16
2. D. Kundur and D. Hatzinakos, "A robust digital image watermarking method using wavelet-based fusion," Proc. IEEE ICIP'97, vol.1, 1997, pp.544-547 17
3. X. G. Xia, C. G. Boncelet and G. R. Arce, "A multiresolution watermark for digital images," Proc. IEEE ICIP'97, Vol.1,1997, pp.548-551 17
4. H. Inoue, A. Miyazaki, A. Yamamoto, etal., "A digital watermark bases on the wavelet transform and its robustness on image compression," Proc. IEEE ICIP'98, Vol.2, 1998, pp.391-423 17
5. R. Dugad, K. Ratakonda and N. Ahuja, "A new wavelet-based scheme for watermarking image," Proc. IEEE ICIP'98, vol.2, 1998, pp.419-423 17
6. W. W. Zhu, Z. X. Xiong and Y. Q. Zhang, "Multiresolution watermarking for images and video: a unified approach," Proc. IEEE ICIP'98, vol.1, 1998, pp.465-468 17

7. H. Inoue, A. Kiomiyazaki and T. Katsura, "An image watermarking method based on the wavelet transform," Proc. IEEE ICIP'99, vol.1, 1999, pp.296-300 17, 19
8. J. R. Kim and Y. S. Moon, "A robust wavelet-based digital watermarking using Level-adaptive thresholding," Proc. IEEE ICIP'99, vol.2, 1999, pp.226-230 17
9. S. K. Bhattacharjee and M. Kutter, "Compression tolerant image authentication", Proc. IEEE ICIP'98, Vol.1,1998 17, 18
10. I. Daubechies, "Ten Lectures on Wavelets," CBMS-NSF conference series in applied mathematics, SIAM Ed. 17
11. J. M. Shapiro, "Embeded image coding using zerotrees of wavelet coefficients," IEEE trans. On Signal Processing, Vol.41, No.12, 1993, pp.3445-3462 17

A Study on Preconditioning Multiwavelet Systems for Image Compression

Wonkoo Kim and Ching-Chung Li

University of Pittsburgh, Dept. of Electrical Engineering
Pittsburgh, PA 15261, USA
wonkoo@home.com
ccl@ee.pitt.edu

Abstract. We present a study on applications of multiwavelet analysis to image compression, where filter coefficients form matrices. As a multiwavelet filter bank has multiple channels of inputs, we investigate the data initialization problem by considering prefilters and postfilters that may give more efficient representations of the decomposed data. The interpolation postfilter and prefilter are formulated, which are capable to provide a better approximate image at each coarser resolution level. A design process is given to obtain both filters having compact supports, if exist. Image compression performances of some multiwavelet systems are studied in comparison to those of single wavelet systems.

1 Nonorthogonal Multiwavelet Subspaces

Let us define a multiresolution analysis of $L^2(\mathbb{R})$ generated by several scaling functions, with an increasing sequence of function subspaces $\{V_j\}_{j\in\mathbb{Z}}$ in $L^2(\mathbb{R})$:

$$\{0\} \subset \ldots \subset V_{-1} \subset V_0 \subset V_1 \subset \ldots \subset L^2(\mathbb{R}). \tag{1}$$

Subspaces V_j are generated by a set of scaling functions $\phi^1, \phi^2, \ldots, \phi^r$ (namely, multiscaling functions) such that

$$V_j := clos_{L^2(\mathbb{R})} < \phi^m_{j,k} \; : \; 1 \leq m \leq r, \, k \in \mathbb{Z} >, \quad \forall j \in \mathbb{Z}, \tag{2}$$

i.e., V_j is the closure of the linear span of $\{\phi^m_{j,k}\}_{1\leq m\leq r,\, k\in\mathbb{Z}}$ in $L^2(\mathbb{R})$, where

$$\phi^m_{j,k}(x) := 2^{j/2}\phi^m(2^j x - k), \quad \forall x \in \mathbb{R}. \tag{3}$$

Then we have a sequence of multiresolution subspaces $\{V_j\}$ generated by a set of multiscaling functions, where the resolution gets finer and finer as j increases.

Let us define inter-spaces $W_j \subset L^2(\mathbb{R})$ such that $V_{j+1} := V_j \dot{+} W_j$, $\forall j \in \mathbb{Z}$, where the plus sign with a dot ($\dot{+}$) denotes a nonorthogonal direct sum. W_j is the complement to V_j in V_{j+1}, and thus W_j and W_l with $j \neq l$ are disjoint but may not be orthogonal to each other. If $W_j \perp W_l$, $\forall j \neq l$, we call them

Y. Y. Tang et al. (Eds.): WAA 2001, LNCS 2251, pp. 22–36, 2001.

semi-orthogonal wavelet spaces [1]. By the nature of construction, subspaces W_j can be generated by r base functions, $\psi^1, \psi^2, \ldots, \psi^r$ that are multiwavelets. The subspace W_j is the closure of the linear span of $\{\psi^m_{j,k}\}_{1\le m\le r,\, k\in\mathbb{Z}}$:

$$W_j := clos_{L^2(\mathbb{R})} < \psi^m_{j,k} \; : 1 \le m \le r,\, k \in \mathbb{Z} >, \quad \forall j \in \mathbb{Z}, \tag{4}$$

where

$$\psi^m_{j,k}(x) := 2^{j/2}\psi^m(2^j x - k), \quad \forall x \in \mathbb{R}. \tag{5}$$

We may express multiscaling functions and multiwavelets as vector functions:

$$\boldsymbol{\phi}(x) := \begin{pmatrix} \phi^1(x) \\ \vdots \\ \phi^r(x) \end{pmatrix}, \quad \boldsymbol{\psi}(x) := \begin{pmatrix} \psi^1(x) \\ \vdots \\ \psi^r(x) \end{pmatrix}, \quad \forall x \in \mathbb{R}. \tag{6}$$

Also, in vector form, let us define

$$\boldsymbol{\phi}_{j,k}(x) := 2^{j/2}\boldsymbol{\phi}(2^j x - k) \quad \text{and} \quad \boldsymbol{\psi}_{j,k}(x) := 2^{j/2}\boldsymbol{\psi}(2^j x - k), \quad \forall x \in \mathbb{R}. \tag{7}$$

Since the multiscaling functions $\phi^m \in V_0$ and the multiwavelets $\psi^m \in W_0$ are all in V_1, and since V_1 is generated by $\{\phi^m_{1,k}(x) = 2^{1/2}\phi^m(2x-k)\}_{1\le m\le r,\, k\in\mathbb{Z}}$, there exist two ℓ^2 matrix sequences $\{H_n\}_{n\in\mathbb{Z}}$ and $\{G_n\}_{n\in\mathbb{Z}}$ such that we have a *two-scale relation* for the multiscaling function $\boldsymbol{\phi}(x)$:

$$\boldsymbol{\phi}(x) = 2\sum_{n\in\mathbb{Z}} H_n \boldsymbol{\phi}(2x - n), \quad x \in \mathbb{R}, \tag{8}$$

which is also called as a *two-scale matrix refinement equation (MRE)*, and for multiwavelet $\boldsymbol{\psi}(x)$:

$$\boldsymbol{\psi}(x) = 2\sum_{n\in\mathbb{Z}} G_n \boldsymbol{\phi}(2x - n), \quad x \in \mathbb{R}, \tag{9}$$

where H_n and G_n are $r \times r$ square matrices. We are interested in finite sequences of H_n and G_n, namely, FIR (Finite Impulse Response) filter pairs.

Using the fractal interpolation, Geronimo, Hardin, and Massopust successfully constructed a very important multiwavelet system [2,3,4] which has two orthogonal multiscaling functions and two orthogonal multiwavelets. Their four matrix coefficients H_n satisfy the MRE for a multiscaling function $\boldsymbol{\phi}(x)$:

$$H_0 = \begin{bmatrix} \frac{3}{10} & \frac{4\sqrt{2}}{10} \\ -\frac{\sqrt{2}}{40} & -\frac{3}{20} \end{bmatrix}, \; H_1 = \begin{bmatrix} \frac{3}{10} & 0 \\ \frac{9\sqrt{2}}{40} & \frac{1}{2} \end{bmatrix}, \; H_2 = \begin{bmatrix} 0 & 0 \\ \frac{9\sqrt{2}}{40} & -\frac{3}{20} \end{bmatrix}, \; H_3 = \begin{bmatrix} 0 & 0 \\ -\frac{\sqrt{2}}{40} & 0 \end{bmatrix}, \tag{10}$$

and other four matrix coefficients G_n generate a multiwavelet $\boldsymbol{\psi}(x)$:

$$G_0 = \begin{bmatrix} -\frac{\sqrt{2}}{40} & -\frac{3}{20} \\ -\frac{1}{20} & -\frac{3\sqrt{2}}{20} \end{bmatrix}, \; G_1 = \begin{bmatrix} \frac{9\sqrt{2}}{40} & -\frac{1}{2} \\ \frac{9}{20} & 0 \end{bmatrix}, \; G_2 = \begin{bmatrix} \frac{9\sqrt{2}}{40} & -\frac{3}{20} \\ -\frac{9}{20} & \frac{3\sqrt{2}}{20} \end{bmatrix}, \; G_3 = \begin{bmatrix} -\frac{\sqrt{2}}{40} & 0 \\ \frac{1}{20} & 0 \end{bmatrix} \tag{11}$$

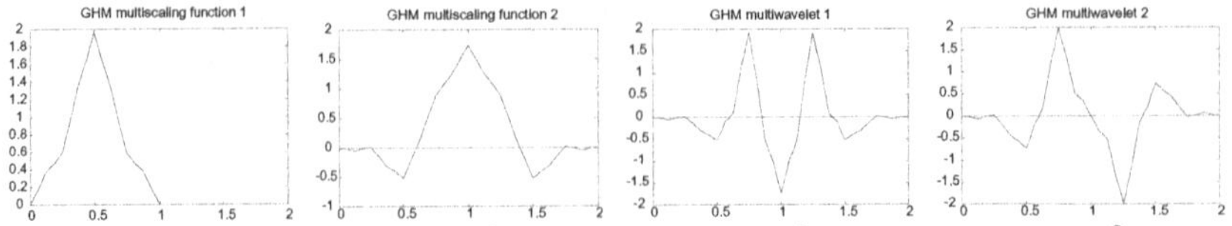

(a) ϕ^1 (b) ϕ^2 (c) ψ^1 (d) ψ^2

Fig. 1. Geronimo-Hardin-Massopust orthogonal multiscaling functions and multiwavelets

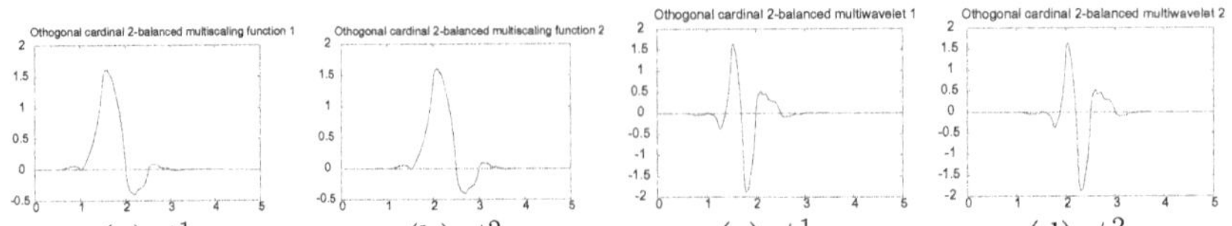

(a) ϕ^1 (b) ϕ^2 (c) ψ^1 (d) ψ^2

Fig. 2. Cardinal 2-balanced orthogonal multiscaling functions and multiwavelets

The GHM (Geronimo-Hardin-Massopust) orthogonal multiscaling functions are shown in Figure 1(a) and (b), and their corresponding orthogonal multiwavelets are shown in (c) and (d). The GHM multiwavelet system has very remarkable properties: its scaling functions and wavelets are orthogonal, very shortly supported, symmetric or antisymmetric, and it has second order approximation so that locally constant and locally linear functions are in V_j.

Another example of orthogonal multiwavelet is shown in Figure 2[5,6,7], where multiscaling functions are shown in figures (a) and (b), and multiwavelet functions are shown in figures (c) and (d), respectively. Two scaling functions in each cardinal balanced multiwavelet system are the same functions up to a half integer shift in time, and also the wavelets are the same up to a half integer shift in time. The approximation orders of the cardinal balanced orthogonal multiwavelet systems are 2 for cardinal 2-balanced, 3 for cardinal 3-balanced, and 4 for cardinal 4-balanced systems. The cardinal 2-balanced orthogonal multiwavelet filters are given by

$$\mathbf{H}(z) = \begin{bmatrix} b(z) & 0.5z^{-1} \\ z^{-5}b(-1/z) & 0.5z^{-2} \end{bmatrix}, \qquad \mathbf{G}(z) = \begin{bmatrix} -b(z) & 0.5z^{-1} \\ -z^{-5}b(-1/z) & 0.5z^{-2} \end{bmatrix}, \tag{12}$$

where $b(z) = 0.015625 + 0.123015364784490z^{-1} + 0.46875z^{-2} - 0.121030729568979z^{-3} + 0.015625z^{-4} - 0.001984635215512z^{-5}$. For more details on cardinal balanced orthogonal multiwavelets, refer to the paper written by I. Selesnick [6].

We should note that a scalar system with one scaling function cannot combine symmetry, orthogonality, and the second order approximation together. Furthermore, the solution of a scalar refinement equation with four coefficients is supported on the interval [0,3], while multiscaling functions with four matrix coefficients can be supported on a shorter interval.

Since all elements of both $\phi(2x)$ and $\phi(2x-1)$ are in V_1 and $V_1 = V_0 \dot{+} W_0$, there exist two ℓ^2 matrix sequences $\{\tilde{H}_n\}_{n\in\mathbb{Z}}$ and $\{\tilde{G}_n\}_{n\in\mathbb{Z}}$ such that

$$\phi(2x-k) = \sum_{n\in\mathbb{Z}} \left[\tilde{H}_{k-2n}^T \phi(x-n) + \tilde{G}_{k-2n}^T \psi(x-n)\right], \quad \forall k \in \mathbb{Z}, \tag{13}$$

which is called the *decomposition relation* of ϕ and ψ.[1]

We have two pairs of sequences $(\{H_n\}, \{G_n\})$ and $(\{\tilde{H}_n\}, \{\tilde{G}_n\})$, which are unique due to the direct sum relationship $V_1 = V_0 \dot{+} W_0$. A carefully chosen pair of sequences $(\{H_n\}, \{G_n\})$ can generate multiscaling functions and multiwavelets and thus multiwavelet subspaces; hence, they can completely characterize a multiwavelet analysis.

2 Multiwavelet Decomposition and Reconstruction

From the formulas (8), (9), and (13), the following signal decomposition and reconstruction algorithms can be derived. Let $v_j \in V_j$ and $w_j \in W_j$ so that

$$v_j(x) := \sum_{k\in\mathbb{Z}} \boldsymbol{c}_{j,k} \cdot \phi(2^j x - k) = \sum_{k\in\mathbb{Z}} \boldsymbol{c}_{j,k}^T \, \phi(2^j x - k); \tag{14}$$

$$w_j(x) := \sum_{k\in\mathbb{Z}} \boldsymbol{d}_{j,k} \cdot \psi(2^j x - k) = \sum_{k\in\mathbb{Z}} \boldsymbol{d}_{j,k}^T \, \psi(2^j x - k), \tag{15}$$

where $\cdot$ denotes a dot product between two vectors and $\cdot^T$ denotes the transpose operator. The scale factor $2^{j/2}$ is not explicitly shown here for simplicity but incorporated into the sequences $\boldsymbol{c}_{j,k}$ and $\boldsymbol{d}_{jk}$. By the relation $V_j = V_{j-1} \dot{+} W_{j-1}$,

$$\begin{aligned} v_j(x) &:= v_{j-1}(x) + w_{j-1}(x) \\ &= \sum_{k\in\mathbb{Z}} \boldsymbol{c}_{j-1,k} \cdot \phi(2^{j-1}x - k) + \sum_{k\in\mathbb{Z}} \boldsymbol{d}_{j-1,k} \cdot \psi(2^{j-1}x - k), \quad \forall j \in \mathbb{Z}. \end{aligned} \tag{16}$$

Thus we have the following recursive *decomposition (analysis) formulas*:

$$\boldsymbol{c}_{j-1,k} = \sum_n \tilde{H}_{n-2k} \, \boldsymbol{c}_{j,n} = \sum_n \tilde{H}_{-n} \, \boldsymbol{c}_{j,2k-n}, \quad \forall j \in \mathbb{Z}; \tag{17}$$

$$\boldsymbol{d}_{j-1,k} = \sum_n \tilde{G}_{n-2k} \, \boldsymbol{c}_{j,n} = \sum_n \tilde{G}_{-n} \, \boldsymbol{c}_{j,2k-n}, \quad \forall j \in \mathbb{Z}. \tag{18}$$

An original data sequence $\boldsymbol{c}_0$ $(=\{\boldsymbol{c}_{0,k}\}_k)$ is decomposed into $\boldsymbol{c}_1$ and $\boldsymbol{d}_1$ data sequences, and the sequence $\boldsymbol{c}_1$ is further decomposed into $\boldsymbol{c}_2$ and $\boldsymbol{d}_2$ sequences, etc.. Keeping this process recursively, the original sequence $\boldsymbol{c}_0$ is decomposed into $\boldsymbol{d}_1, \boldsymbol{d}_2, \boldsymbol{d}_3, \ldots$. Note that this process continuously reduces the data size by half for each decomposed sequence but it conserves the total data size.

[1] We here intentionally transposed the matrices of $\tilde{H}$ and $\tilde{G}$ and reversed indexing instead of $2n-k$, for some convenience in representing formulas of dual relationship.

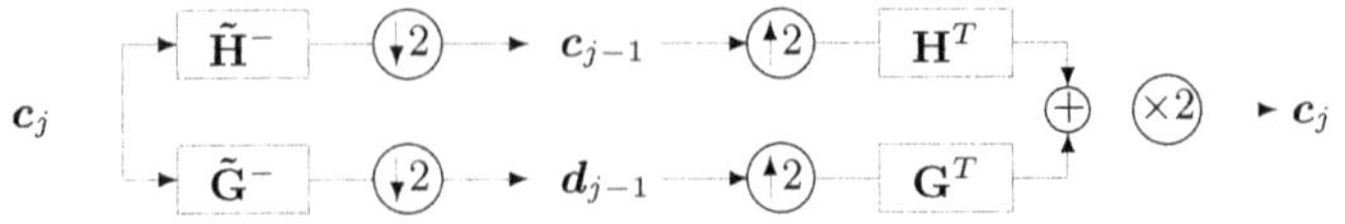

(a) Filterbanks derived from multiwavelet analysis

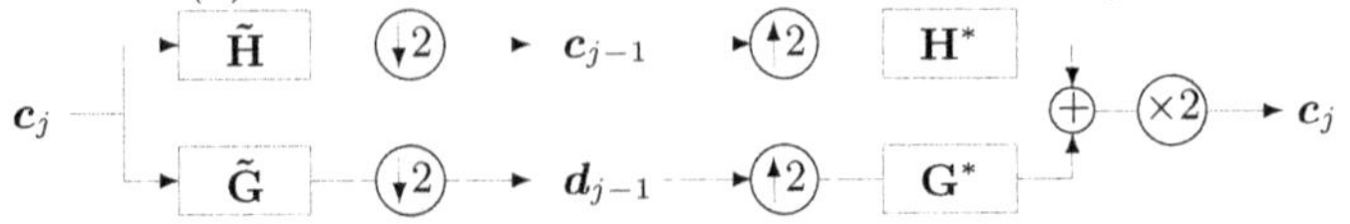

(b) Multiwavelet filterbanks by reverse indexing

Fig. 3. The multiwavelet transform filter banks. Filters are $r \times r$ matrices and data paths are r lines, where $r = 2$ in our examples. The multiwavelet systems (a) and (b) are equivalent, except that filter indices are all reversed between the two systems

Let $\mathbf{D}_K$, $K \geq 1$, be the *subsampling (downsampling) operator* defined by

$$(\mathbf{D}_K \boldsymbol{x})[n] := \boldsymbol{x}[Kn], \tag{19}$$

where K is a subsampling rate and $\boldsymbol{x}$ is a sequence of vector-valued samples. The *decomposition formulas* can be rewritten in the Z-transform domain as

$$\boldsymbol{c}_{j-1}(z) = \mathbf{D}_2 \tilde{\mathbf{H}}^{-}(z) \boldsymbol{c}_j(z), \tag{20}$$
$$\boldsymbol{d}_{j-1}(z) = \mathbf{D}_2 \tilde{\mathbf{G}}^{-}(z) \boldsymbol{c}_j(z), \tag{21}$$

where the superscript $^-$ denotes reverse indexing, i.e., $\mathbf{H}^- := \mathbf{H}^{*^T}$.

From the two-scale relations (8), (9) and from (14), (15), we have the following recursive *reconstruction (synthesis) formula*:

$$\boldsymbol{c}_{j,k} = 2 \sum_n \left(H_{k-2n}^T \, \boldsymbol{c}_{j-1,n} + G_{k-2n}^T \, \boldsymbol{d}_{j-1,n} \right). \tag{22}$$

Let $\mathbf{U}_K$, $K \geq 1$, be the *upsampling operator* defined by

$$(\mathbf{U}_K \boldsymbol{x})[n] := \begin{cases} \boldsymbol{x}[\frac{n}{K}], & \text{if } \frac{n}{K} \text{ is an integer;} \\ 0, & \text{otherwise,} \end{cases} \tag{23}$$

where K is an upsampling rate and $\boldsymbol{x}$ is a sequence of vector-valued samples. Then the *reconstruction formula* can be rewritten in the Z-transform domain as

$$\boldsymbol{c}_j(z) = 2 \left[\mathbf{H}^T(z) \mathbf{U}_2 \boldsymbol{c}_{j-1}(z) + \mathbf{G}^T(z) \mathbf{U}_2 \boldsymbol{d}_{j-1}(z) \right] \tag{24}$$

The decomposition and reconstruction systems implemented by multiwavelet filterbanks are shown in Figure 3, where the system (a) is the exact implementation of our equations derived. If we take reverse indexing for all filters, we have the system (b), and the *multiwavelet decomposition formulas* become

$$\boldsymbol{c}_{j-1}(z) = \mathbf{D}_2 \tilde{\mathbf{H}}(z) \boldsymbol{c}_j(z), \tag{25}$$
$$\boldsymbol{d}_{j-1}(z) = \mathbf{D}_2 \tilde{\mathbf{G}}(z) \boldsymbol{c}_j(z), \tag{26}$$

and the *reconstruction formula* becomes

$$\boldsymbol{c}_j(z) = 2\left[\mathbf{H}^*(z)\mathbf{U}_2\boldsymbol{c}_{j-1}(z) + \mathbf{G}^*(z)\mathbf{U}_2\boldsymbol{d}_{j-1}(z)\right]. \tag{27}$$

Note that the input data $\boldsymbol{c}_j$ is a sequence of vector-valued data, every data path has r lines, and filters are $r \times r$ matrices. We restrict $r = 2$ in this study. Constructing a vector-valued sequence $\boldsymbol{c}_j$ from a signal or an image is nontrivial. As an 1-D input signal is vectorized, the direction of filter indexing will affect the reconstructed signal in an undesirable way, if the vectorization scheme does not match with filter indexing. This effect does not happen in a scalar wavelet system, whose filters are not matrices. As we do not take reverse indexing for data sequences, we will take the system (a) of Figure 3 in our implementation. A prefilter for the chosen input scheme will be designed later in Section 5.

3 Biorthogonality and Perfect Reconstruction Condition

From the two-scale dilation equations (8), (9), and the decomposition relation (13), we have the following biorthogonality conditions:

$$\mathbf{H}(z)\tilde{\mathbf{H}}^*(z) + \mathbf{H}(-z)\tilde{\mathbf{H}}^*(-z) = \mathbf{I}_r; \tag{28}$$
$$\mathbf{H}(z)\tilde{\mathbf{G}}^*(z) + \mathbf{H}(-z)\tilde{\mathbf{G}}^*(-z) = \mathbf{0}_r; \tag{29}$$
$$\mathbf{G}(z)\tilde{\mathbf{H}}^*(z) + \mathbf{G}(-z)\tilde{\mathbf{H}}^*(-z) = \mathbf{0}_r; \tag{30}$$
$$\mathbf{G}(z)\tilde{\mathbf{G}}^*(z) + \mathbf{G}(-z)\tilde{\mathbf{G}}^*(-z) = \mathbf{I}_r, \tag{31}$$

which completely characterize the biorthogonality between the analysis filter pair $(\tilde{\mathbf{H}}, \tilde{\mathbf{G}})$ and the synthesis filter pair $(\mathbf{H}, \mathbf{G})$. (Namely, $\mathbf{H} \perp \tilde{\mathbf{G}}$ and $\tilde{\mathbf{H}} \perp \mathbf{G}$.)

Let $\mathbf{H_m}(z)$ denote the modulation matrix[2] of $(\mathbf{H}, \mathbf{G})$ as defined by

$$\mathbf{H_m}(z) := \begin{bmatrix} \mathbf{H}(z) & \mathbf{H}(-z) \\ \mathbf{G}(z) & \mathbf{G}(-z) \end{bmatrix}, \tag{32}$$

and $\tilde{\mathbf{H}}_\mathbf{m}(z)$ denote the modulation matrix of $(\tilde{\mathbf{H}}, \tilde{\mathbf{G}})$ similarly defined, then the above biorthogonality condition becomes

$$\mathbf{H_m}(z)\tilde{\mathbf{H}}^*_\mathbf{m}(z) = \begin{bmatrix} \mathbf{H}(z) & \mathbf{H}(-z) \\ \mathbf{G}(z) & \mathbf{G}(-z) \end{bmatrix} \begin{bmatrix} \tilde{\mathbf{H}}^*(z) & \tilde{\mathbf{G}}^*(z) \\ \tilde{\mathbf{H}}^*(-z) & \tilde{\mathbf{G}}^*(-z) \end{bmatrix} = \begin{bmatrix} \mathbf{I}_r & \mathbf{0} \\ \mathbf{0} & \mathbf{I}_r \end{bmatrix} = \mathbf{I}_{2r}. \tag{33}$$

From the decomposition and reconstruction formulas (20), (21) and (24), we have the following perfect reconstruction (PR) condition:

$$\tilde{\mathbf{H}}^*_\mathbf{m}(z)\mathbf{H_m}(z) = c\,\mathbf{I}_{2r}, \tag{34}$$

where c is a non-zero constant (a scale change in the reconstructed signal is allowed).

[2] The modulation matrix is also called as the AC (alias component) matrix[8].

Theorem 1. *For two matrix filter pairs* $(\mathbf{H}, \mathbf{G})$ *and* $(\tilde{\mathbf{H}}, \tilde{\mathbf{G}})$, *the modulation matrices* $\mathbf{H_m}(z)$ *and* $\tilde{\mathbf{H}}_\mathbf{m}(z)$ *are defined by*

$$\mathbf{H_m}(z) := \begin{bmatrix} \mathbf{H}(z) & \mathbf{H}(-z) \\ \mathbf{G}(z) & \mathbf{G}(-z) \end{bmatrix}, \quad \tilde{\mathbf{H}}_\mathbf{m}(z) := \begin{bmatrix} \tilde{\mathbf{H}}(z) & \tilde{\mathbf{H}}(-z) \\ \tilde{\mathbf{G}}(z) & \tilde{\mathbf{G}}(-z) \end{bmatrix}. \tag{35}$$

Then

$$\mathbf{H_m}(z)\tilde{\mathbf{H}}^*_\mathbf{m}(z) = \tilde{\mathbf{H}}^*_\mathbf{m}(z)\mathbf{H_m}(z) = c\,\mathbf{I}_{2r}, \tag{36}$$

where c *is a nonzero constant, is the necessary and sufficient condition for the two matrix filter pairs* $(\mathbf{H}, \mathbf{G})$ *and* $(\tilde{\mathbf{H}}, \tilde{\mathbf{G}})$ *to be biorthogonal and to ensure the perfect reconstruction. If these filter pairs generate multiscaling functions and multiwavelets, then they are biorthogonal.*

For orthogonal filter pairs, we have $\tilde{\mathbf{H}} = \mathbf{H}$ and $\tilde{\mathbf{G}} = \mathbf{G}$, and then

$$\mathbf{H_m}(z)\mathbf{H}^*_\mathbf{m}(z) = \mathbf{H}^*_\mathbf{m}(z)\mathbf{H_m}(z) = c\mathbf{I}_{2r}. \tag{37}$$

Hence, $\mathbf{H_m}(z)$ is paraunitary (lossless), i.e., unitary for all z on the unit circle.

4 Construction of Biorthogonal Multiwavelets

Plonka and Strela constructed biorthogonal Hermite cubic (piecewise cubic polynomial) multiscaling functions and multiwavelets using the cofactor method [9,10]. The coefficient matrix

$$\mathbf{H}(z) = \frac{1}{16}\begin{bmatrix} 4(1+z^{-1})^2 & -2(1-z^{-1})(1+z^{-1}) \\ 3(1-z^{-1})(1+z^{-1}) & -1+4z^{-1}-z^{-2} \end{bmatrix} \tag{38}$$

generates Hermite cubic multiscaling functions, where $\det \mathbf{H}(z) = (1+z^{-1})^4/128$. A possible choice of $\tilde{\mathbf{H}}$ for dual functions is

$$\tilde{\mathbf{H}}(z) = \frac{1}{32}\begin{bmatrix} z-8+18z^{-1}-8^{-2}+z^{-3} & -3z+12-12z^{-2}+3z^{-3} \\ 2z-8+8z^{-2}-2z^{-3} & -4z+8+24z^{-1}+8z^{2}-4z^{-3} \end{bmatrix} \tag{39}$$

By the biorthogonality conditions, we have

$$\tilde{\mathbf{G}}(z) = \frac{z^{-1}}{16}\begin{bmatrix} -4(1-z^{-1})^2 & 6(1-z^{-1})(1+z^{-1}) \\ -(1-z^{-1})(1+z^{-1}) & 1+4z^{-1}+z^{-2} \end{bmatrix} \tag{40}$$

and by cofactor method,

$$\mathbf{G}(z) = \frac{1}{32}\begin{bmatrix} 1+8z^{-1}+18z^{-2}+8z^{-3}+z^{-4} & -1-4z^{-1}+4z^{-3}+z^{-4} \\ 6+24z^{-1}-24z^{-3}-6z^{-4} & -4-8z^{-1}+24z^{-2}-8z^{-3}-4z^{-4} \end{bmatrix}. \tag{41}$$

The Hermite cubic multiscaling functions and multiwavelets generated by $\mathbf{H}$ and $\mathbf{G}$ are shown in Figure 4 (a)–(d). Their corresponding biorthogonal multiscaling functions and multiwavelets are shown in Figure 4 (e)–(h).

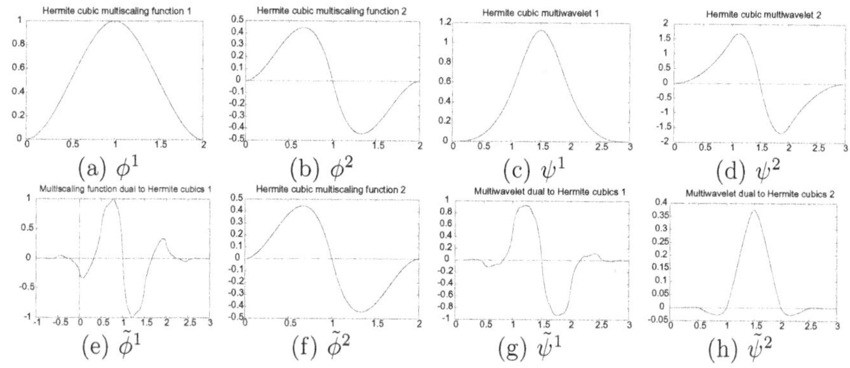

Fig. 4. Hermite cubics and their dual multiwavelets

5 Preconditioning Multiwavelet Systems

In this section we consider multiwavelet systems that analyze discrete data, and investigate how to precondition a multiwavelet system by prefiltering input data, which is not necessary for the case of single (or scalar) wavelet systems.

5.1 Prefilters and Postfilters

Consider the multiwavelet series expansion:

$$f_j(t) := \sum_k \boldsymbol{c}_{j,k}^T \boldsymbol{\phi}(2^j t - k) \tag{42}$$

From a given 1-D signal $x[n]$, construct a vector-valued sequence $\boldsymbol{x}[n]$ by

$$\boldsymbol{x}[n] := \begin{bmatrix} x[nr] \\ \vdots \\ x[nr + r - 1] \end{bmatrix}, \quad r \geq 1 \tag{43}$$

Let us define a *prefilter* $\mathbf{Q}(z)$, which maps a vector-valued sequence space onto itself, such that the coefficient vector sequence $\boldsymbol{c}_{0,k}$ is obtained by filtering $\boldsymbol{x}[n]$:

$$\boldsymbol{c}_0(z) = \mathbf{Q}(z)\boldsymbol{x}(z) \tag{44}$$

For any $j \leq 0$, $\boldsymbol{c}_{j,k}$ is decomposed to $\{\boldsymbol{c}_{j-1,k}, \boldsymbol{d}_{j-1,k}\}$ by a layer of multiwavelet decomposition. Recursive multiwavelet decompositions down to a resolution level $J < 0$ give us a set of decomposed data sequences $\boldsymbol{c}_{J,k}$ and $\{\boldsymbol{d}_{j,k}\}_{J \leq j < 0}$. Recursive multiwavelet reconstruction from the decomposed data set gives the original coefficient vector $\boldsymbol{c}_{0,k}$. Then $\boldsymbol{x}(z)$ is reconstructed by applying a *postfilter* $\mathbf{P}(z)$:

$$\boldsymbol{x}(z) = \mathbf{P}(z)\boldsymbol{c}_0(z) \tag{45}$$

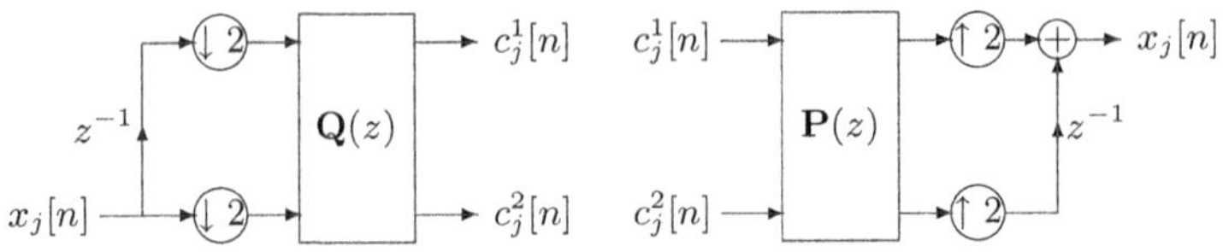

(a) Prefilter (b) Postfilter

Fig. 5. Prefilter and postfilter blocks. A unit delay and downsampling in a prefilter block (a) vectorize the 1-D input data sequence $x_j[n]$ to a vector-valued sequence, where the prefilter output $[c_j^1[n]\ c_j^2[n]]$ is the input to multiwavelet decomposition filter banks. A unit delay and upsampling in a postfilter block (b) serialize the two-channel postfilter output vector sequence to the 1-D output signal $x_j[n]$, where $[c_j^1[n]\ c_j^2[n]]$ are from the outputs of multiwavelet reconstruction filter banks

The postfilter $\mathbf{P}$ must be an inverse of the prefilter $\mathbf{Q}$ up to some unit delays for the perfect reconstruction:

$$\mathbf{P}(z)\mathbf{Q}(z) = z^{-l}\mathbf{I}, \quad \text{for some integer } l. \tag{46}$$

We may assume $l = 0$ (no delay) for convenience.

Define

$$\boldsymbol{x}_0(z) := \boldsymbol{x}(z) \quad \text{and} \quad \boldsymbol{x}_j(z) := \mathbf{P}(z)\boldsymbol{c}_j(z). \tag{47}$$

Then $\{\boldsymbol{x}_j\}_{j<0}$ are the projections of $\boldsymbol{x}$ into (discrete-time) multiscaling spaces at lower resolutions. This implies that a postfilter should be applied to a coefficient vector $\boldsymbol{c}_j$ if we want to see a decomposed signal at the resolution level $j < 0$.

For an r-channel multiwavelet system, the construction of a vector-valued input sequence from an 1-D signal can be implemented in a prefilter block by serial-to-parallel conversion (vectorization) by using $r-1$ unit delays and then downsampling each channel at the rate r. The block diagrams of a prefilter and a postfilter blocks for a 2-channel multiwavelet system are shown in Figure 5.

5.2 Interpolation Prefilter and Postfilter

In the multiwavelet case, in order to avoid the undesirable visual effect, we need a prefilter that computes multiscaling coefficient sequence $\mathbf{c}_{0,k}$ from a discrete-time input signal before starting the multiwavelet decomposition[11,12,13]. In this section, we develop a process of finding a pair of prefilter and postfilter such that

$$f_0(t) := \sum_k \boldsymbol{c}_{0,k}^T \boldsymbol{\phi}(t-k) \tag{48}$$

interpolates an original signal $x_0[n]$. Since we have r scaling functions, a continuous-time signal $f_0(t)$ is sampled at the interval of $1/r$ at the 0-th resolution level:

$$f_0(\frac{n}{r}) = \sum_{k\in\mathbb{Z}} \boldsymbol{c}_{0,k}^T\, \boldsymbol{\phi}(\frac{n}{r}-k) = \sum_{k\in\mathbb{Z}} \boldsymbol{\phi}(\frac{n}{r}-k)^T\, \mathbf{c}_{0,k}, \tag{49}$$

and we impose an interpolation property by $f_0(\frac{n}{r}) = x_0[n]$. We construct vector-valued sequences $\boldsymbol{f}_0[n]$ and $\boldsymbol{x}_0[n]$ from the sampled sequence $f_0(n/r)$ and the 1-D signal $x_0[n]$, respectively:

$$\boldsymbol{f}_0[n] := \begin{bmatrix} f_0(n) \\ f_0(n+\frac{1}{r}) \\ \vdots \\ f_0(n+\frac{r-1}{r}) \end{bmatrix}, \qquad \boldsymbol{x}_0[n] := \begin{bmatrix} x_0[nr] \\ x_0[nr+1] \\ \vdots \\ x_0[nr+r-1] \end{bmatrix}, \tag{50}$$

then the interpolation condition $f_0(n/r) = x_0[n]$ gives the following relation:

$$\boldsymbol{f}_0[n] = \boldsymbol{x}_0[n] = \sum_{k\in\mathbb{Z}} P_{n-k}\, \boldsymbol{c}_0[k] = \sum_{k\in\mathbb{Z}} P_k\, \boldsymbol{c}_0[n-k], \tag{51}$$

where P_n is an $r \times r$ matrix sequence and defined by

$$P_n := \begin{bmatrix} \boldsymbol{\phi}(n)^T \\ \boldsymbol{\phi}(n+\frac{1}{r})^T \\ \vdots \\ \boldsymbol{\phi}(n+\frac{r-1}{r})^T \end{bmatrix}. \tag{52}$$

This is an interpolation postfilter that maps the space of scaling coefficients $\boldsymbol{c}_j[k]$ to the space of sampled signals $\boldsymbol{f}_j[n]$. At any resolution level j, a decomposed signal can be obtained by filtering scaling coefficients $\boldsymbol{c}_j[k]$ by the postfilter P_n:

$$\boldsymbol{x}_j[n] = \sum_{k\in\mathbb{Z}} P_{n-k}\, \boldsymbol{c}_j[k] = \sum_{k\in\mathbb{Z}} P_k\, \boldsymbol{c}_j[n-k]. \tag{53}$$

This relation is expressed in the Z-transform domain as

$$\boldsymbol{x}_j(z) = \mathbf{P}(z)\boldsymbol{c}_j(z), \tag{54}$$

where $\mathbf{P}(z) := \sum_n P_n z^{-n}$. By (52), P_n is a finite sequence (FIR filter) if the scaling vector function $\boldsymbol{\phi}$ is compactly supported.

We define a prefilter $\mathbf{Q}(z)$ such that

$$\mathbf{Q}(z)\mathbf{P}(z) = \mathbf{P}(z)\mathbf{Q}(z) = \mathbf{I}_r. \tag{55}$$

Then the scaling coefficient $\boldsymbol{c}_j(z)$ is obtained by filtering the signal $\boldsymbol{x}_j(z)$:

$$\boldsymbol{c}_j(z) = \mathbf{Q}(z)\boldsymbol{x}_j(z). \tag{56}$$

To have an FIR solution to the above condition (55), $\det(\mathbf{P}(z))$ must have the form of $\det(\mathbf{P}(z)) = \alpha z^{-l}$, where α is a constant and l is an integer.

For the GHM orthogonal multiwavelet system, an interpolation postfilter $\mathbf{P}$ is obtained from the GHM scaling functions (Figure 1(a) & (b)):

$$\mathbf{P}(z) = \begin{bmatrix} 0 & 1.73210618015z^{-1} \\ 1.95965444133 & -0.519631854046 - 0.519631854046z^{-1} \end{bmatrix} \tag{57}$$

The corresponding prefilter $\mathbf{Q}$ is computed from the condition $\mathbf{P}(z)\mathbf{Q}(z) = \mathbf{I}$,

$$\mathbf{Q}(z) = \mathbf{P}^{-1}(z) = \begin{bmatrix} 0.1530923245z + 0.1530923245 & 0.5103077369 \\ 0.5773517497z & 0. \end{bmatrix} \tag{58}$$

For the cardinal 2-balanced (also 3-balanced or 4-balanced) orthogonal multiwavelet system, we obtain the postfilter and prefilter as

$$\mathbf{P}(z) = \begin{bmatrix} 0 & \sqrt{2}z^{-2} \\ \sqrt{2}z^{-1} & 0 \end{bmatrix}, \quad \mathbf{Q}(z) = \begin{bmatrix} 0 & z/\sqrt{2} \\ z^2/\sqrt{2} & 0 \end{bmatrix}. \tag{59}$$

The biorthogonal Hermite cubic multiwavelet system does not give a stable prefilter for an interpolation postfilter. In this case, we need to design a different pair of prefilter and postfilter for those systems. One possible solution is to design an orthogonal prefilter.

5.3 Orthogonal Prefilter

A prefilter $\mathbf{Q}(z) := \sum_n Q_n z^{-n}$ is said to be *orthogonal* if

$$\|Q * \boldsymbol{c}\| = \|\boldsymbol{c}\| \tag{60}$$

for all $\boldsymbol{c} \in \ell^2(\mathbb{Z})^r$, where Q is an impulse response (a sequence of $r \times r$ matrices) of $\mathbf{Q}(z)$ and $*$ denotes a discrete (matrix) convolution operator. The above condition $\|\mathbf{Q}(z)\boldsymbol{c}(z)\| = \|\boldsymbol{c}(z)\|$ is equivalent to the paraunitary condition of $\mathbf{Q}(z)$:

$$\mathbf{Q}(z)\mathbf{Q}(z^{-1})^T = \mathbf{I}. \tag{61}$$

An FIR filter $\mathbf{Q}(z)$ is paraunitary if and only if it is of the form

$$\mathbf{Q}(z) = \mathbf{Q}(1) \prod_{i=1}^{N} (\mathbf{I} - P_i + P_i z^{\epsilon_i}), \tag{62}$$

where $\mathbf{Q}(1)$ is an orthogonal (unitary) matrix, $\epsilon_i = \pm 1$, and P_i for $i = 1, ..., N$ are orthogonal (unitary) matrices [8]. Higher approximation orders will give quite complex relations, so here we consider a prefilter only up to the approximation order 2. Then, for a minimal filter length ($N = 2$), we need to find P_1 and P_2 such that $\mathbf{Q}(z) = \mathbf{Q}(1)(\mathbf{I} - P_1 + P_1 z)(\mathbf{I} - P_2 + P_2 z)$ satisfies the above orthogonality condition. A delay factor z^{-2} may be introduced to make $\mathbf{Q}(z)$ causal. An example of orthogonal prefilter of approximation order 2 for the GHM orthogonal multiwavelet system is given by $\mathbf{Q}(z) := Q_0 + Q_1 z^{-1}$, where

$$\begin{aligned} Q_0 &= \begin{bmatrix} 0.11942337067748 & 0.99158171438258 \\ 0.04967860804828 & -0.00598315472909 \end{bmatrix}, \\ Q_1 &= \begin{bmatrix} -0.00598315472909 & -0.04967860804828 \\ 0.9915817143825 & -0.11942337067748 \end{bmatrix}. \end{aligned} \tag{63}$$

Table 1. Compression performances of wavelet systems

	PSNR [dB]						
	Multiwavelets				Single Wavelets		
	Orthogonal			Biorth.	Orthogonal		Biorth.
CR	GHM (i)	GHM (o)	CardBal2	H-Cubics	D4	D6	Bin9-7
2	48.929	47.933	48.317	44.262	47.410	48.232	49.162
4	41.012	40.500	41.327	36.964	39.483	41.233	42.388
8	36.126	35.717	37.041	32.212	34.762	36.874	38.626
16	32.259	31.887	32.922	29.004	30.956	32.568	35.481
32	28.786	28.348	29.296	26.730	27.617	28.810	31.799
64	26.031	25.590	26.070	23.847	24.991	25.532	27.535
128	23.379	23.036	23.381	21.414	22.688	23.106	24.121
256	20.566	20.572	20.785	20.453	20.559	20.672	20.712
Prefilter	Inter.	Orth.	Inter.	Orth.	N/A		

6 Compression Performances

Multiwavelet systems have been explored for applications to data compression and image processing [5,13,14,15,16]. With the prefilers and postfilters that we have designed for multiwavelet systems, we have investigated the applications of these systems to image compression and examined their compression performances. Experimental studies are performed on the level of compression performances of three multiwavelet systems (two orthogonal multiwavelets, GHM and cardinal balanced, and one biorthogonal multiwavelet, Hermite cubics) in comparison to some single wavelet systems (Daubechies' D4 and D6 orthogonal wavelets and binary 9-7 biorthogonal wavelet). We consider a simple compression scheme with a uniform quantizer, which removes a certain number of small values from highpassed subimages but keeps the larger values to achieve a specified compression ratio (CR). We used six test images (5125128-bit) of Lena, Airplane, Baboon, Peppers, Sailboat, and Wavy in our experiments. Our experiments suggested that wavelet decomposition up to the 3rd or 4th level would give a reasonably high compression ratio and a good reconstruction.

To describe the image fidelity, PSNR (peak signal-to-noise ratio) is defined by

$$\text{PSNR [dB]} := 20 \log \left(255 / \sqrt{\frac{1}{MN} \sum_{i=1}^{M} \sum_{j=1}^{M} (f[i,j] - s[i,j])^2} \right), \qquad (64)$$

where f is a $M \times N$ noisy or distorted image (decompressed or reconstructed image) and s is the $M \times N$ original image. The PSNR values shown in Table 1 are the average values taken from the experimental results for the six test images at each given compression ratio. The image compression performances of orthogonal wavelet systems are shown in Figure 6(b) and some biorthogonal wavelet systems in Figure 6(a). In orthoronal wavelet systems, multiwavelet systems perform better than single wavelet systems with comparable support lengths.

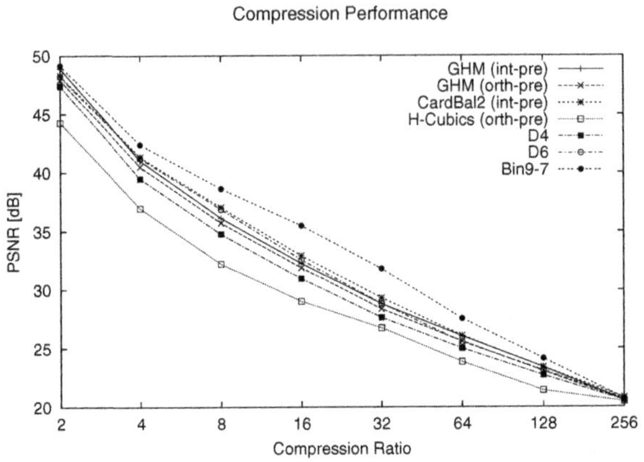

(a) Biorthogonal systems

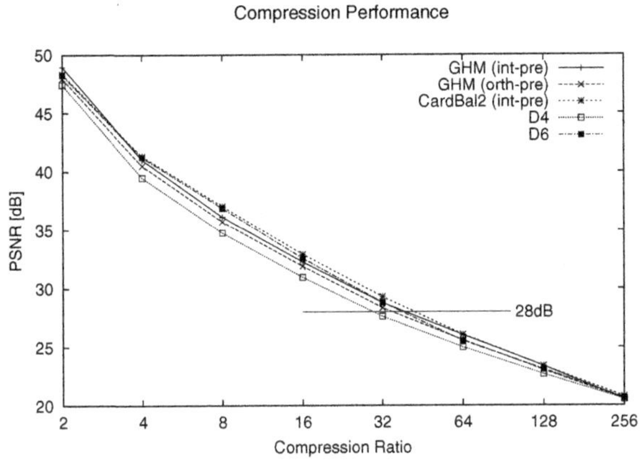

(b) Orthogonal systems

Fig. 6. Compression performances of wavelet systems

However, the binary 9-7 biorthogonal single wavelet system significantly outperforms other wavelet systems, because it has a higher order of approximation and symmetric functions. The biorthogonal Hermite cubic multiwavelet system with an orthogonal prefilter of approximation order of 2 did not give a desirable compression performance. The reason is that this orthogonal prefilter is not a good approximation to an interpolation prefilter because of its lower approximation order (only 2) while the Hermite cubics have the 4th order approximation. We have yet to find a good biorthogonal multiwavelet filters and prefilters.

7 Conclusion

In this paper, multiwavelet systems are applied to image compression. Each line of image data is vectorized for r channel inputs of a multiwavelet system. A general method of prefiltering the inputs has been formulated to provide data to the multiwavelet filter bank, which should enable the reconstruction of the original data after postfiltering. A design process for interpolation prefilter-postfilter, if exist, has been developed, which will provide a better approximation image at each coarser resolution level. These filters must be of the finite impulse response type, or else, an orthogonal prefilter of some approximation order can be designed. The prefilters and postfilters have been designed for 3 multiwavelet systems (GHM, cardinal balanced, and Hermite cubics). Using these filters, image compression performances of orthogonal multiwavelet systems have been shown to be better than those of the scalar orthogonal wavelet systems.

References

1. Chui, C. K.: An Introduction to Wavelets. Volume 1 of Wavelet Analysis and Its Applications. Academic Press (1992) 23
2. Geronimo, J. S., Hardin, D. P., Massopust, P. R.: Fractal functions and wavelet expansions based on several scaling functions. Journal of Approximation Theory **78** (1994) 373–401 23
3. Donovan, G. C., Geronimo, J., Hardin, D. P.: Intertwining multiresolution analyses and the construction of piecewise polynomial wavelets. SIAM Journal of Mathematical Analysis **27** (1996) 1791–1815 23
4. Donovan, G., Geronimo, J. S., Hardin, D. P., Massopust, P. R.: Construction of orthogonal wavelets using fractal interpolation functions. SIAM Journal of Mathematical Analysis **27** (1996) 1158–1192 23
5. Strela, V., Walden, A. T.: Orthogonal and biorthogonal multiwavelets for signal denoising and image compression. SPIE Proc. 3391 AeroSense 98, Orlando, Florida, April 1998 (1998) 24, 33
6. Selesnick, I. W.: Interpolating multiwavelet bases and the sampling theorem. IEEE Trans. on Signal Processing **47** (1999) 1615–1621 24
7. Chui, C. K., Lian, J.: A study on orthonormal multi-wavelets. J. Appl. Numer. Math. **20** (1996) 273–298 24
8. Vaidyanathan, P. P.: Multirate Systems and Filter Banks. Prentice-Hall, New Jersey (1993) 27, 32
9. Strela, V.: Multiwavelets: Theory and Applications. PhD thesis, Massachusetts Institute of Technology, Cambridge, Mass. (1996) 28
10. Plonka, G., Strela, V.: Construction of multiscaling functions with approximation and symmetry. SIAM Journal of Mathematical Analysis **29** (1998) 481–510 28
11. Xia, X. G., Geronimo, J. S., Hardin, D. P., Suter, B. W.: Design of prefilters for discrete multiwavelet transforms. IEEE Trans. on Signal Processing **44** (1996) 25–35 30
12. Xia, X. G.: A new prefilter design for discrete multiwavelet transforms. IEEE Trans. on Signal Processing **46** (1998) 1558–1570 30
13. Miller, J. T., Li, C. C.: Adaptive multiwavelet initialization. IEEE Trans. on Signal Processing **46** (1998) 3282–3291 30, 33

14. Xia, T., Jiang, Q.: Optimal multifilter banks: design, related symmetric extension transform and application to image compression. IEEE Trans. on Signal Processing **47** (1999) 1878–1889 33
15. Jiang, Q.: On the design of multifilter banks and orthogonal multiwavelet bases. IEEE Trans. on Signal Processing **46** (1998) 3292–3303 33
16. Strela, V., Heller, P., Strang, G., Topiwala, P., Heil, C.: The application of multiwavelet filter banks to image processing. IEEE Trans. on Image Processing **8** (1999) 548–563 33

Reduction of Blocking Artifacts in Both Spatial Domain and Transformed Domain

Wing-kuen Ling and P. K. S. Tam

Department of Electronic and Information Engineering
The Hong Kong Polytechnic University
Hung Hom, Kowloon, Hong Kong
Hong Kong Special Administrative Region, China
Tel: (852) 2766-6238, Fax: (852) 2362-8439
Email: bingo@encserver.eie.polyu.edu.hk

Abstract. In this paper, we propose a bi-domain technique to reduce the blocking artifacts commonly incurred in image processing. Some pixels are sampled in the shifted image block and some high frequency components of the corresponding transformed block are discarded. By solving for the remaining unknown pixel values and the transformed coefficients, a less blocky image is obtained. Simulation results using the Discrete Cosine Transform and the Slant Transform show that the proposed algorithm gives a better quantitative result and image quality than that of the existing methods.

1 Introduction

Many images are very large in size, and so it typically requires an extensive computation to process a whole image. Hence, dividing an image into a number of small blocks with size *8x8* for processing is very common in practice, such as that employed in the JPEG, MPEG-1/2, H.261/263 standards. However, the block-based coded images suffer from a kind of distortion, called blocking artifacts, especially when the compression ratio is high. There are boundaries among the blocks and these boundaries are very disturbing.

Several techniques have been developed to reduce the blocking artifacts: The theory of projection onto convex set (POCS) has been proposed [1], but it requires a large number of iterations for convergence. Methods using interleaved image blocks before the encoding were also suggested [2], but they are not in conformity with the coding standards. Lowpass filtering over the block-based coded images were also proposed [3], but it may cause serious distortions when the image contains high frequency components. Some adaptive filtering approaches have been proposed [4], but the cost is too high.

In this paper, we propose a bi-domain de-blocking technique, which samples the shifted image block at certain fixed locations and discards some high frequency components of the corresponding transformed block. By solving for the remaining unknown pixel values and the transformed coefficients, a less blocky image is obtained. This idea has been carried out for the Discrete Cosine Transform (DCT) and the Slant Transform. It is found that the proposed bi-domain de-blocking technique reduces blocking artifacts effectively.

Y. Y. Tang et al. (Eds.): WAA 2001, LNCS 2251, pp. 37-43, 2001.

2 Vector Representation of an Image Transform

In the application of transform techniques to image processing, a linear separable orthonormal block transform with block size *8x8* can be expressed as $\mathbf{Y}=\mathbf{F}\cdot\mathbf{X}\cdot\mathbf{F}^T$, where **X** is the *8x8* image block, **Y** is the *8x8* transformed block and **F** is the *8x8* transformed matrix. Every element in the matrix **Y** can be expressed as:

$$y_{pq} = \sum_{m=1}^{8}\sum_{n=1}^{8} f_{pk}\cdot x_{km}\cdot f_{qm}, \tag{1}$$

where x_{km} is at the kth row and m^{th} column of matrix **X**, f_{ij} is at the ith row and j^{th} column of matrix **F** and y_{pq} is at the pth row and q^{th} column of matrix **Y**.

Since y_{pq} is a linear combination of x_{km}, we can express it as follows:

$$\begin{bmatrix} y_{11} \\ \vdots \\ y_{pq} \\ \vdots \\ y_{88} \end{bmatrix} = \begin{bmatrix} f_{11}\cdot f_{11} & \cdots & f_{1m}\cdot f_{1n} & \cdots & f_{18}\cdot f_{18} \\ \vdots & \ddots & \vdots & \cdot^{\cdot^{\cdot}} & \vdots \\ f_{p1}\cdot f_{q1} & \cdots & f_{pm}\cdot f_{qn} & \cdots & f_{p8}\cdot f_{q8} \\ \vdots & \cdot^{\cdot^{\cdot}} & \vdots & \ddots & \vdots \\ f_{81}\cdot f_{81} & \cdots & f_{8m}\cdot f_{8n} & \cdots & f_{88}\cdot f_{88} \end{bmatrix} \cdot \begin{bmatrix} x_{11} \\ \vdots \\ x_{mn} \\ \vdots \\ x_{88} \end{bmatrix}. \tag{2}$$

The above equation is a vector representation of an image transform from an image vector **x** to a transformed vector **y** with $\mathbf{y}=\mathbf{C}\cdot\mathbf{x}$, where **C** is a *64x64* matrix and **x** is a *64x1* column vector. Similarly, the inverse transform can be represented as $\mathbf{x}=\mathbf{T}\cdot\mathbf{y}$ with $\mathbf{T}=\mathbf{C}^{-1}$, where **T** is a *64x64* matrix and **y** is a *64x1* column vector.

3 Blocking Effect Model

By shifting the block-based coded image four pixels both horizontally and vertically, the visible edge is at the middle of the shifted image block. This shifted image block can be modeled using four *4x4* matrices [5] as follows:

$$X = \begin{bmatrix} a_{4x4} & b_{4x4} \\ c_{4x4} & d_{4x4} \end{bmatrix}. \tag{3}$$

If the compression ratio of the block-based coder is too high that all the AC coefficients of the block-based coded image are quantized to zero, then the four matrices of the shifted image block become four constant matrices. Consequently, we only have four different pixel values (a, b, c and d) in the shifted coded image block.

4 Bi-domain De-blocking Algorithm

Let x_{ij}^{old} be the ith row and j^{th} column of the unprocessed shifted image block and y_{ij}^{old} be the ith row and j^{th} column of the corresponding transformed block. Similarly, let x_{ij}^{new} be the ith row and j^{th} column of the processed shifted image block and y_{ij}^{new} be the ith row and j^{th} column of the corresponding transformed block.

For the ideal terrace image block, that is, there are only four different pixel values in the shifted image block, the minimum number of the sampling points in the pixel domain is four. In order to reduce the blocking artifacts, those sampling points should be sampled as far to the block edge as possible. Hence, we sample at the corners of the shifted image block and we have $x_{11}^{old}=x_{11}^{new}$, $x_{18}^{old}=x_{18}^{new}$, $x_{81}^{old}=x_{81}^{new}$, $x_{88}^{old}=x_{88}^{new}$. For the other pixel values, we let them to be unknown at this stage and to be determined in the next stage.

In the transformed domain, some coefficients, especially the high frequency components, suffer from the blocking artifacts because the block edge always contains high frequency components. Hence, in order to reduce the blocking artifacts, we should discard the high frequency components, that is, setting $y_{ij}^{new}=0$ for some i, j. For the remaining transformed coefficients, we let them to be unknown at this stage.

According to the sampling scheme mentioned above, we can pick up the corresponding rows in equation (2) and break down the matrix multiplication into a sum of two matrix multiplications, as follows:

$$\begin{bmatrix} x_{11} \\ x_{18} \\ x_{81} \\ x_{88} \end{bmatrix} = \begin{bmatrix} \cdots & t_{1,8\cdot(p-1)+q} & \cdots \\ \cdots & t_{8,8\cdot(p-1)+q} & \cdots \\ \cdots & t_{57,8\cdot(p-1)+q} & \cdots \\ \cdots & t_{64,8\cdot(p-1)+q} & \cdots \end{bmatrix} \cdot \begin{bmatrix} \vdots \\ y_{pq} \\ \vdots \end{bmatrix} + \begin{bmatrix} \cdots & t_{1,8\cdot(m-1)+n} & \cdots \\ \cdots & t_{8,8\cdot(m-1)+n} & \cdots \\ \cdots & t_{57,8\cdot(m-1)+n} & \cdots \\ \cdots & t_{64,8\cdot(m-1)+n} & \cdots \end{bmatrix} \cdot \begin{bmatrix} \vdots \\ y_{mn} \\ \vdots \end{bmatrix}, \tag{4}$$

where $t_{i,j}$ is at the ith row and j^{th} column of the matrix $\mathbf{T}$. The above equation can be expressed in the form of $\mathbf{x}_1=\mathbf{S}_1\cdot\mathbf{y}_1+\mathbf{S}_2\cdot\mathbf{y}_2$, where $\mathbf{y}_1$ is a vector containing the low frequency components, $\mathbf{y}_2$ is a vector containing the high frequency components, and $\mathbf{x}_1$ is the vector containing the four corners of the shifted image block. $\mathbf{S}_1$ and $\mathbf{S}_2$ are the corresponding matrices.

Since we keep four pixel values unchanged after processing, we can set up four linear independent equations in the spatial domain. In order to find out the unknown pixel values, we need to set sixty transformed coefficients to zero ($\mathbf{y}_2^{\mathbf{new}}=\mathbf{0}$) and find out the remaining four transformed coefficients ($\mathbf{y}_1^{\mathbf{new}}$). The detail procedure is as follows:

Since $\mathbf{x}_1$ remains unchanged after processing, we have $\mathbf{x}_1^{\mathbf{new}}=\mathbf{x}_1^{\mathbf{old}}$. As we discard the high frequency components of the processed shifted image block, so we have $\mathbf{y}_2^{\mathbf{new}}=\mathbf{0}$. This implies that $\mathbf{x}_1^{\mathbf{new}}=\mathbf{x}_1^{\mathbf{old}}=\mathbf{S}_1\cdot\mathbf{y}_1^{\mathbf{new}}=\mathbf{S}_2\cdot\mathbf{y}_1^{\mathbf{old}}+\mathbf{S}_2\cdot\mathbf{y}_2^{\mathbf{old}} \Rightarrow \mathbf{y}_1^{\mathbf{new}}=\mathbf{S}_1^{-1}\cdot\mathbf{x}_1^{\mathbf{old}}$. Figure 1 shows the block diagram of the proposed algorithm.

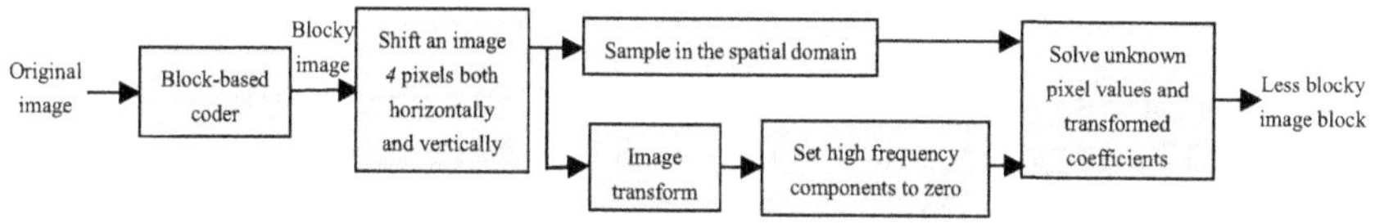

Fig. 1. Block diagram of the proposed de-blocking technique

The idea is applied to the Discrete Cosine Transform (DCT) and the Slant Transform (ST) as follows:

4.1 Discrete Cosine Transform (DCT)

It can be shown that the only non-zero DCT coefficients for the ideal terrace image block are located at the positions (p,q) for $p=1,2,4,6,8$ and $q=1,2,4,6,8$. The lowest four frequency components are at the positions $(1,1)$, $(1,2)$, $(2,1)$ and $(2,2)$. Hence, we set $\mathbf{y}_1^{\mathbf{new}}=[y_{11}^{new}\ y_{12}^{new}\ y_{21}^{new}\ y_{22}^{new}]^{\mathrm{T}}$. As $\mathbf{x}_1^{\mathbf{new}}=\mathbf{x}_1^{\mathbf{old}}=[a\ b\ c\ d]^{\mathrm{T}}$ and $\mathbf{y}_1^{\mathbf{new}}=\mathbf{S}_1^{-1}\cdot\mathbf{x}_1^{\mathbf{old}}$. It can be shown that $y_{11}^{new}=2*(a+b+c+d)$, $y_{12}^{new}=1.4419*(a-b+c-d)$, $y_{21}^{new}=1.4419*(a+b-c-d)$ and $y_{22}^{new}=1.0396*(a-b-c+d)$. By doing the IDCT, that is, $\mathbf{X}^{\mathbf{new}}=\mathbf{F}^{\mathrm{T}}\cdot\mathbf{Y}^{\mathbf{new}}\cdot\mathbf{F}$, where $\mathbf{F}$ is the DCT matrix, and expanding this IDCT equation, it can be shown that the pixel values in the reconstructed image block is

$x_{km}^{new}=2*(a+b+c+d)*f_{1m}*f_{1k}+1.4419*(a-b+c-d)*f_{2m}*f_{1k}+1.4419*(a+b-c-d)*f_{1m}*f_{2k}+1.0396*(a-b-c+d)*f_{2m}*f_{2k}$, for $k=1,2,...,8$ and $m=1,2,...,8$. This equation can be expressed in the form of $\mathbf{X}^{\mathbf{new}}=a*\mathbf{A}+b*\mathbf{B}+c*\mathbf{C}+d*\mathbf{D}$, where **A**, **B**, **C** and **D** are constant matrices and image independent and are shown in figure 2 diagrammatically. These four matrices can be viewed as the interpolation matrices. Since these interpolation matrices are smooth, so the reconstructed image block is also smooth and the blocking artifacts are reduced.

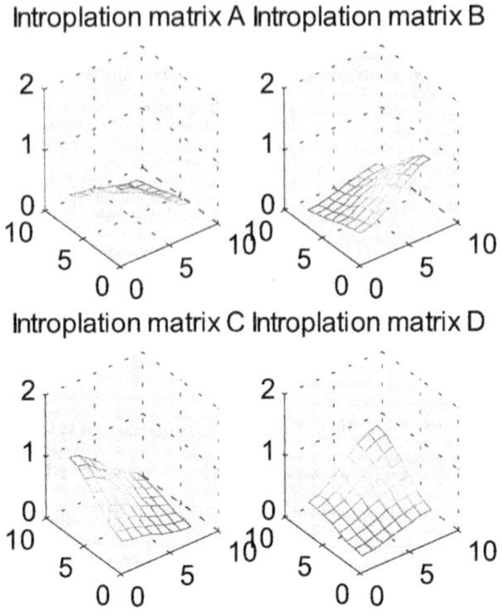

Fig. 2. Interpolation matrices for DCT de-blocking technique

4.2 Slant Transform (ST)

Similarly, it can be shown that the transformed coefficients for the ideal terrace image block are non-zero only at the positions (p,q) for $p=1,2,6$ and $q=1,2,6$. Hence, we can set $\mathbf{y}_1^{\mathbf{new}}=[y_{11}^{new}\ y_{12}^{new}\ y_{21}^{new}\ y_{22}^{new}]^T$ and set the remaining sixty coefficients to zero ($\mathbf{y}_2^{\mathbf{new}}=\mathbf{0}$), and use the same method as before to find $\mathbf{X}^{\mathbf{new}}$. It can be shown that $y_{11}^{new}=2*(a+b+c+d)$, $y_{12}^{new}=1.3093*(a-b+c-d)$, $y_{21}^{new}=1.3093*(a+b-c-d)$ and $y_{22}^{new}=6/7*(a-b-c+d)$. The reconstructed image block ($\mathbf{X}^{\mathbf{new}}$) is now interpolated by four new constant matrices **A'**, **B'**, **C'** and **D'** as shown in figure 3.

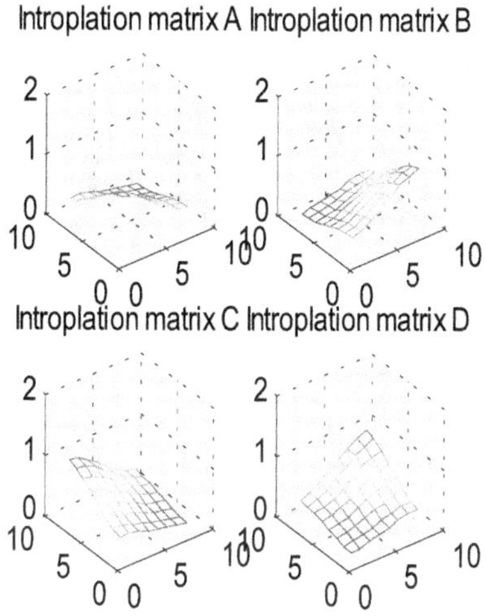

Fig. 3. Interpolation matrices for Slant de-blocking technique

5 Simulation Results

The DCT de-blocking technique and the Slant de-blocking technique are applied to the JPEG-coded images "Tiffany", "Cancer" and "Woman" of size *512x512* adaptively. The effectiveness of the proposed algorithm is estimated by both quantitative measurement and qualitative evaluation.

For the quantitative measurement, the blocking artifact is mainly due to the grid noise in the monotone areas. The intensity of the monotone areas of a natural image usually changes very slowly, but there is a tendency for the intensity in the block-based coded image to change abruptly from one block to another as modeled in the section III. Therefore, we propose the following methodology to measure the quantitative result:

If the four neighbor *8x8* image blocks are all DC blocks, that is, all the pixel values in the individual block are constant, then we sum up the error square in these four blocks, and compute the mean square error (*MSE*) of all these blocks as follows:

$$MSE = \frac{1}{N} \cdot \sum_{(i,j)\in Q} [R(i,j) - O(i,j)]^2 , \quad (5)$$

where **O** is the original image, **R** is the reconstructed image, **Q** is the region where there are four neighbor *8x8* DC blocks and *N* is the total number of pixels in **Q**.

Table 1 shows the comparison of the results of applying some common existing methods and our proposed de-blocking techniques. It can be seen from table 1 that our proposed algorithm gives better quantitative results than that of the existing methods. The qualitative results shown in the figure 4 also show that our proposed algorithm gives a better image quality than that of the existing methods.

Fig. 4. Simulation results of the comparison of the existing methods and our proposed algorithms

	Tiffany(0.238bpp)	Cancer(0.139bpp)	Woman(0.223bpp)
JPEG coded image	32.6421	22.6819	24.8619
DCT zero-masking technique [5]	30.4607	19.2036	21.6293
DCT coefficient weighting technique [5]	29.4858	19.0579	20.9516
Bi-domain DCT de-blocking technique	26.9824	16.7301	19.908
Bi-domain Slant de-blocking technique	27.0419	16.4105	20.5486

Table 1. Simulation results of calculated MSE by applying some common existing methods and our proposed algorithms

6 Concluding Remarks

In this paper, we propose a bi-domain de-blocking technique, which samples some pixel values in the shifted image block and discards some high frequency components in the corresponding transformed block. By solving for the remaining unknown pixel values and the transformed coefficients, we obtain a less blocky image.

The proposed algorithm can be applied to the enhancement of very high compression ratio block-based coded images. The given image can be first compressed to a very high compression ratio image through the block-based coder, and then the blocky image is enhanced by the proposed algorithm. The simulation results using the Discrete Cosine Transform and the Slant Transform show that the blocking artifacts are reduced significantly.

Further research work will study the effect on the image quality of the number of sampling points and the number of discarded transformed coefficients. The positions of the sampling points and the transformed coefficients will also be considered.

Acknowledgement

The work described in this paper was substantially supported by a grant from the Hong Kong Polytechnic University with account number G-V968.

References

1. Zakhor A.: Iterative procedures for reduction of blocking effects in transform image coding. IEEE Transactions on Circuits and System for Video Technology, Vol. 2, No. 1. (1992) 91-95.
2. Malvar H. S. and Staelin D. H.: The LOT: Transform coding without blocking effects. IEEE Transactions on Acoustics, Speech, and Signal Processing, Vol. 37, No. 4. (1989) 553-559.
3. Reeve H. C. and Lim J. S.: Reduction of blocking effects in image coding. Optical Engineering, Vol. 23, No. 1. (1984) 34-37.
4. Lee Y. L., Kim H. C. and Park H. W.: Blocking effect reduction of JPEG images by signal adaptive filtering. IEEE Transactions on Image Processing, Vol. 7, No. 2. (1998) 229-234.
5. Ling W. K. and Zeng B.: A novel method for blocking effect reduction in DCT-coded images. Proceedings of the 1999 IEEE International Symposium on Circuit and System (ISCAS), Vol. 4. (1999) 46-49.

Simple and Fast Subband De-blocking Technique by Discarding the High Band Signals

Wing-kuen Ling and P. K. S. Tam

Department of Electronic and Information Engineering
The Hong Kong Polytechnic University
Hung Hom, Kowloon, Hong Kong
Hong Kong Special Administrative Region, China
Tel: (852) 2766-6238, Fax: (852) 2362-8439
Email: bingo@encserver.eie.polyu.edu.hk

Abstract. In this paper, we propose a simple and fast post-processing de-blocking technique to reduce blocking artifacts. The block-based coded image is first decomposed into several subbands. Only the low frequency subband signals are retained and the high frequency subband signals are discarded. The remaining subband signals are then reconstructed to obtain a less blocky image. The ideas are demonstrated by a cosine filter bank and a modulated sine filter bank. The simulation result shows that the proposed algorithm is effective in the reduction of blocking artifacts.

1 Introduction

Transform codecs, such as those based on the Discrete Cosine Transform (DCT), are simple codecs widely applied in the industry. However, they usually produce undesirable blocking artifacts at high compression ratios. This is because each block in an image is transformed independently, and the correlation among adjacent blocks is not exploited. Thus, at a high compression ratio, quantization errors lead to blocking artifacts.

In order to tackle this problem, the lapped transform before encoding was proposed to capture the correlation information among the adjacent blocks [8]. However, this pre-processing technique requires a decrease of compression ratio and so it is not adopted in the international standard. Some subband de-blocking techniques [2, 3, 4] have also been proposed, but they are too complex in terms of implementation and computation.

In this paper, we propose a simple and fast post-processing subband de-blocking technique, which discards some high band signals and retains the remaining low band signals. The algorithm is tested by the cosine filter banks and the modulated sine filter banks. The simulation results show that this algorithm can suppress the blocking artifact effectively in both the quantitative measurement and the qualitative evaluation.

2 De-blocking System

Since a block-based transform and a lapped transform can be viewed as a discrete time linear time periodic varying system,

Y. Y. Tang et al. (Eds.): WAA 2001, LNCS 2251, pp. 44-48, 2001.

it can be realized by a filter bank structure [1]. Due to the fact that block edges always contain high frequency components [5], we propose to retain the low frequency band signals and discard the high band frequency signals.

The more low band signals are retained, the more block boundaries will be captured in the reconstructed image. However, the image details will be destroyed if we only keep a very little subset of the subband signals. We have conducted an intensive simulation and found that the best performance corresponds to retain two subband signals and discard the remaining high band signals.

The block diagram of the subband de-blocking system is shown in figure 1. There are many ways to select the analysis filters, $h_j[n]$, for $j=0,1,..,M$, and the synthesis filters, $f_j[n]$, for $j=0,1,..,M$, where the quantizers are designed as $Q_j(x)=x$, for $j=0,1$, and $Q_j(x)=0$, for $j=2,3,..,M$. The design of the filters should give a perfect reconstruction system when the quantizers are removed. This is because the error introduced due to the filter bank structure is illuminated in the perfect reconstruction system. In this paper, a cosine filter bank [6] and a modulated sine filter bank [7] are selected to demonstrate this idea.

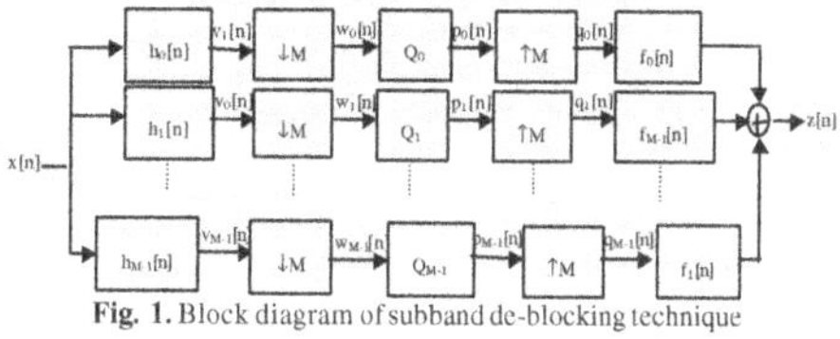

Fig. 1. Block diagram of subband de-blocking technique

2.1 Cosine Filter Bank

The impulse responses of the synthesis filters, $f_j[n]$, for $j=0,1,..,7$, are the transform basis functions of the DCT and the impulse responses of the analysis filters, $h_j[n]$, for $j=0,1,..,7$, are equal to the time-reversed basis functions [6] as follows:

$$h_j[n]=\alpha_j \cdot \cos\left(\frac{\pi \cdot j \cdot (2 \cdot n+1)}{16}\right), \tag{1}$$

for $j=0,1,..,7$ and for $n=0,1,..,7$, where:

$$\alpha_j = \begin{cases} \frac{1}{\sqrt{8}} & ; j=0, \\ \frac{1}{2} & ;\text{otherwise}, \end{cases}$$

$$f_j[n]=\alpha_j \cdot \cos\left(\frac{\pi \cdot j \cdot (15-2 \cdot n)}{16}\right), \tag{2}$$

for $j=0,1,..,7$ and for $n=0,1,..,7$.

2.2 Modulated Sine Filter Bank

The modulated sine filter bank is similar to the cosine filter bank except that the impulse responses of the synthesis filters, $f_j[n]$, for $j=0,1,..,7$, are the transform basis functions of the modulated sine transform and the impulse responses of the analysis filters, $h_j[n]$, for $j=0,1,...,7$, are equal to its time-reversed basis functions [7] as follows:

$$h_j[n]=\frac{1}{2} \cdot \sin\left(\frac{\pi}{16} \cdot \left(\frac{15}{2}-n\right)\right) \cdot \cos\left(\frac{\pi}{8} \cdot \left(j+\frac{1}{2}\right)\left(\frac{23}{2}-n\right)\right), \tag{3}$$

for $j=0,1,..,7$ and for $n=0,1,..,7$,

$$f_j[n]=\frac{1}{2}\cdot\sin\left(\frac{\pi}{16}\cdot\left(n+\frac{1}{2}\right)\right)\cdot\cos\left(\frac{\pi}{8}\cdot\left(j+\frac{1}{2}\right)\cdot\left(n+\frac{9}{2}\right)\right), \tag{4}$$

for $j=0,1,..,7$ and for $n=0,1,..,7$.

3 Simulation Results

The proposed de-blocking technique is applied to the JPEG-coded image "Cancer" of size *512x512* adaptively. The effectiveness of the proposed algorithm can be estimated by both the quantitative measurement and the qualitative evaluation.

For the quantitative measurement, the blocking artifact is mainly due to the grid noise in the monotone areas. Since the intensity of the monotone areas of most natural image change very slowly, but there is a tendency for the intensity in the block-based coded image to change abruptly from one block to another, we propose the following methodology to measure this effect:

If the four neighbor *8x8* image blocks are all DC blocks, that is, all the pixel values in the individual blocks are constant, then we sum up the error square in these four blocks, and finally we compute the mean square error (*MSE*) of all these blocks as follows:

$$MSE=\frac{1}{N}\cdot\sum_{(i,j)\in Q}[R(i,j)-O(i,j)]^2, \tag{5}$$

where **O** is the original image, **R** is the reconstructed image, **Q** is the region where there are four neighbor *8x8* DC blocks and *N* is the total number of pixels in **Q**.

Table 1 shows the comparison of the results of applying existing methods and our proposed de-blocking technique. It can be seen from table 1 that our proposed algorithm gives better quantitative results than that of the existing methods. The qualitative results shown in figure 2 also demonstrates that our proposed algorithm gives a better image quality than that of the existing methods.

	Cancer(0.139bpp)
JPEG coded image	22.6819
DCT zero-masking technique [5]	19.2036
DCT coefficient weighting technique [5]	19.0579
Cosine de-blocking technique	17.9483
Modulated sine de-blocking technique	18.4175

Table 1. Simulation results calculated by MSE of applying existing methods and our proposed algorithms

4 Concluding Remarks

In this paper, we have proposed a simple and fast post-processing subband de-blocking technique, which discards the high band signals and only retains the lowest two low band signals. This algorithm is tested by a cosine filter bank and a modulated sine filter bank. The simulation results show that our proposed method is very effective.

Since it adopts the existing transform codec and do not affect the compression ratio, the proposed algorithm can be applied to the enhancement of very high compression ratio block-based coded images. The given image can be first compressed to a very high compression ratio image through the block-based coder, and then the blocky image is enhanced by the proposed algorithm.

Further research work will focus on the finding of the best filter bank that gives the highest coding gain.

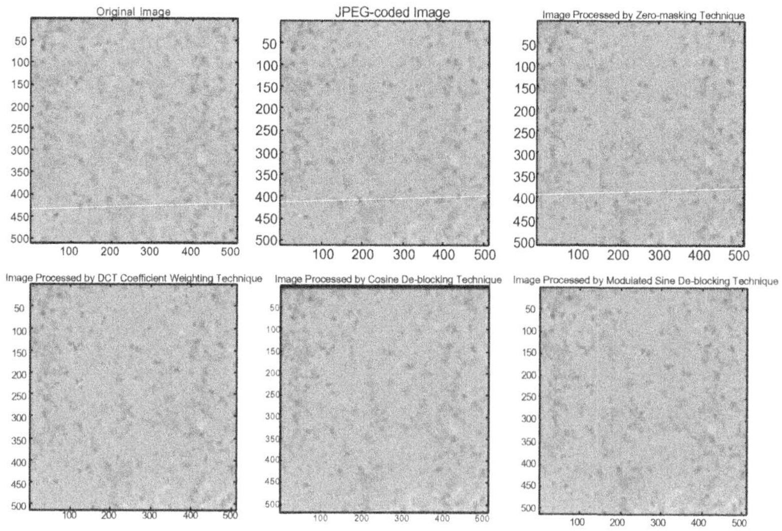

Fig. 2. Simulation results of the comparison of the existing methods and our proposed algorithms

Acknowledgement

The work described in this paper was substantially supported by a grant from the Hong Kong Polytechnic University with account number G-V968.

References

1. Malvar H. S.: Extended Lapped Transforms: Properties, Applications, and Fast Algorithms. IEEE Transactions on Signal Processing, Vol. 40, No. 11. (1992) 2703-2714.
2. Sung W. H., Chan Y. H. and Siu W. C.: Subband Adaptive Regularization Method for Removing Blocking Effect. in Proc. ICIP, Vol. 2. (1995) 523-526.
3. Rabiee H. R. and Kashyap R. L.: Image De-blocking with Wavelet-Based Multiresolution Analysis and Spatially Variant OS Filters. in Proc. ICIP, Vol. 1. (1997) 318-321.
4. Hsung T. C., Lun P. K. and Siu W. C.: A Deblocking Technique for JPEG Decoded Image Using Wavelet Transform Modulus Maxima Representation. in Proc. ICIP, Vol. 1. (1996) 561-564.
5. Ling W. K. and Zeng B.: A Novel Method for Blocking Effect Reduction in DCT-Coded Images. in Proc. ISCAS, Vol. 4. (1999) 46-49.
6. Malvar H. S.: Lapped Transforms for Efficient Transform/ Subband Coding. IEEE Transactions on Acoustics, Speech, and

Signal Processing, Vol. 38, No. 6. (1990) 969-978.

7. Malvar H. S.: Efficient Signal Coding with Hierarchical Lapped Transforms. in Proc. ICASSP, Vol. 3. (1990) 1519-1522.

8. Malvar H. S. and Staelin D. H.: The LOT: Transform coding without blocking effects. IEEE Transactions on Acoustics, Speech, and Signal processing, Vol. 37. (1989) 553-559.

A Method with Scattered Data Spline and Wavelets for Image Compression*

Guan Lütai and Lu Feng

Department of Scientific Computing and Computer Applications
Zhongshan University, Guangzhou 510275, P. R. China

Abstract. This paper presents a method for image compression. First, selecting some scattered data points on some lines of a plane to construct an interpolating spline surface approach to the image, then, one kind of wavelets for this spline function is given. By different codes to spline and wavelets, an image compression finished.

1 Introduction

In [1], we discussed spline-wavelets of plane scattered data for data compression. The basic idea is using spline interpolation first, then, by spline-wavelets for data compression.

From [3]-[7], some different multivariate spline interpolation for scattered data were given. Think of image data be in some lines, we can simplify the local support multivariate splines to scattered data in *Hilbert* space of [1] in this case.

In this paper, a method for image compression is presented. To select some scattered data points on some lines to construct an interpolating spline approach surface for the image first, then to give a spline-wavelets decomposition for the spline function using different codes to the spline and wavelets to finish image compression.

2 Polynomial Natural Spline Local Basis Interpolation for Large Scattered Data on Some Lines

Problem I: Given scattered data points on some lines of a plane $(x_i, y_{i,j}), i = 1, 2, \dots n_0; j = 1, 2, \dots m_i$ and real numbers $z_{i,j}, i = 1, 2, \dots n_0; j = 1, 2, \dots m_i$, find a function $f(x, y) \in H^{mn}(R) \bigcap D_z$ satisfying

$$J_1(f) = \min_{u \in H^{mn}(R) \bigcap D_z} J_1(u)$$

$$where : H^{mn}(R) = \{u(x, y) | \frac{\partial^{m+n}(u)}{\partial x^m \partial y^n} \in L_2(R), \frac{\partial^{\alpha+\beta}(u)}{\partial x^\alpha \partial y^\beta}$$

* This work is supported by Natural Science Foundation of Guangdong(9902275), Foundation of Zhongshan University Advanced Research Centre.

Y. Y. Tang et al. (Eds.): WAA 2001, LNCS 2251, pp. 49–53, 2001.

is an absolutely continuous function,

$$\alpha = 0, \ldots, m-1, \beta = 0, \ldots, n-1; (x,y) \in R = [a,b] \times [c,d]\},$$

$$D_z = \{u(x,y) | u(x_i, y_{ij}) = z_{ij}, i = 1, \ldots n_0, j = 1, \ldots m_i\},$$

Let: $u^{(m,n)}(x,y) = \frac{\partial^{m+n} u(x,y)}{\partial x^m \partial y^n}$,

$$J_1(u) = \iint\limits_{R^2} (u^{(m,n)}(x,y))^2 dx dy + \int_a^b \sum_{\nu=0}^{n-1} (u^{(m,\nu)}(x,c))^2 dx$$
$$+ \int_c^d \sum_{\mu=0}^{m-1} (u^{(\mu,n)}(a,y))^2 dy$$

We call the solution of this problem I natural interpolation spline function for scattered data on some lines.

Theorem 1 *A natural interpolation spline function for scattered data on some lines $f(x,y)$ has the following explicit and closed-form expression:*

$$f(x,y) = \sum_{i=1}^{n_0} \sum_{j=1}^{m_i} \alpha_{ij} G_1(x, x_i) G_2(y, y_{i,j}) + \sum_{i=0}^{m-1} \sum_{j=0}^{n-1} c_{ij} x^i y^j$$

where:

$$G_1(x,t) = \frac{(t-x)_+^{2m-1}}{(2m-1)!} + \sum_{\mu=0}^{m-1} \frac{(x-a)^\mu}{\mu!} \{ \frac{(a-t)^{2m-\mu-1}}{(2m-\mu-1)!} + (-1)^{m-\mu} \frac{(a-t)^\mu}{\mu!} \}$$

$$G_2(y,\tau) = \frac{(\tau-y)_+^{2n-1}}{(2n-1)!} + \sum_{\nu=0}^{n-1} \frac{(y-c)^\nu}{\nu!} \{ \frac{(c-\tau)^{2n-\nu-1}}{(2n-\nu-1)!} + (-1)^{n-\nu} \frac{(c-\tau)^\nu}{\nu!} \}$$

Let

$$B_{ij}(y) = \begin{vmatrix} 1 & y_{i,j} & \cdots & y_{i,j}^{2n-1} & G_2(y, y_{i,j}) \\ 1 & y_{i,j+1} & \cdots & y_{i,j+1}^{2n-1} & G_2(y, y_{i,j+1}) \\ \cdots & \cdots & \cdots & \cdots & \cdots \\ 1 & y_{i,j+2n} & \cdots & y_{i,j+2n}^{2n-1} & G_2(y, y_{i,j+2n}) \end{vmatrix},$$

if $j + 2n > m_i$, then to $0 < \varepsilon_1 < \varepsilon_2 < \cdots < \varepsilon_{2n}$, let $y_{i,m_i+1} = y_{i,m_i} + \varepsilon_1, \cdots, y_{i,m_i+2n} = y_{i,m_i} + \varepsilon_{2n}$.

We can prove that $B_{i,j}(y)$ is B-spline with knots $y_{i,j}, y_{i,j+1}, \cdots, y_{i,j+2n}$

Theorem 2 *A natural interpolation spline function for scattered data on some lines $f(x,y)$ has the following local basis explicit and closed-form expression:*

$$f(x,y) = \sum_{i=1}^{n_0} \sum_{j=1}^{m_i} \alpha_{ij} G_1(x, x_i) B_{ij}(y) + \sum_{i=0}^{m-1} \sum_{j=0}^{n-1} c_{ij} x^i y^j$$

Theorem 3 *The coefficients* $\{\alpha_{ij}\}_{i=1,\cdots,n_0,j=1,\cdots,m_i}$ *and* $\{c_{ij}\}_{i=0,\cdots,m-1,j=0,\cdots,n-1}$ *of a natural interpolation spline function for scattered data on some lines* $f(x,y)$ *can be solved by the following linear system:*

$$\begin{bmatrix} A & B \\ B^T & 0 \end{bmatrix} \begin{bmatrix} \Lambda \\ 0 \end{bmatrix} = \begin{bmatrix} Z \\ 0 \end{bmatrix}$$

Z *is a given real number set* $\{z_{ij}\}, i = 1,\cdots,n_0; j = 1,\cdots,m_i$,$\Lambda$ *is a unknown coefficient vector* $(\alpha_{ij})_{i=\overline{1,n_0};j=\overline{1,m_i}}$ *and* $C = (c_{ij})_{i=\overline{0,m-1};j=\overline{0,n-1}}$ *Elements of matrix* B *are* $b_{ij,\mu\nu} = (x_i^\mu y_{ij}^\nu)_{\mu=0,\cdots,m-1;\nu=0,\cdots,n-1;i=1,\cdots,n_0;j=1,\cdots,m_i}$;
Elements of matrix A *are* $a_{ij}^{\alpha\beta} = G_1(x_\alpha, x_i)B_{ij}(y_{\alpha,\beta}), i, \alpha = 1,\cdots,m-1; \beta = 1,\cdots,m_\alpha; j = 1,\cdots,m_i$. *and* 0 *is a zero matrix.*

3 Algorithm for Polynomial Natural Interpolation Splines on Some Lines

An algorithm for an interpolating spline surface approach to the image on some lines is given as follows:

1) Select suitable points on some lines: $(x_i, y_{ij}), i = 1,\cdots,n_0; j = 1,\cdots,m_i$. To an image with k rows and l columns ($k < l$),we use the even rows as the lines. To every line,we find the image points with suddenly changing color as our suitable points,then adding one to two points in the two image points that there exist similar colors in.
2) To $m = n = 1$ or $m = 1, n = 2$ or $m = n = 2$,compute non-zero numbers of matrix A and matrix B.
   when $m = 1, G_1(x_\alpha, x_i) = (x_i - x_\alpha)_+ + a - x_i - 1$;
   when $m = 2, G_1(x_\alpha, x_i) = \frac{(x_i - x_\alpha)_+^3}{6} + \frac{(a-x_i)^3}{6} + 1 + (x_\alpha - a)[\frac{(a-x_i)^2}{2} - a + x_i]$
   B_{ij} is a B-spline,when $n = 1$,it is a piece-wise polynomial with one degree;when $n = 2$,it is a piece-wise polynomial with three degree. We can use the following formula:

$$B_{ij}(y) = B_{ij}^n(y) = \frac{y - y_{ij}}{y_{i,j+n} - y_{i,j}} B_{ij}^{n-1}(y) + \frac{y_{i,j+n+1} - y}{y_{i,j+n+1} - y_{i,j+1}} B_{i,j+1}^{n-1}(y)$$

$$B_{ij}^0(y) = \begin{cases} 1 & y \in [y_{ij}, y_{i,j+1}] \\ 0 & otherwise \end{cases}$$

3) Using gerneralized conjugate gradient acceleration of iteration method to find the solution of the spline interpolation problem. (Theorem 3)

4 Wavelets for the Polynomial Natural Interpolation Splines on Some Lines

Just like one variate case [2] and two variate cases [1],we can define a multiresolution analysis to the polynomial natural interpolation spline on some lines. Then a theorem to the polynomial natural interpolation spline basis on some lines being the basis of scale function space is given,and its dimension is discussed.

Note $S_{2m,2n}$ be a spline space with natural spline local basis on some lines of$(2m, 2n)$ order,$S^0_{2m,2n}$ be a subspace of $S_{2m,2n}$ with zero condition on the refinement points on some lines. A similar theorem of (m, n) order differential operator $D^{m,n}$ operating $S^0_{2m,2n}$ onto the wavelet space for local basis polynomial natural spline on some lines is obtained. Then using *Lagrange* interpolation method,a wavelet basis of this wavelet space is constructed.

5 Image Compression Algorithm

By local basis polynomial natural splines on some lines and wavelets for this local basis polynomial natural splines method, an algorithm for image compression is shown as follows:

1) Using the algorithm in 3, selecting suitable points on some lines constructing local basis polynomial natural spline interpolating on some lines to approach this image data.
2) To threat these data by wavelet decomposition,these wavelets are wavelets for the local basis polynomial natural spline on some lines in 4.
3) Using different coding method to compress these data.

6 Conclusion

Notice the local support properties of B-spline,the coefficient matrix is sparse matrix,and the zero elements in the matrix are more and more than it in [1].

Acknowledgements

This work is supported by the Foundation of Zhongshan University Advanced Research Centre. This work is supported by Natural Science Foundation of Guangdong (9902275).

References

1. Lütai Guan, *Spline-wavelet of plane scattered data for data compression,* in "ICMI'99 Proceedings", Hongkong Baptist University (1999), VI-127-VI-131 (ISBN 962-85415-2-8).
2. Lütai Guan, *Spline-wavelets of free knots for signal processing,* in "Proceeding of ICSP'96", eds. Yuan Baozhong & Tang Xiaofang, IEEE Press, (1996) 311-314.
3. Chui, C. K. and L. T. Guan, Multivariate polynomial natural splines for interpolation of scattered data and other applications, in "Workshop on Computational Geometry" eds. A. Conte et al, World Scientific (1993) 77-96.

4. Lütai Guan, Bivarate polynomial natural spline for smoothing or generalized interpolation of the scattered data, Chinese J. of Num. Math & Appl. 16:1 (1994) 1-14.
5. Dong Hong, Recent progress on multivariate spline, in "Approximation Theory: in memory of A. K.Varme" eds. N. K.Govil, Marcel Dekker Inc. N. Y., 265-291 (1998).

{lilst4,jocst4}@pitt.edu,
ccl@ee.pitt.edu

chuang@cs.pitt.edu

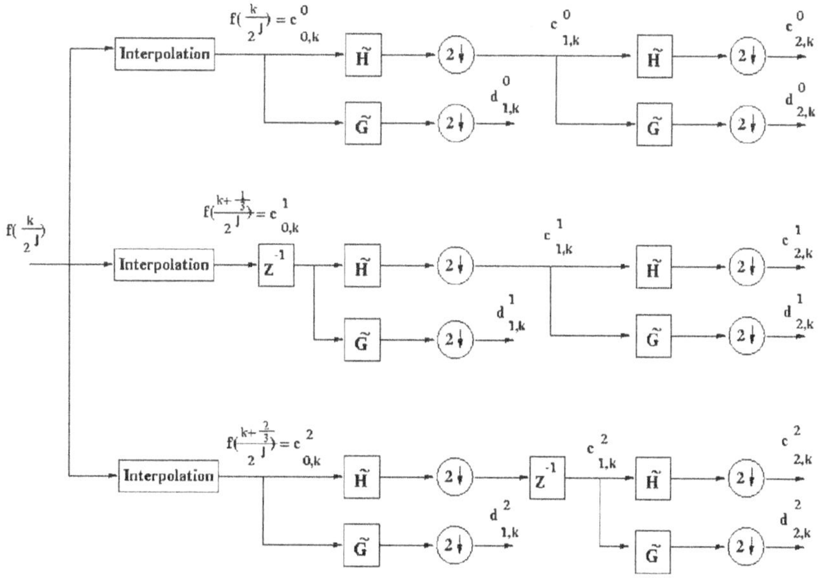

f(k/2^J)
Interpolation
f(k/2^J) = c^0_{0,k}
H̃
G̃
2↓
c^0_{1,k}
d^0_{1,k}
c^0_{2,k}
d^0_{2,k}
Interpolation
f((k+1/3)/2^J) = c^1_{0,k}
Z^-1
H̃
G̃
2↓
c^1_{1,k}
d^1_{1,k}
c^1_{2,k}
d^1_{2,k}
Interpolation
f((k+2/3)/2^J) = c^2_{0,k}
H̃
G̃
2↓
Z^-1
c^2_{1,k}
d^2_{1,k}
c^2_{2,k}
d^2_{2,k}

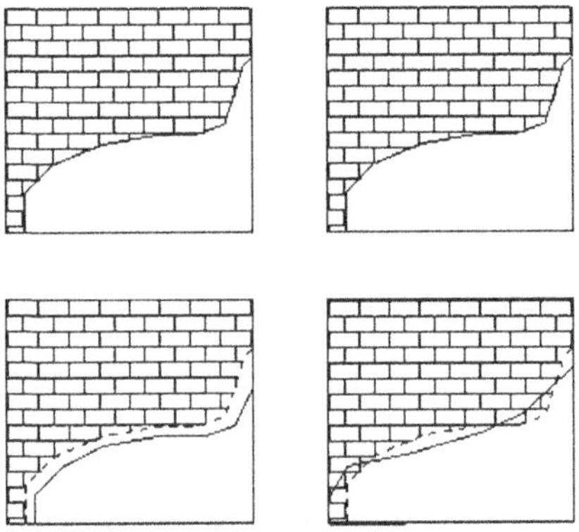

σ_i

$\sigma_i)$

= – – –

= – – –

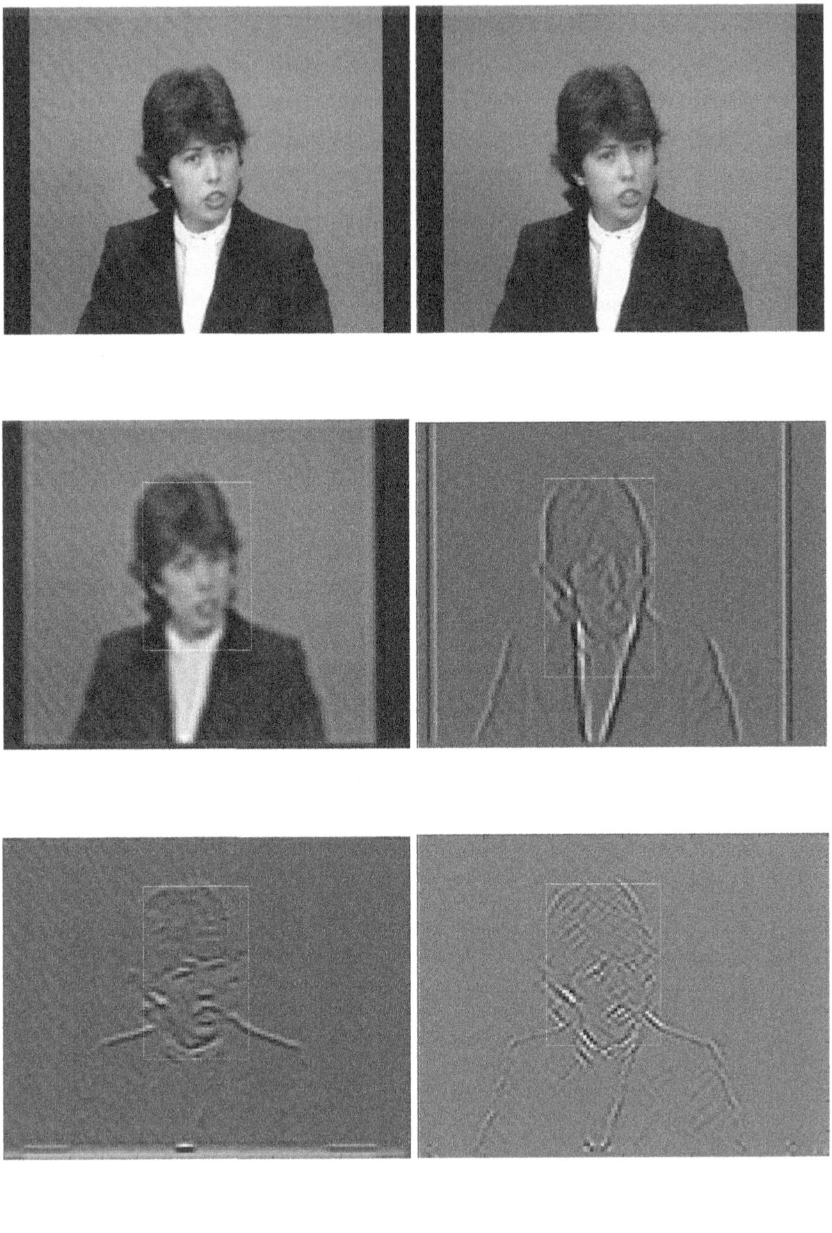

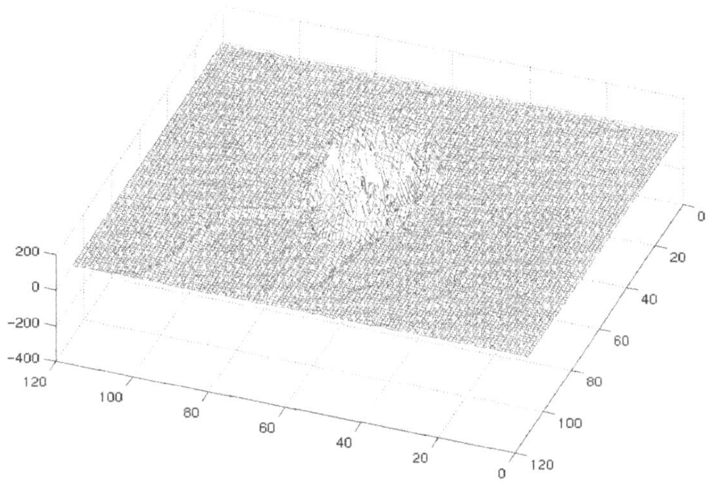
200
0
-200
-400
120
100
80
60
40
20
0
0
20
40
60
80
100
120

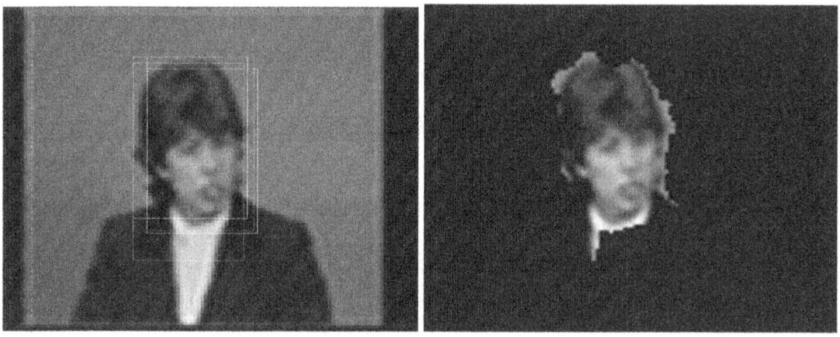

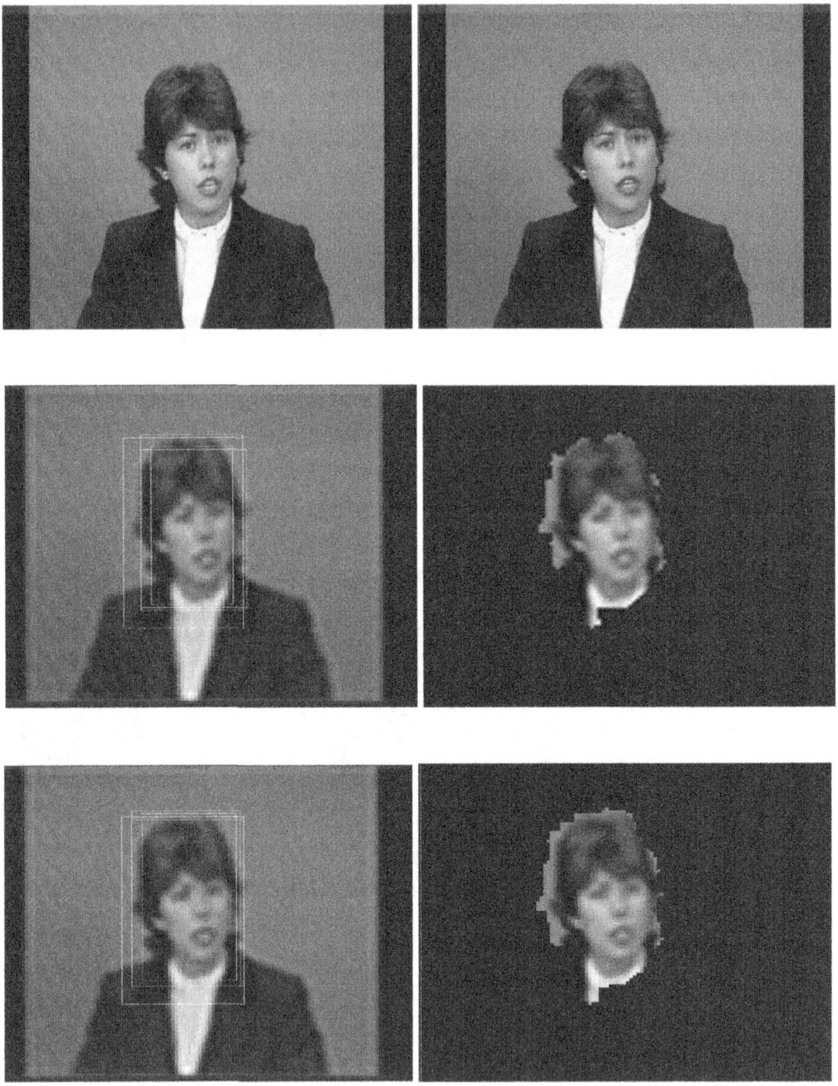

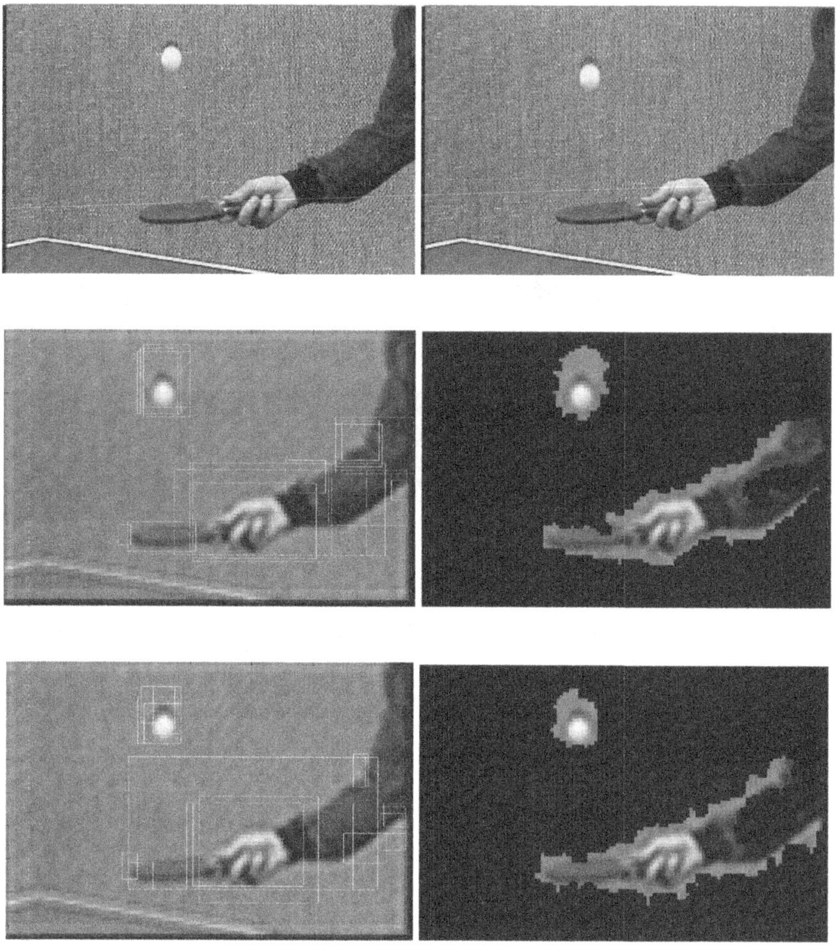

Embedded Zerotree Wavelet Coding of Image Sequence

Mbainaibeye Jérôme and Noureddine Ellouze

Laboratoire de Système et Traitement du Signal (LSTS)
Ecole Nationale d'Ingénieurs de Tunis
BP 37, Tunis le Belvédère 1002, Tél :874 700
Jerome.mbai@enit.ru.tn
N.Ellouze@enit.rnu.tn

Abstract. In this paper we present an image sequence coding system based on Embedded Zerotree Wavelet algorithm (EZW). Difference between the image in the coder and the reconstructed previous image in the decoder is used as technique for removing the temporal redundancies. The first image is encoded in intra-mode by EZW algorithm and a specific binary codebook CB1. The subsequent images in the sequence are encoded by performing the difference between the reconstructed previous image in the decoder and the current image in the coder; this difference (residual image) is then encoded by EZW algorithm and a specific binary codebook CB2. Simulations are operated on Claire and Alexis sequences. The results show that the system can provides best reconstruction quality as well objectively as subjectively for a minimum given bit rate. Progressive transmission, rate control for constant bit-rate and rate scalability are the main characteristics of this system.

1. Introduction

In multimedia applications, digital image compression is generally used for storage and transmission. MPEG1, MPEG2 and H263 are standards used in moving images coding. MPEG2 uses DCT applied in blocks of 8 x 8 pixels where motion estimation and compensation are performed. H263 uses also DCT for low bit rate. Due to the fact that image is split in blocks then the above standards produce image quality affected by block effects at low bit rate. Shapiro proposed Embedded Zerotree Wavelet algorithm (EZW) for image compression [1], which uses dependencies among wavelet subbands [2]-[6]. This coder outperforms today JPEG standard, ranging from low bit-rate to high bit-rate. Since, many developments in image compression using wavelet transform are performed [7]-[16]; improvements have been obtained by modification of EZW [10, 12, 16]. JPEG2000 is a new standard based on wavelet transform [17]. B.J.Kim and *al.* extended the Set Partitioning in Hierarchical Three (SPIHT) for video sequences [18] which exploits the energy clustering property of 3D subband/wavelet coefficients. Despite of the realization of MPEG4 and MPEG7 standards, the adoption of wavelet to video coding constitutes a special challenge. One can apply 2D wavelet coding in combination with motion compensation to temporal prediction, or one can consider the sequence as a three-dimensional array of data and making compression with 3D-wavelet analysis. These approaches present

Y. Y. Tang et al. (Eds.): WAA 2001, LNCS 2251, pp. 65-75, 2001.

some difficulties that arise from the fundamental property of discrete wavelet transform which is space-varying operator.
In this paper, we present an image sequence coding system based on EZW algorithm. Image sequences are characterized by great similarities between consecutive images (the term image in this paper is used to design frame). These similarities are known as temporal redundancies. The removing of these temporal redundancies is the key technique, which improves the compression performances. In some standards such as MPEG1-2 and H.263, this is performed by motion estimation and compensation where displaced blocks are searched and encoded to predict the current image from the previous one. In our approach, the temporal redundancies removing process is operated by calculating the difference between the consecutive images in the sequence. Discrete wavelet transform is applied on the residual images. EZW algorithm is used for the encoding process. This paper is organized as follows: in section 2, we present a short description of EZW algorithm; section 3 presents the proposed image sequence coding system. Results and discussions are presented in section 4; the conclusion is finally presented in section 5.

2. Embedded Zerotree Wavelet Coding

The EZW algorithm encodes images in embedded fashion from their dyadic wavelet representations. The goal of embedded coding is to generate a single encoded bit stream that allows achieving any desired bit rate while giving the best reconstruction quality at that rate. In wavelet domain, image is represented by approximation coefficients (called DC subband) and detail coefficients (called AC subbands). These coefficients are represented in trees. The trees are structured according to a rule such that a parent coefficient in AC subband is related to four children in the next finer AC subband for the same orientation and same spatial location. Only the parent coefficient in DC subband is related to three children, one in each of the three coarsest AC subbands. The EZW algorithm encodes these coefficients by using a sequence of thresholds. The initial value of the threshold T_0 is defined such that $C < 2T_0$ where C is the maximum wavelet coefficient. A coefficient X_i is significant if $X_i \geq T$. Significance map consists of scanning the wavelet coefficient matrix to decide whether or not a coefficient is significant and it is generated in each bit plane. Two passes are performed for each threshold value: the dominant pass and the subordinate pass. All significant coefficients found in dominant pass are encoded by four symbols which are ZTR (Zerotree Root), IZ (Isolated Zero), POS (significant Positive) and NEG (significant Negative). ZTR symbol is generated for an insignificant coefficient, which has no significant children. IZ symbol is generated for an insignificant coefficient, which has at least one significant child. POS and NEG are generated for significant coefficients which are positive and negative respectively. In finer AC subbands where the coefficients have no children, IZ and ZTR are merged into Z (zero) symbol. The subordinate pass refines the quantized coefficients to obtain the best approximation of wavelet coefficients.
EZW algorithm is particularly interesting for applications such as rate and quality scalabilities since encoder and decoder can terminate the encoding and decoding process at any time and gives a target rate or target distortion.

3. Coding of Image Sequences

A general structure of image sequence coding system is composed by encoder and decoder (figure 1). We shortly describe this system, referring to MPEG2 where the orthogonal transform is the DCT and the entropy coding is the variable length coding (VLC). In the encoder, blocks of the first image in the sequence are encoded in intra-mode without any reference. In fact, DCT is applied in blocks of 8 x 8 pixels. The quantized coefficients are then encoded using VLC coding to produce the bit stream. The subsequent images are encoded by prediction from the previous images using motion estimation and compensation technique. The motion estimation process tries to detect the displaced blocks between the current image and the previous image. These blocks are then encoded to predict the current image. Of course, for constant bit rate applications, bit rate control algorithm is used to prevent the underflow or overflow. However, MPEG does not specify the way to search the displaced blocks; this is the detail that the system designer can choose to implement in one or many possible ways. This is also the case of bit rate control algorithm where complexity versus quality issues need not to be addressed relative to individual application. The decoder performs the inverse operations accomplished by the encoder.

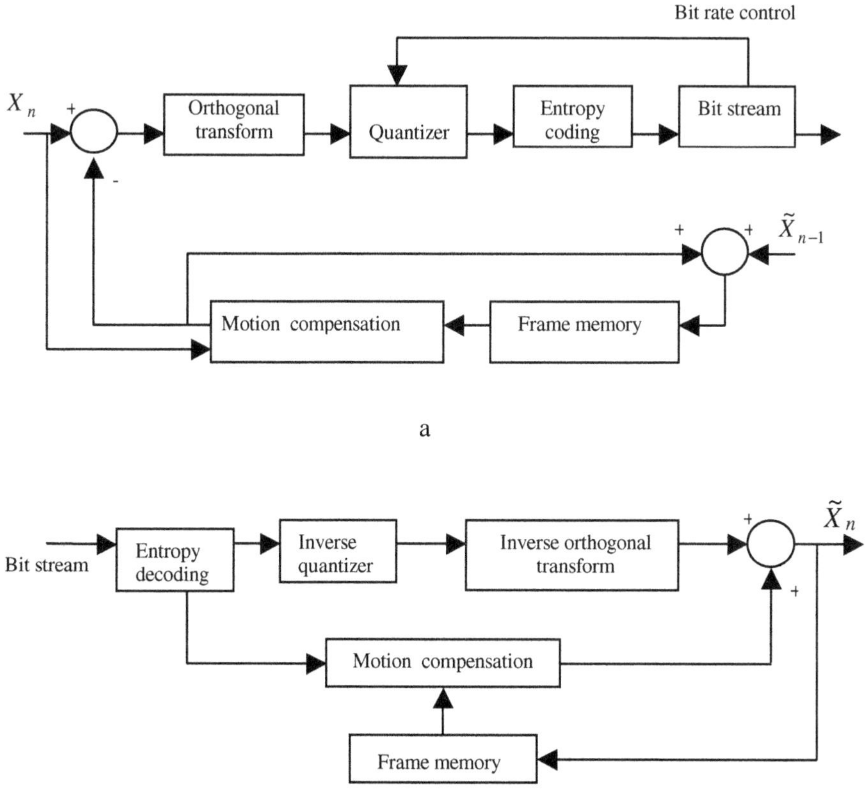

Fig. 1. General structure of image sequence coding system: a) Encoder, b) Decoder

3.1 Proposed System

Figure 2 shows the structure of the proposed image sequence coding. Compared to figure 1, our system differs by the following considerations:
- The image is not split in blocks;
- Wavelet transform is applied on the whole image;
-Temporal redundancy removing process is operated on the whole image and not on blocks;
-The encoding is realized for limited channel or in the other words for a given level. The encoder contains three components:
-The Discrete Wavelet Transform (DWT), which represents the image in the wavelet domain.
-The Embedded Zerotree Wavelet Quantization (EZWQ) which quantizes the wavelet coefficients in embedded fashion and produces the EZW symbols;
-The Binary coding which encodes the produced EZW symbols by a specific defined binary codes [19]. The decoder performs the inverse of the encoder's operations.

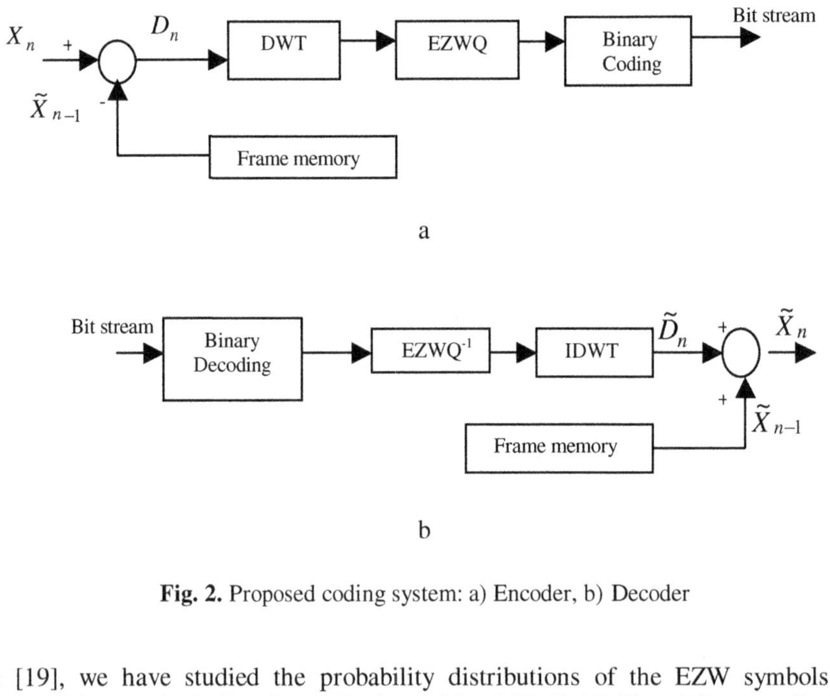

Fig. 2. Proposed coding system: a) Encoder, b) Decoder

In [19], we have studied the probability distributions of the EZW symbols for standard images including Lena, Barbara, Mandrill, Goldhill and Peppers. In fact, these images are decomposed in wavelet domain using the Daubechies biorthogonal wavelet 9/7-tap filter bank [22]. Five scales are performed [20, 21]. EZW algorithm is used to generate the different symbols described in the section 2. The probability distributions of these symbols are estimated. From these distributions, we have defined binary codes for each symbol in each subband. A specific codebook which we called CB1 is built. Using CB1 in still image coding, the obtained results outperform the Flexible Zerotree Codec [21]. According to this performance, we have extended the probability distributions analysis of EZW symbols to image sequences. So, differences between consecutive images in the image sequences are calculated. Some image sequences including Alexis, Claire, Mother & daughter, Salesman are used in the experimentation. These differences are then decomposed in wavelet domain. Similarly, EZW algorithm is used to generate the different symbols where their probability distributions are estimated. From these distributions, we have defined the binary codes for each symbol in each subband. A specific binary codebook for image difference, which we called CB2, is then built. Since the first image in the sequence is considered as the still image, it is encoded without any reference. The subsequent images are encoded by prediction. In our system, the first image is encoded by using CB1 and the subsequent images in the sequence are encoded by CB2.

3.2 Encoding Protocol

Two configurations are analysed in the terms of objective and subjective reconstruction qualities. For the first configuration, the following steps are performed:
1. The first image in the sequence (designed by X_0) is decomposed in wavelet domain and encoded by using EZW algorithm and CB1. The produced bit stream is considered as a reference bit stream;

2. The difference between the current image X_n and the previous image X_{n-1} in the encoder frame memory is calculated to remove temporal redundancies.
3. The obtained residual image D_n is decomposed in wavelet domain and encoded by using EZW and CB2. The bit stream produced in this case is the residual bit stream;

4. The current image is reconstructed by adding the residual image $\tilde{D}_n$ and the previous image $\tilde{X}_{n-1}$ in the decoder.

The following expressions summarize image difference calculating, and reconstruction process:

$$D_n = X_n - X_{n-1} \tag{3.1}$$

$$\tilde{X}_n = \tilde{X}_{n-1} + \tilde{D}_n \tag{3.2}$$

The expression 3.1 provides the difference D_n between the current image X_n and the previous image X_{n-1}. The expression 3.2 gives the reconstruction of the current image $\tilde{X}_n$ from the residual image $\tilde{D}_n$ and the reconstructed previous image $\tilde{X}_{n-1}$.
For the second configuration, only the step 2 is changed where difference is calculated between the current image and the reconstructed previous in the decoder. Expression 3.1 becomes:

$$D_n = X_n - \tilde{X}_{n-1} \tag{3.3}$$

4. Experimental Results and Discussions

Simulations are operated on Claire and Alexis sequences. Decomposition is performed using 9/7 filter bank [27]. The image size is rescaled to 256 x 256 pixels before decomposition. To reproduce an image from the received binary symbols, the output bit stream includes seven bytes as header information: four bytes for horizontal and vertical dimensions of the image, one byte for the filter bank, one byte for the decomposition levels and one byte for the initial threshold. Since the horizontal and vertical dimensions, the filter bank, and the number of decomposition levels are the same for the residual image, only initial threshold can change; so, one byte header is included in the residual bit stream to inform the decoder to update the initial threshold. Figure 3 shows PSNR versus image number for Claire sequence at 56 Kbits/s where only the 36 first images are reconstructed. The curve labelled "Serie1" is the result of the second configuration and the curve labelled "Série 2" is the result for the first configuration.

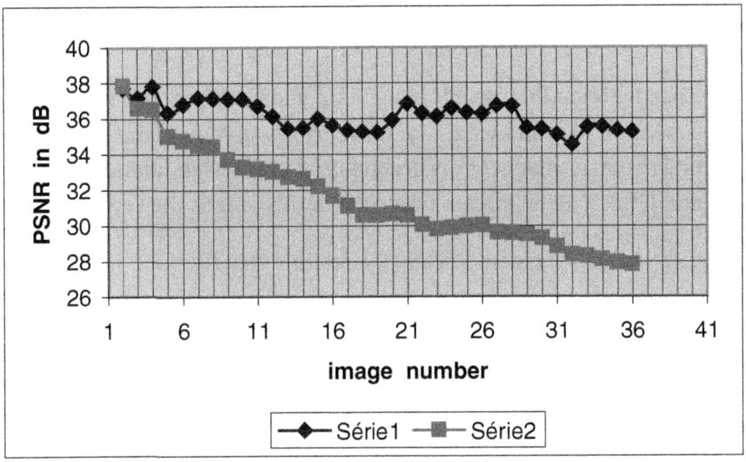

Fig. 3. Claire sequence at 56 Kbits/s and 10 fps
Serie1: second configuration result
Serie2: first configuration result

We observe in figure 3 that the reconstruction quality in the case of the first configuration decreases where the coding is performed by using the difference between the current image X_n and the previous image X_{n-1} (which is assumed to be transmitted without any loss and any quantization error). The reality is that the previous image in the decoder is affected by the quantization error. The difference between X_n and X_{n-1} can not cover information which exists between these consecutive images. Since the reconstruction is performed by using expression 3.2 and the bit rate is limited to 56 Kbits/s, there is not enough bits to improve the reconstruction quality. This is the main reason which explains the observed degradation quality.

We then repeat the encoding process by using the second configuration. Then Claire and Alexis sequences are encoded at 56 Kbits/s and 10 fps. Figure 4 and figure 5 show PSNR versus image number.

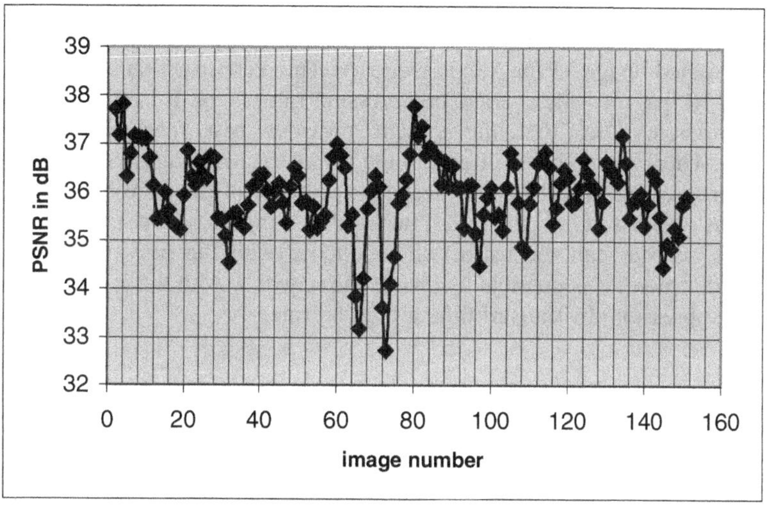

Fig. 4. Claire sequence at 56 Kbits/s and 10 fps
Mean PSNR: 35.98 dB

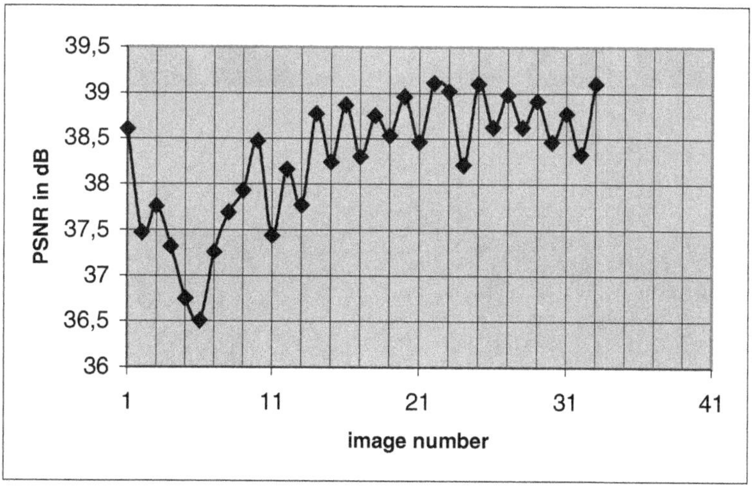

Fig. 5. Alexis sequence at 56 Kbits/s and 10 fps
Mean PSNR: 38.28 dB

It is shown in figure 4 and figure 5 that the system provides good reconstruction quality objectively where average PSNR of 35.98 dB and 38.28 dB are reached respectively for Claire and Alexis sequences. Figure 6 and figure 7 show the original and reconstructed images. Figure 6 A and B are respectively the original and reconstructed images 61, figure 6 C and D are respectively the original and reconstructed image 134 for Claire sequence. Figure 7 E and F are respectively the original and reconstructed image 20, figure 7 G and H are respectively the original and reconstructed image 33 for Alexis sequence. The reconstruction is operated at 56 Kbits/s and 10 fps, then the average compression ratio is 94. It is shown, despite of this compression ratio, that the system provides best reconstruction quality subjectively. There are no block effects in the reconstructed images. Since the system keeps the progressive encoding and decoding property of the EZW algorithm, it is robust against the loss of information. It means that if the encoder ceases the encoding process, the decoder can reconstruct the sequence with the previous received bit stream. Furthermore, it is possible to encode the sequence with the maximum quality (loss less compression) by transmitting at high bit rate.

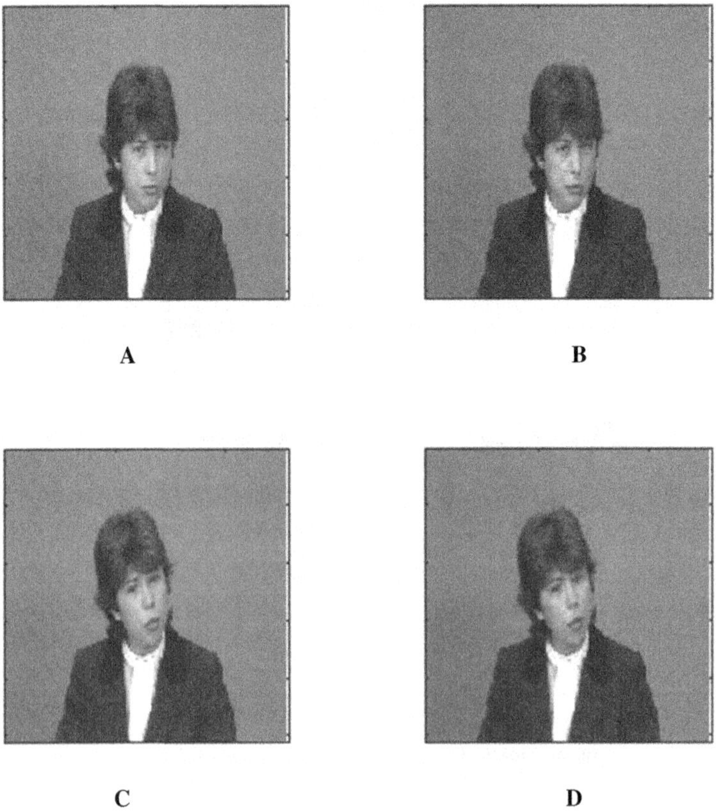

Fig. 6. Reconstruction results of Claire sequence at 56 Kbits/s and 10 fps
A: original image 61, B: reconstructed image 61
C: original image 134, D: reconstructed image 134

E

F

G

H

Fig. 7. Reconstruction results of Alexis sequence at 56 Kbits/s and 10 fps
E: original image 20, F: reconstructed image 20
G: original image 33, H: reconstructed image 33

5. Conclusion

In this paper, we have presented an image sequence coding system based on EZW algorithm and binary coding. Difference between the current image in the coder and the reconstructed image in the decoder is used as technique for removing temporal redundancies. The residual image is then decomposed in wavelet domain and encoded. Specific binary codebooks are built and used in the encoding process.

Experimental results show that the system provides best reconstruction quality as well objectively as subjectively. What explains the performance of our system is the fact that temporal redundancies are removed between the current image and the reconstructed previous image in the decoder. This enables the encoder to minimize the overall distortion due to the quantization error and improves the reconstruction quality.

References

[1] J.M.Shapiro, "Embedded image coding using zerotree of wavelet coefficients", IEEE Trans. on Signal Processing, Vol.41, No.12, pp.3445-3462, Dec.1993.
[2] I.Daubechies, "Orthonormal bases of compactly supported wavelets", Communication on Pure and Applied Mathematics, V.41, pp.909-996, Nov.1988.
[3] S.Mallat, "Atheory for multi-resolution signal decomposition: the wavelet representation", IEEE Trans. on Pattern Analysis and Machine Intelligence, Vo.11, pp.674-693, July 1989.
[4] I. Daubechies, Ten Lectures on Wavelets, SIAM, Philadelphia, PA, 1992.
[5] J.D.Villasenor, B.Belzer, and J.Lio, "Wavelet filter evaluation for image compression", IEEE Trans. on Image Processing, Vol.4, No.8, pp.1053-1060, Aug.1995.
[6] G.Strang, and T. Nguyen, Wavelets and Filter Banks, Wallesley-Cambridge Press, Wellesley, MA, 1996.
[7] A.Zandi, J.D.Allen, E.L.Schwartz, and M.Boliek, "CREW: Compression with Reversible Embedded wavelet", IEEE Data Compression Conference, pp.212-221, Snowbird, Mar.1995.
[8] A.Said, and W.A.Pearlman, "An image multi-resolution representation for loss less and lossy compression", IEEE Trans. on Image Processing, Vol.5, No.9, pp.1303-1310, Sep.1996.
[9] Y.Chen, and W.A.Pearlman, "Three-dimensional subband coding of video using zerotree method", Proc. SPIE, Visual Communications and Image Processing, pp.1302-1309, Orlando, Mar. 1996.
[10] A.Said, and W.A.Pearlman, "A new fast and efficient image codec based on set partitioning in hierarchical trees", IEEE Trans. on Circuits and Systems for Video Technology, Vol.6, No.3, pp.243-250, Jun. 1996.
[11] S.A.Martucci, I.Sodagar, T.H.Chiang, and Y.Q.Zhang, " A zerotree wavelet coder", IEEE Trans. on Circuits and Systems for Video Technology, Vol.7, No.1, pp.109-118, Feb. 1997.
[12] J.Li, P.Cheng, and C.Kuo, " On the improvement of embedded zerotree wavelet coding", Proc. SPIE, Visual Communications and Image Processing, pp.1490-1501, Orlando, Apr. 1995.
[13] H.Man, F.Kossentini, and M.Smith,"Robust EZW image coding for noisy channels", IEEE Signal Processing Letters, Vol.4, No.8, pp.227-229, Aug. 1997.
[14] C.D.Creusere, "A new method for robust image compression based on the embedded zerotree wavelet algorithm", IEEE Trans. on Image Processing, Vol.6, No.10, pp.1436-1442, Oct. 1997.
[15] J.K.Rogers, and P.C.Cosman, "Wavelet zerotree image compression with packetization", IEEE Signal Processing Letters, Vol.5, No.5, pp.105-107, May 1998.
[16] S.Joo, H.Kikuchi, S.Sasaki, and J.Shin, "Flexible Zerotree coding of Wavelet coefficeints", IEICE Trans. Fundamentals, Vol.E82-A, No.4, Apr. 1999.
[17] Michael W. Marcellin, Michael J.Gormish, Ali Bilgin, and Martin P.Boliek, " An overview of JPEG-2000", Proc. IEEE Data Compression Conference, pp.523-541, 2000.
[18] Beong-Jo Kim, and W.A.Pearlman, "An embedded wavelet video coder using three-dimensional set partitioning in hierarchical trees (SPIHT)", Proc. DCC'97, IEEE Data Compression Conference, pp.251-260, Snowbird, UT, Mar. 1997.

[19] M. Jérôme, "Optimal Image Coding based on Probability Distribution of Embedded Zerotree Wavelet Symbols", Tunisian-German Conference on Smart Systems and Devices SSD, pp.666-671, Hammamet, Tunisia, March 27-30, 2001.
[20] M. Jérôme et N. Ellouze, "Etude énergétique de l'analyse multi-résolution d'images par ondelette , Proc. in JTEA'2000, Tome1, pp.103-109, 24-25 Mar. 2000 Hammamet, Tunisia.
[21] M. Jérôme and N. Ellouze, "Image Wavelet Coefficients Quantization by Embedded Zerotree Wavelet Algorithm", Proc. in ACIDCA'2000, *International conference on Artificial and Computational Intelligence for Decision, Control and Automation in Engineering and Industrial Applications,* pp.1-5, *Monastir*, 22-24 March 2000.
[22] M.Antoni, M.Barlaud, P.Mathieu, and I.Daiubechies, "Image coding using wavelet transform", IEEE Trans. on Image Processing, Vol.1, No.2, pp.205-220, Apr. 1992.

Wavelet-Based Video Compression Using Long-Term Memory Motion-Compensated Prediction and Context-Based Adaptive Arithmetic Coding

Detlev Marpe[1], Thomas Wiegand[1], and Hans L. Cycon[2]

[1] Image Processing Department
Heinrich-Hertz-Institute (HHI) for Communication Technology
Einsteinufer 37, 10587 Berlin, Germany
{marpe,wiegand}@hhi.de
[2] University of Applied Sciences (FHTW Berlin)
Allee der Kosmonauten 20–22, 10315 Berlin, Germany
hcycon@fhtw-berlin.de

Abstract. In this paper, we present a novel design of a wavelet-based video coding algorithm within a conventional hybrid framework of temporal motion-compensated prediction and transform coding. Our proposed algorithm involves the incorporation of multi-frame motion compensation as an effective means of improving the quality of the temporal prediction. In addition, we follow the rate-distortion optimizing strategy of using a Lagrangian cost function to discriminate between different decisions in the video encoding process. Finally, we demonstrate that context-based adaptive arithmetic coding is a key element for fast adaptation and high coding efficiency. The combination of overlapped block motion compensation and frame-based transform coding enables blocking-artifact free and hence subjectively more pleasing video. In comparison with a highly optimized MPEG-4 (Version 2) coder, our proposed scheme provides significant performance gains in objective quality of 2.0–3.5 dB PSNR.

1 Introduction

Multi-frame prediction [11] and variable block size motion compensation in a rate-distortion optimized motion estimation and mode selection process [12,10] are powerful tools to improve the coding efficiency of today's video coding standards. In this paper, we present the design of a video coder, dubbed *DVC*, which demonstrates how most elements of the state-of-the-art in video coding as currently implemented in the test model long-term [2] (TML8) of the ITU-T H.26L standardization project can be successfully integrated in a blocking-artifact free video coding environment. In addition, we provide a solution for an efficient macroblock based intra coding mode within a frame-based residual coding method, which is extremely beneficial for improving the subjective quality as well as the error robustness.

Y. Y. Tang et al. (Eds.): WAA 2001, LNCS 2251, pp. 76–86, 2001.

We further explain how appropriately designed entropy coding tools, which have already been introduced in some of our previous publications [6,7] and which, in some modified form [5], are now part of TML8, help to improve the efficiency of a wavelet-based residual coder.

In our experiments, we compared our proposed wavelet-based DVC coder against an improved MPEG-4 coder [10], where both codecs were operated using a fixed frame rate, fixed quantization step sizes and a search range of ± 32 pels. We obtained coding results for various sequences showing that our proposed video coding system yields a coding gain of 2.0–3.5 dB PSNR relative to MPEG-4. Correspondingly, the visual quality provided by the DVC coder compared to that of the block-based coding approach of MPEG-4 is much improved, especially at very low bit rates.

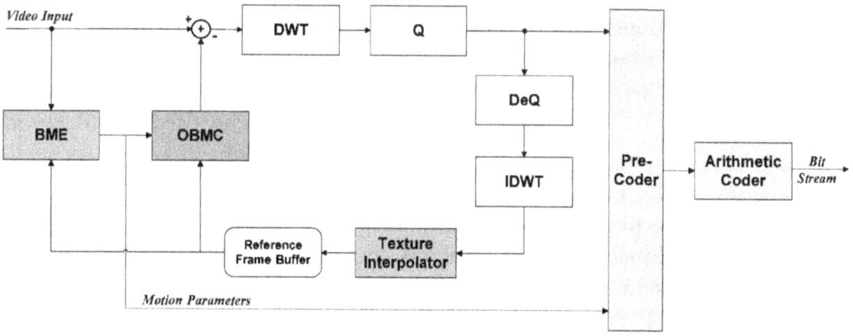

Fig. 1. Block diagram of the proposed coding scheme

2 Overview of the DVC Scheme

Fig. 1 shows a block diagram of the proposed DVC coder. As a hybrid system, it consists of a temporal predictive loop along with a spatial transform coder. Temporal prediction is performed by using a *block motion estimation* (BME) and an *overlapped block motion compensation* (OBMC), such that the reference of each predicted block can be obtained from a long-term *reference frame memory*. Coding of the motion compensated *P*-frames as well as of the initial intra (*I*) frame is performed by first applying a *discrete wavelet transform* (DWT) to an entire frame. Uniform scalar *quantization* (Q) with a central dead-zone around zero similar to that designed for H.263 is then used to map the dynamic range of the wavelet coefficients to a reduced alphabet of decision levels. Prior to the final *arithmetic coding* stage, the *pre-coder* further exploits redundancies of the quantized wavelet coefficients in a 3-stage process of partitioning, aggregation and conditional coding.

Table 1. Macroblock partition modes

Mode	Block Size	Partition
1	16×16	Leave MB as a whole
2	16×8	Split MB into 2 sub-blocks
3	8×16	Split MB into 2 sub-blocks
4	8×8	Split MB into 4 sub-blocks

3 Motion-Compensated Prediction

3.1 Motion Model

As already stated above, the motion model we used is very similar to that of the H.26L TML8 design [2]. In essence it relies on a simple model of block displacements with variable bock sizes. Given a partition of a frame into macroblocks (MB) of size 16×16 pels, each macroblock can be further sub-divided into smaller blocks, where each sub-block has its own displacement vector. Our model supports 4 different partition modes, as shown in Table 1.

Each macroblock may use a different reference picture out of a long-term frame memory. In addition to the predictive modes represented by the 4 different MB partition modes in Table 1, we allow for an additional macroblock-based intra coding mode in P-frames. This local intra mode is realized by computing the DC for each 8×8 sub-block of each spectral component (Y,U,V) in a macroblock and by embedding the DC-corrected sub-blocks into the residual frame in a way, which is further described in the following section.

3.2 Motion Estimation and Compensation

Block motion estimation is performed by an exhaustive search over all integer pel positions within a pre-defined search window around the motion vector predictor, which is obtained from previously estimated sub-blocks in the same way as in TML8 [2]. In a number of subsequent steps, the best integer pel motion vector is refined to the final $\frac{1}{4}$-pel accuracy by searching in a 3×3 sub-pel window around the refined candidate vector. All search positions are evaluated by using a Lagrangian cost function, which involves a rate and distortion term coupled by a Lagrangian multiplier. For all fractional-pel displacements, distortion in the transform domain is estimated by using the Walsh-Hadamard transform, while the rate of the motion vector candidates is estimated by using a fixed, pre-calculated table. This search process takes place for each of the 4 macroblock partitions and each reference frame, and the cost of the overall best motion vector candidate(s) of all 4 macroblock modes is finally compared against the cost of the intra mode decision to choose the macroblock mode with minimum cost.

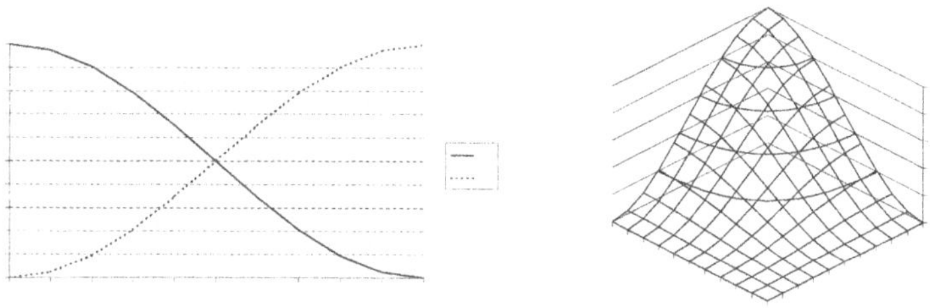

(a) (b)

Fig. 2. (a) 1-D profile of 2-D weighting functions along the horizontal or vertical axes of two neighboring overlapping blocks. (b) 2-D weighting function

The prediction error luminance (chrominance) signal is formed by the weighted sum of the differences between all 16 × 16 (8 × 8) overlapping blocks from the current frame and their related overlapping blocks with displaced locations in the reference frame, which have been estimated in the BME stage for the corresponding core blocks. In the case of an intra macroblock, we compute the weighted sum of the differences between the overlapping blocks of the current intra blocks and its related DC-values. As a weighting function w, we used the 'raised cosine', as shown in Fig. 2. For a support of $N \times N$ pels, it is given by

$$w(n,m) = w_n \cdot w_m, \quad w_n = \frac{1}{2}\left[1 - \cos\frac{2\pi n}{N}\right] \quad \text{for } n = 0, \ldots, N. \tag{1}$$

In our presented approach, we choose $N = 16$ ($N = 8$) for the luminance (chrominance, resp.) in Eq. (1), which results in a 16×16 (8×8) pixel support centered over a "core" block of size 8 × 8 (4 × 4) pels for the luminance (chrominance, resp.). For the texture interpolation of sub-pel positions, the same filters as specified in TML8 [2] have been used.

4 Wavelet Transform

In wavelet-based image compression, the so-called 9/7-wavelet with compact support [3] is the most popular choice. Our proposed coding scheme, however, utilizes a class of biorthogonal wavelet bases associated with infinite impulse response (IIR) filters, which was recently constructed by Petukhov [9]. His approach relies on the construction of a dual pair of rational solutions of the matrix equation

$$M(z)\tilde{M}^T(z^{-1}) = 2I, \tag{2}$$

where I is the identity matrix, and

$$M(z) = \begin{pmatrix} h(z) & h(-z) \\ g(z) & g(-z) \end{pmatrix}, \quad \tilde{M}(z) = \begin{pmatrix} \tilde{h}(z) & \tilde{h}(-z) \\ \tilde{g}(z) & \tilde{g}(-z) \end{pmatrix}$$

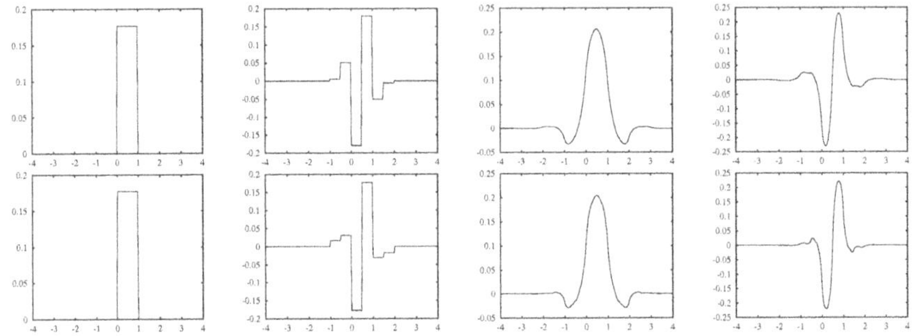

Fig. 3. *From left to right*: scaling function of analysis, analyzing wavelet, scaling function of synthesis, and synthesizing wavelet used for I-frame coding (*top row*) and P-frame coding (*bottom row*)

are so-called 'modulation matrices'.

In [9], a one-parametric family of filters h_a, g_a, $\tilde{h}_a$ and $\tilde{g}_a$ satisfying Eq. (2) was constructed:[1]

$$h_a(z) = \frac{1}{\sqrt{2}}(1+z), \tag{3}$$

$$\tilde{h}_a(z) = \frac{(2+a)(z^{-1}+3+3z+z^2)(z^{-1}+b+z)}{4\sqrt{2}(2+b)(z^{-2}+a+z^2)}, \tag{4}$$

$$g_a(z) = \frac{(2+a)(z^{-1}-3+3z-z^2)(-z^{-1}+b-z)}{4\sqrt{2}(2+b)}, \tag{5}$$

$$\tilde{g}_a(z) = \frac{1}{\sqrt{2}} \frac{1-z^{-1}}{z^{-2}+a+z^2}, \tag{6}$$

where $b = \frac{4a-8}{6-a}$, $|a| > 2$, $a \neq 6$.

To adapt the choice of the wavelet basis to the nature and statistics of the different frame types of intra and inter mode, we performed a numerical simulation on this one-parametric family of IIR filter banks yielding the optimal value of $a = 8$ for intra frame mode and $a = 25$ for inter frame mode in Eqs. (3)–(6). Graphs of these optimal basis functions are presented in Fig. 3. Note that the corresponding wavelet transforms are efficiently realized with a composition of recursive filters [9].

5 Pre-coding of Wavelet Coefficients

For encoding the quantized wavelet coefficients, we follow the conceptual ideas initially presented in [6] and later refined in [7]. Next, we give a brief review of the involved techniques. For more details, the readers are referred to [6,7].

[1] h_a and g_a denote low-pass and high-pass filters of the decomposition algorithm, respectively, while $\tilde{h}_a$ and $\tilde{g}_a$ denote the corresponding filters for reconstruction.

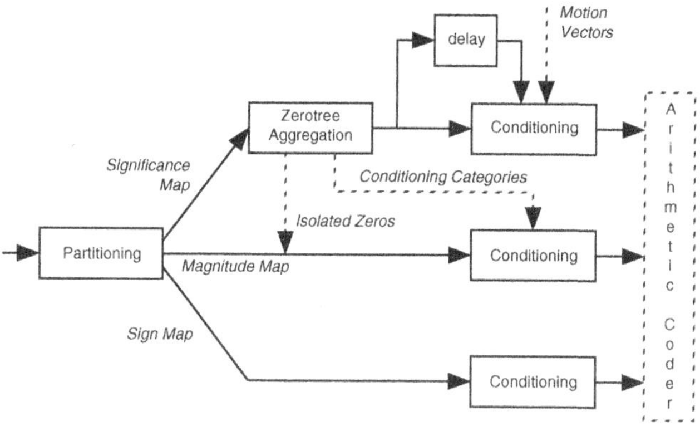

Fig. 4. Schematic representation of the pre-coder used for encoding the quantized wavelet coefficients

5.1 Partitioning

As shown in the block diagram of Fig. 4, an initial 'partitioning' stage divides each frame of quantized coefficients into three sub-sources: a significance map, indicating the position of significant coefficients, a magnitude map holding the absolute values of significant coefficients, and a sign map with the phase information of the wavelet coefficients. Note that all three sub-sources inherit the subband structure from the quantized wavelet decomposition, so that there is another partition of each sub-source according to the given subband structure.

5.2 Zerotree Aggregation

In a second stage, the pre-coder performs an 'aggregation' of insignificant coefficients using a quad-tree related data structure. These so-called *zerotrees* [4,6] connect insignificant coefficients, which share the same spatial location along the multiresolution pyramid. However, we do not consider zero-tree roots in bands below the maximum decomposition level. In inter-frame mode, coding efficiency is further improved by connecting the zerotree root symbols of all three lowest high-frequency bands to a so-called 'integrated' zerotree root which resides in the LL-band.

5.3 Conditional Coding

The final 'conditioning' part of the pre-coding stage supplies the elements of each source with a 'context', *i.e.*, an appropriate model for the actual coding process in the arithmetic coder. Fig. 5 (a) shows the prototype template used for conditioning of elements of the significance map. In the first part, it consists of a causal neighborhood of the actual coding event C, which depends on the

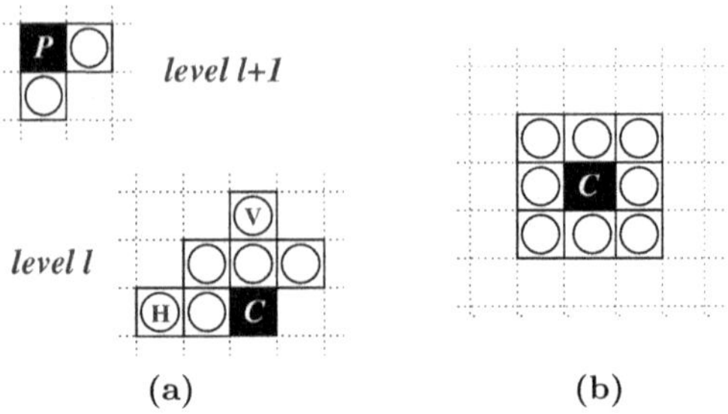

Fig. 5. (a) Two-scale template (*white circles*) with an orientation dependent design for conditional coding of an event C of the significance map; V, H: additional element used for vertical and horizontal oriented bands, respectively. (b) 8-neighborhood of significance used for conditioning of a given magnitude C

scale and orientation of a given band. Except for the lowest frequency bands, the template uses additional information of the next upper level (lower resolution) represented by the neighbors of the parent P of C, thus allowing a 'prediction' of the non-causal neighborhood of C. The processing of the lowest frequency band depends on the intra/inter decision. In intra mode, mostly non-zero coefficients are expected in the LL-band, so there is no need for coding a significance map. For P-frames, however, we indicate the significance of a coefficient in the LL-band by using the four-element kernel of our prototype template (Fig. 5 (a)), which is extended by the related entry of the significance map belonging to the previous P-frame.

The processing of subbands is performed band-wise in the order from lowest to highest frequency bands and the partitioned data of each band is processed such that the significance information is coded (and decoded) first. This allows the construction of special conditioning categories for the coding of magnitudes using the local significance information. Thus, the actual conditioning of magnitudes is performed by classifying magnitudes of significant coefficients according to the local variance estimated by the significance of their 8-neighborhood (cf. Fig. 5 (b)). For the conditional coding of sign maps, we are using a context built of two preceding signs with respect to the orientation of a given band [7].

For coding of the LL-band of I-frames, the proposed scheme uses a DPCM-like procedure with a spatially adaptive predictor and a backward driven classification of the prediction error using a six-state model.

6 Binarization and Adaptive Binary Arithmetic Coding

All symbols generated by the pre-coder are encoded using an adaptive binary arithmetic coder, where non-binary symbols like magnitudes of coefficients or motion vector components are first mapped to a sequence of binary symbols by means of the unary code tree. Each element of the resulting "intermediate" codeword given by this so-called *binarization* will then be encoded in the subsequent process of binary arithmetic coding.

At the beginning of the overall encoding process, the probability models associated with all different contexts are initialized with a pre-computed start distribution. For each symbol to encode the frequency count of the related binary decision is updated, thus providing a new probability estimate for the next coding decision. However, when the total number of occurrences of symbols related to a given model exceeds a pre-defined threshold, the frequency counts will be scaled down. This periodical rescaling exponentially weighs down past observations and helps to adapt to non-stationarities of a source.

For intra and inter frame coding we use separate models. Consecutive P-frames as well as consecutive motion vector fields are encoded using the updated related models of the previous P-frame and motion vector field, respectively. The binary arithmetic coding engine used in our presented approach is a straightforward implementation similar to that given in [13].

7 Experimental Results

7.1 Test Conditions

To illustrate the effectiveness of our proposed coding scheme, we used an improved MPEG-4 coder [10] as a reference system. This coder follows a rate-distortion (R-D) optimized encoding strategy by using a Lagrangian cost function, and it generates bit-streams compliant with MPEG-4, Version 2 [1]. Most remarkable is the fact that this encoder provides PSNR gains in the range from 1.0–3.0 dB, when compared to the MoMuSys reference encoder (VM17) [10]. For our experiments, we used the following encoding options of the improved MPEG-4 reference coder:

- $\frac{1}{4}$-pel motion vector accuracy enabled
- Global motion compensation enabled
- Search range of ± 32 pels
- 2 B-frames inserted ($IBBPBBP\ldots$)
- MPEG-2 quantization matrix used

For our proposed scheme, we have chosen the following settings:

- $\frac{1}{4}$-pel motion vector accuracy
- Search range of ± 32 pels around the motion vector predictor
- No B-frames used ($IPPP\ldots$)
- Five reference pictures were used for all sequences except for the 'News'-sequence (see discussion of results below)

Coding experiments were performed by using the test sequences 'Foreman' and 'News' both in QCIF resolution and with 100 frames at a frame rate of 10 Hz. Only the first frame was encoded as an I-frame; all subsequent frames were encoded as P-frames or B-frames. For each run of a sequence, a set of quantizer parameters according to the different frame types (I,P,B) was fixed.

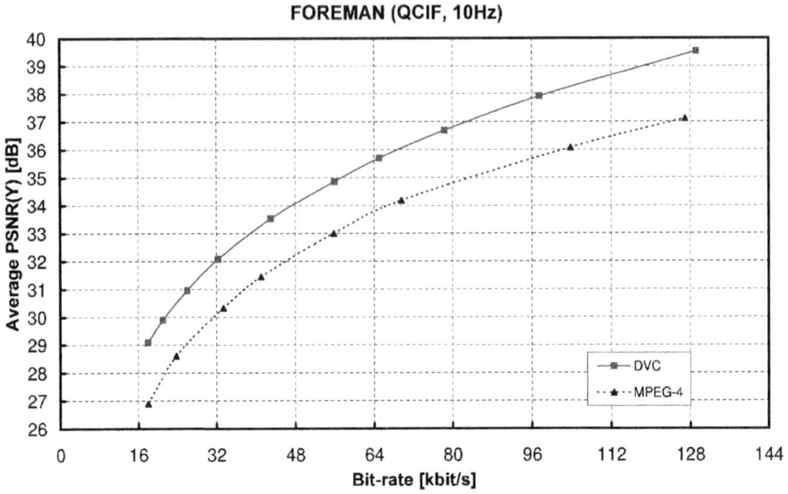

Fig. 6. Average Y-PSNR against bit-rate using the QCIF test sequence 'Foreman' at a frame rate of 10 Hz

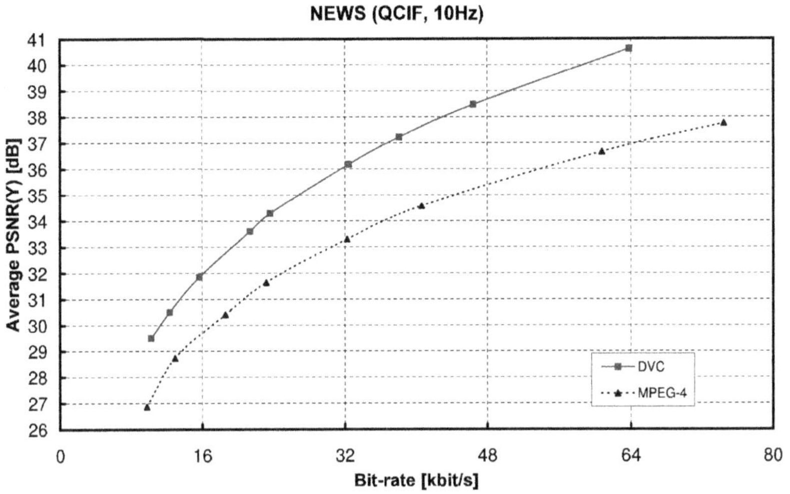

Fig. 7. Average Y-PSNR against bit-rate using the QCIF test sequence 'News' at a frame rate of 10 Hz

(a) (b)

Fig. 8. Comparison of subjective reconstruction quality: Frame no. 22 of 'Foreman' at 32 kbit/s. (a) DVC reconstruction (b) MPEG-4 reconstruction. Note that the MPEG-4 reconstruction has been obtained by using a de-blocking filter

7.2 Test Results

Figs. 6–7 show the average PSNR gains obtained by our proposed DVC scheme relative to the MPEG-4 coder for the test sequences 'Foreman' and 'News', respectively. For the 'Foreman'-sequence, significant PSNR gains of 2.0–2.5 dB on the luminance component have been achieved (cf. Fig. 6). Figure 8 shows a comparison of the visual quality for a sample reconstruction at 32 kbit/s. The results we obtained for the "News"-sequence show dramatic PSNR improvements of about 2.5–3.5 dB. To demonstrate the ability of using some *a priori* knowledge about the scene content, we checked for this particular sequence in addition to the five most recent reference frames one additional reference frame 50 frames back in the past according to the repetition of parts of the scene content. By using the additional reference frame memory for this particular test case, we achieved an additional gain of about 1.5 dB PSNR on the average compared to the case where the reference frame buffer was restricted to the five most recent reference frames only.

8 Conclusions and Future Research

The coding strategy of DVC has proven to be very efficient. PSNR gains of up to 3.5 dB relative to an highly optimized MPEG-4 coder have been achieved. However, it should be noted that in contrast to the MPEG-4 coding system, no B-frames were used in the DVC scheme, although it can be expected that DVC will benefit from the usage of B-frames in the same manner as the MPEG-4 coder, *i.e.*, depending on the test material, additional PSNR improvements of up to 2 dB might be achievable. Another important point to note, when comparing the coding results of our proposed scheme to that of the highly R-D optimized MPEG-4 encoder used for our experiments, is the fact that up to now, we did not incorporate any kind of high-complexity R-D optimization method. We even did not optimize the motion estimation process with respect to the overlapped motion compensation, although it is well known, that conventional block motion estimation is far from being optimal in an OBMC framework [8]. Furthermore, we believe that the performance of our zerotree-based wavelet coder can be further improved by using a R-D cost function for a joint optimization of the quantizer and the zerotree-based encoder. Thus, we expect another significant gain by exploiting the full potential of encoder optimizations inherently present in our DVC design. This topic will be a subject of our future research.

References

1. ISO/IEC JTC1SC29 14496-2 MPEG-4 Visual, Version 2. 83
2. Bjontegaard, G. (ed.): H.26L Test Model Long Term Number 8 (TML8), ITU-T SG 16 Doc. VCEG-N10 (2001) 76, 78, 79
3. Cohen, A., Daubechies, I., Feauveau, J.-C.: Biorthogonal Bases of Compactly Supported Wavelets, Comm. on Pure and Appl. Math., Vol. 45 (1992) 485–560 79
4. Lewis, A., Knowles, G.: Image Compression Using the 2D Wavelet Transform, IEEE Trans. on Image Processing, Vol. 1, No. 2 (1992) 244–250 81
5. Marpe, D., Blättermann, G., Wiegand, T.: Adaptive Codes for H.26L, ITU-T SG 16 Doc. VCEG-L13 (2001) 77
6. Marpe, D., Cycon, H. L.: Efficient Pre-Coding Techniques for Wavelet-Based Image Compression, *Proceedings Picture Coding Symposium 1997*, 45–50 77, 80, 81
7. Marpe, D., Cycon, H. L.: Very Low Bit-Rate Video Coding Using Wavelet-Based Techniques, IEEE Trans. on Circuits and Systems for Video Technology, Vol. 9, No. 1 (1999) 85–94 77, 80, 82
8. Orchard, M. T., Sullivan, G. J.: Overlapped Block Motion Compensation: An Estimation-Theoretic Approach, IEEE Trans. on Image Processing, Vol. 3, No. 5 (1994) 693–699 86
9. Petukhov, A. P.: Recursive Wavelets and Image Compression, Proceedings International Congress of Mathematicians 1998 79, 80
10. Schwarz, H., Wiegand, T.: An Improved MPEG-4 Coder Using Lagrangian Coder Control, ITU-T SG 16 Doc. VCEG-M49 (2001) 76, 77, 83
11. Wiegand, T., Zhang, X., Girod, B.: Long-Term Memory Motion-Compensated Prediction, IEEE Trans. on Circuits and Systems for Video Technology, Vol. 9, No. 1 (1999) 70–84 76
12. Wiegand, T., Lightstone, M., Mukherjee, D., Campbell, T. G., Mitra, S. K.: Rate-Distortion Optimized Mode Selection for Very Low Bit Rate Video Coding and the Emerging H.263 Standard, IEEE Trans. on Circuits and Systems for Video Technology, Vol. 6, No. 2 (1996) 182-190 76
13. Witten, I., Neal, R., Cleary, J.: Arithmetic Coding for Data Compression, Communications of the ACM, Vol. 30 (1987) 520–540 83

ueno@iss.soka.ac.jp

( )

( )

$$(+) = - \quad / \quad - \qquad ( ) \quad + \quad =$$

$$(+) \; + \; / = - \quad / \quad - \qquad ( ) \quad + \quad =$$

$$\phi = \phi - \quad =$$

$$\varphi = \varphi - \quad =$$

$\phi \neq \varphi \neq \leq$

Υ

$$\gamma = \frac{= \quad + - \quad + +/- \quad +/}{\sqrt{= \quad + -}\;\sqrt{= \quad + +/- \quad +/}}$$

Υ

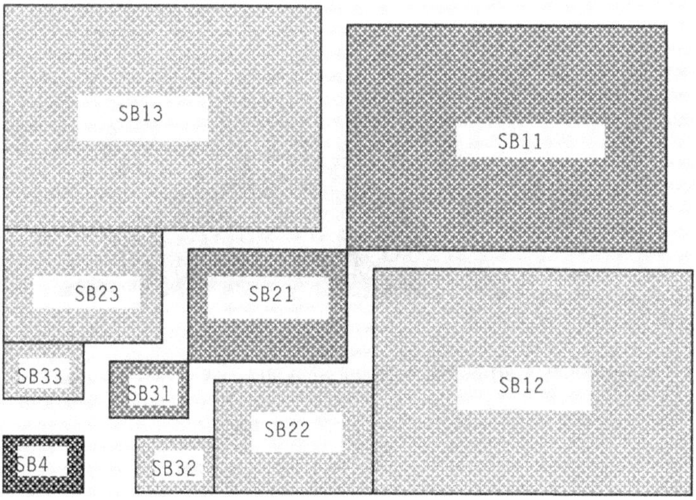

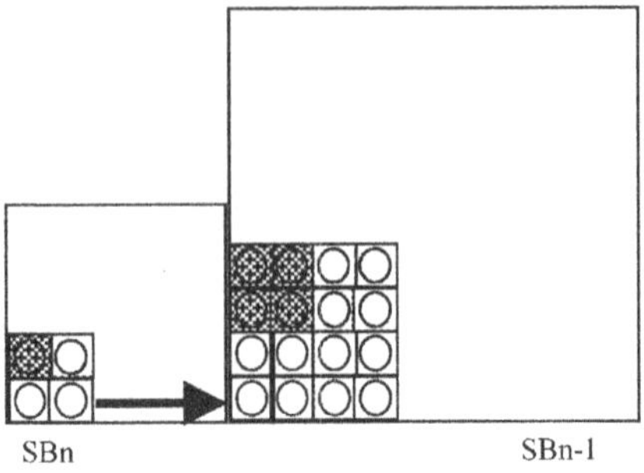

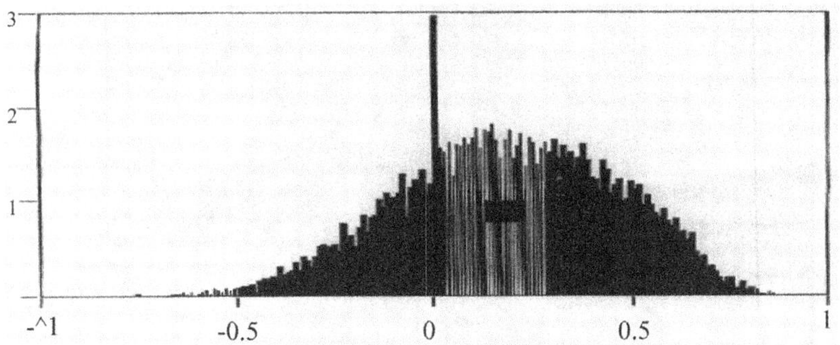

Correlation coefficients between SBn and SBn-1 (Lena Image)

$$- \quad = \quad \times$$

$$< \Big|^{-} \quad + \quad \Big|$$

$\sigma \geq$

σ

$\times$

$\times$

$\times$

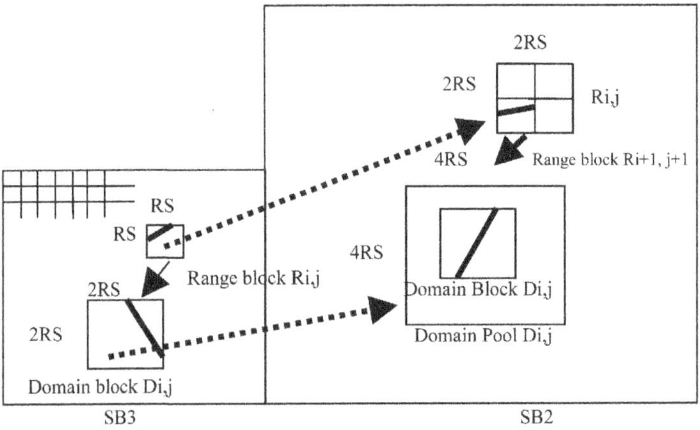

$$\cong \times \alpha \times \quad +$$

α

$$\supset \quad + \quad + \qquad \supset$$

$$+ \quad + \cong \times \alpha \times \times \quad + \Delta$$

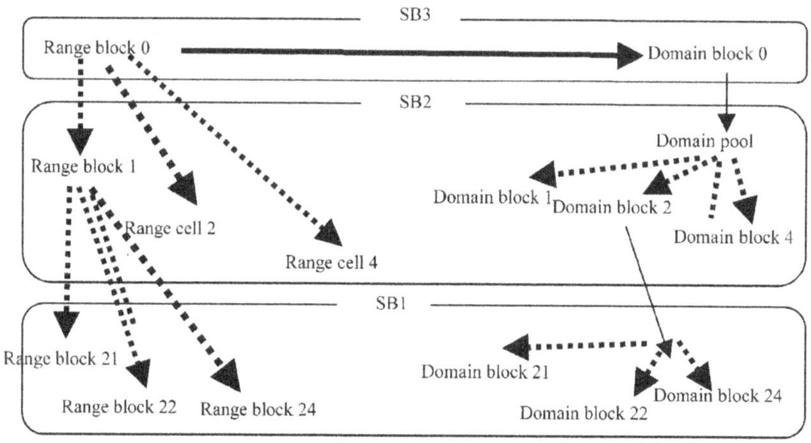

$$= \qquad \overline{\left(\quad - \quad \right)/\gamma}$$

$$\left(\quad - \quad \right)/\gamma$$

$\gamma \qquad \gamma \quad \gamma$

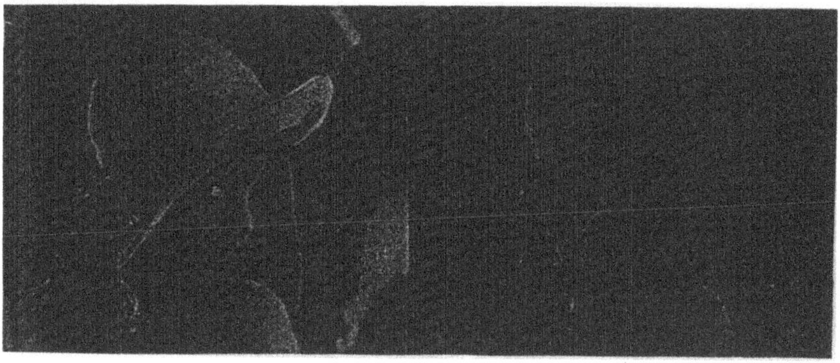

× ×

×

×

×

×

× ×

$\times\{\ \times\ \ +\ \times\ \ \times(\ \ +\ \ )\}=$

/

$\alpha( )$

$$\cong \overline{} = \times\left\{ \times \quad_{\alpha( )}\right\} + \quad \times$$

α

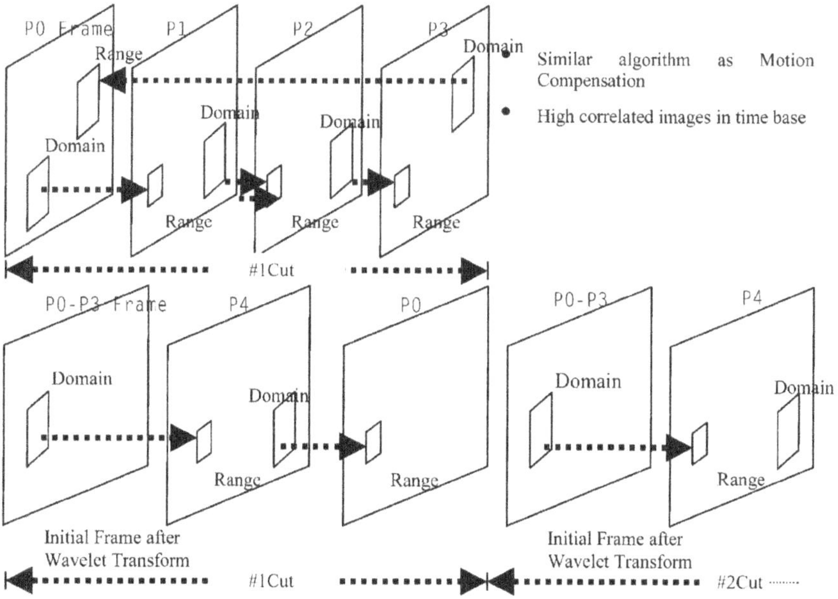
P0 Frame
P1
P2
P3
Range
Domain
Domain
Domain
Domain
Range
Range
Range
Similar algorithm as Motion Compensation
High correlated images in time base
#1Cut
P0-P3 Frame
P4
P0
P0-P3
P4
Domain
Domain
Domain
Domain
Range
Range
Range
Initial Frame after Wavelet Transform
Initial Frame after Wavelet Transform
#1Cut
#2Cut

Integration of Multivariate Haar Wavelet Series*

Stefan Heinrich[1], Fred J. Hickernell[2], and Rong-Xian Yue[3]

[1] FB Informatik, Universität Kaiserslautern
PF 3049, D-67653 Kaiserslautern, Germany
heinrich@informatik.uni-kl.de
[2] Department of Mathematics, Hong Kong Baptist University
Kowloon Tong, Hong Kong SAR, China
fred@hkbu.edu.hk
[3] College of Mathematical Science, Shanghai Normal University
100 Guilin Road, Shanghai 200234, China
rxyue@online.sh.cn

Abstract. This article considers the error of integrating multivariate Haar wavelet series by quasi-Monte Carlo rules using scrambled digital nets. Both the worst-case and random-case errors are analyzed. It is shown that scrambled net quadrature has optimal order. Moreover, there is a simple formula for the worst-case error.

1 Introduction

Digital (t, m, s)-nets and (t, s)-sequences are popular low discrepancy point sets used for quasi-Monte Carlo multidimensional quadrature [8,11]. In recent years it has been shown that these sets are especially effective for integrating multivariate Haar wavelet series. The convergence rate depends on the decay rates of the wavelet series coefficients. This article reports recent results by the authors and others. For proofs the reader is referred to the references cited.

The following section defines the Hilbert space of multivariate Haar wavelet series, $\mathcal{H}_{\text{wav}}$. Section 3 describes constructions of digital nets and sequences, and Section 4 defines the integration problem to be studied. The main results are described in Section 5.

2 Function Spaces Spanned by Haar Wavelets

The space, $\mathcal{H}_{\text{wav}}$, of multivariate Haar wavelets studied here was defined in [15]. The domain of interest is the unit cube, $[0, 1)^s$, where the dimension s is any

* This work was partially supported by a Hong Kong Research Grants Council grant HKBU/2030/99P, by Hong Kong Baptist University grant FRG/97-98/II-99, by Shanghai NSF Grant 00JC14057, and by a Shanghai Higher Education STF Grant.

Y. Y. Tang et al. (Eds.): WAA 2001, LNCS 2251, pp. 99–106, 2001.

positive integer. Let b be an integer greater than one, and define the univariate basic wavelet functions as

$$\psi_\gamma(x) = b^{1/2} 1_{\lfloor bx \rfloor = \gamma} - b^{-1/2} 1_{\lfloor x \rfloor = 0}, \quad \gamma = 0, 1, \ldots, b-1,$$

where $1_{\{\cdot\}}$ denotes the characteristic function, and $\lfloor x \rfloor$ denotes the greatest integer less than x. For each subset u of the coordinate axes $\{1, \ldots, s\}$, let $|u|$ denote the cardinality of u. For each $r \in u$ let κ_r, τ_r and γ_r be integers with $\kappa_r \geq 0$, $0 \leq \tau_r < b^{\kappa_r}$ and $0 \leq \gamma_r < b$. Define the vectors $\boldsymbol{\kappa} = (\kappa_r)_{r \in u}$, $\boldsymbol{\tau} = (\tau_r)_{r \in u}$, and $\boldsymbol{\gamma} = (\gamma_r)_{r \in u}$. Let $\psi_{u\boldsymbol{\kappa}\boldsymbol{\tau}\boldsymbol{\gamma}}$ be a product over $r \in u$ of the dilated and translated wavelets, i.e.,

$$\psi_{u\boldsymbol{\kappa}\boldsymbol{\tau}\boldsymbol{\gamma}}(\mathbf{x}) := b^{(|\boldsymbol{\kappa}| - |u|)/2} \prod_{r \in u} (b 1_{\lfloor b^{\kappa_r + 1} x_r \rfloor = b\tau_r + \gamma_r} - 1_{\lfloor b^{\kappa_r} x_r \rfloor = \tau_r}), \tag{1}$$

where $|\boldsymbol{\kappa}| = \sum_{r \in u} \kappa_r$. For $u = \emptyset$ we take by convention $\psi_{u\boldsymbol{\kappa}\boldsymbol{\tau}\boldsymbol{\gamma}}(\mathbf{x}) = \psi_\emptyset(\mathbf{x}) = 1$. The wavelets defined above are not orthogonal nor linearly independent, but they are nearly so. As observed in [15],

$$\sum_{\gamma_r = 0}^{b-1} \psi_{u\boldsymbol{\kappa}\boldsymbol{\tau}\boldsymbol{\gamma}}(\mathbf{x}) = 0, \quad \forall r \in u, \ \forall u, \boldsymbol{\kappa}, \boldsymbol{\tau}, \boldsymbol{\gamma}_{u - \{r\}},$$

$$\int_{[0,1)^s} \psi_{u\boldsymbol{\kappa}\boldsymbol{\tau}\boldsymbol{\gamma}}(\mathbf{x}) \psi_{u'\boldsymbol{\kappa}'\boldsymbol{\tau}'\boldsymbol{\gamma}'}(\mathbf{x}) \, d\mathbf{x} = \delta_{uu'} \delta_{\boldsymbol{\kappa}\boldsymbol{\kappa}'} \delta_{\boldsymbol{\tau}\boldsymbol{\tau}'} \prod_{r \in u} (\delta_{\gamma_r \gamma_r'} - b^{-1}),$$

where δ is the Kronecker delta function.

The space of integrands, $\mathcal{H}_{\text{wav}}$, consists of all series of wavelet functions (1) whose coefficients converge to zero quickly enough. Let

$$\omega_{u\boldsymbol{\kappa}} = \beta_u b^{-2\alpha|\boldsymbol{\kappa}|}, \qquad \text{and} \qquad \beta_u = \prod_{r \in u} \beta_r$$

for some $\alpha > 0$ and $\beta_r > 0$ for $r = 1, \ldots, s$. Then define the scaled wavelets as $\psi^\omega_{u\boldsymbol{\kappa}\boldsymbol{\tau}\boldsymbol{\gamma}}(\mathbf{x}) = \omega^{1/2}_{u\boldsymbol{\kappa}} \psi_{u\boldsymbol{\kappa}\boldsymbol{\tau}\boldsymbol{\gamma}}(\mathbf{x})$. The space of multivariate Haar wavelets may then be defined as:

$$\mathcal{H}_{\text{wav}} = \Bigg\{ f(\mathbf{x}) = \sum_{u, \boldsymbol{\kappa}, \boldsymbol{\tau}, \boldsymbol{\gamma}} \hat{f}^\omega_{u\boldsymbol{\kappa}\boldsymbol{\tau}\boldsymbol{\gamma}} \psi^\omega_{u\boldsymbol{\kappa}\boldsymbol{\tau}\boldsymbol{\gamma}}(\mathbf{x}) = \hat{\mathbf{f}}^T_\omega \boldsymbol{\psi}_\omega(\mathbf{x}) : \\ \|\hat{\mathbf{f}}_\omega\|_2 < \infty \ \& \sum_{\gamma_r = 0}^{b-1} \hat{f}^\omega_{u\boldsymbol{\kappa}\boldsymbol{\tau}\boldsymbol{\gamma}} = 0, \ \forall r \in u, \ \forall u, \boldsymbol{\kappa}, \boldsymbol{\tau}, \boldsymbol{\gamma}_{u - \{r\}} \Bigg\}.$$

where $\hat{\mathbf{f}}_\omega$ is a column vector of the coefficients $\hat{f}^\omega_{u\boldsymbol{\kappa}\boldsymbol{\tau}\boldsymbol{\gamma}}$, and $\boldsymbol{\psi}_\omega$ is a column vector of the basis functions $\psi^\omega_{u\boldsymbol{\kappa}\boldsymbol{\tau}\boldsymbol{\gamma}}$. Because the wavelets are not linearly independent, the condition on the sum of the series coefficients is required to insure that the series expression for $f \in \mathcal{H}_{\text{wav}}$ is unique. The inner product and norm for $\mathcal{H}_{\text{wav}}$ are defined in terms of the scalar product and $\mathcal{L}_2$-norm of the coefficient vectors:

$$\langle f, g \rangle_{\mathcal{H}_{\text{wav}}} = \left\langle \hat{\mathbf{f}}_\omega, \hat{\mathbf{g}}_\omega \right\rangle_2 = \hat{\mathbf{f}}^T_\omega \hat{\mathbf{g}}_\omega, \quad \|f\|_{\mathcal{H}_{\text{wav}}} = \left\| \hat{\mathbf{f}}_\omega \right\|_2 = (\hat{\mathbf{f}}^T_\omega \hat{\mathbf{f}}_\omega)^{1/2}.$$

3 Digital Sequences

One important family of low discrepancy sequences is the (t, s)-sequences in base b [11]. Moreover, all general constructions of such sequences [1,10,12,16] use the digital method [8,11,13]. Owen [14] proposed a method for randomly scrambling (t, s)-sequences so that they are still (t, s)-sequences with probability one. This random scrambling has been implemented by [7]. The following definition describes the construction of a randomly scrambled digital sequence in a prime base. A similar construction is possible for prime power bases.

Definition 1 *[7] Let $b \geq 2$ be a prime number, and $\mathbf{Z}_b = \{0, 1, \ldots, b-1\}$. Let the following $\infty \times \infty$ matrices and $\infty \times 1$ vectors all have elements in $\mathbf{Z}_b$: predetermined* generator matrices $\mathbf{C}_1, \ldots, \mathbf{C}_s$, *lower triangular* scrambling matrices $\mathbf{L}_1, \ldots, \mathbf{L}_s$ *with nonzero diagonal elements, and* shift vectors $\mathbf{e}_1, \ldots, \mathbf{e}_s$. *For any non-negative integer $i = \cdots i_3 i_2 i_1 (\text{base}\, b)$, define the $\infty \times 1$ vector $\boldsymbol{\Upsilon}(i)$ as the vector of its digits, i.e., $\boldsymbol{\Upsilon}(i) = (i_1, i_2, \ldots)^T$. For any point $z = 0.z_1 z_2 \cdots (\text{base}\, b) \in [0, 1)$, let $\boldsymbol{\phi}(z) = (z_1, z_2, \ldots)^T$ denote the $\infty \times 1$ vector of the digits of z. Then the* scrambled digital sequence *in base b is $\{\mathbf{x}_0, \mathbf{x}_1, \mathbf{x}_2, \ldots\}$, where each $\mathbf{x}_i = (x_{i1}, \ldots, x_{is}) \in [0, 1)^s$ is defined by*

$$\boldsymbol{\phi}(x_{ir}) = \mathbf{L}_r \mathbf{C}_r \boldsymbol{\Upsilon}(i) + \mathbf{e}_r, \quad r = 1, \ldots, s, \ i = 0, 1, \ldots, \tag{2}$$

where all arithmetic operations in the above formula take place using arithmetic modulo b.

The basic non-scrambled digital sequence takes $\mathbf{L}_1 = \cdots = \mathbf{L}_s = \mathbf{I}$, and $\mathbf{e}_1 = \ldots = \mathbf{e}_s = \mathbf{0}$. Owen's randomly scrambled sequence chooses the elements of $\mathbf{L}_1, \ldots, \mathbf{L}_s, \mathbf{e}_1, \ldots, \mathbf{e}_s$ randomly, independently and uniformly over their possible values. The function $\boldsymbol{\phi}$ gives proper b-ary expansions of its arguments, i.e., $\boldsymbol{\phi}(z)$ cannot end in an infinite trail of $b-1$s. Thus, the right side of (2) should not give a vector ending in an infinite trail of $b-1$s almost surely. To insure this, it is assumed that any linear combination of columns of any $\mathbf{C}_r$ cannot be a vector ending in an infinite trail of $b-1$s.

The quality of a digital sequence is often measured by its t-value, which is related to the generator matrices. Smaller values of t imply a better sequence. The lemma below describes how to find the t-value for a digital sequence.

Lemma 1. *[8,9,11] Let $\{\mathbf{x}_0, \mathbf{x}_1, \mathbf{x}_2, \ldots\}$ be a digital sequence in base b with generator matrices $\mathbf{C}_1, \ldots, \mathbf{C}_s$. For any positive integer m let $\mathbf{c}_{rmk}^T$ be the row vector containing the first m columns of the kth row of $\mathbf{C}_r$. Let t be an integer, $0 \leq t \leq m$, such that for all non-negative integers $\boldsymbol{\kappa} = (\kappa_1, \ldots, \kappa_s)$ with $|\boldsymbol{\kappa}| = m - t$ the vectors $\mathbf{c}_{rmk}$, $k = 1, \ldots, \kappa_r$, $r = 1, \ldots, s$, are linearly independent over $\mathbf{Z}_b$. Then for any non-negative integer ν and any $\lambda = 0, \ldots, b-1$ with $\lambda \leq b - (\nu \bmod b)$, the set $\{\mathbf{x}_{\nu b^m}, \ldots, \mathbf{x}_{(\nu+\lambda)b^m - 1}\}$, is a (λ, t, m, s)-net in base b. (Note that a $(1, t, m, s)$-net is the same as a (t, m, s)-net.) If the same value of t holds for all non-negative integers m, then the digital sequence is a (t, s)-sequence.*

4 Problem Formulation

The integration problem studied here is integration over the unit cube:

$$I(f) = \int_{[0,1)^s} f(\mathbf{x}) d\mathbf{x}.$$

Quadrature rules to approximate this integral take the form:

$$Q(f; P, \{w_i\}) = \sum_{i=0}^{n-1} w_i f(\mathbf{x}_i)$$

for some set of nodes $P = \{\mathbf{x}_0, \ldots, \mathbf{x}_{n-1}\} \subset [0,1)^s$ and some set of weights $\{w_i\} = \{w_0, \ldots, w_{n-1}\}$. Quasi-Monte Carlo quadrature methods choose P to be a set of points evenly distributed over the integration domain and $w_i = n^{-1}$ for all i.

The quality of a quadrature rule can be assessed by a worst-case or random-case analysis [5]. Let $\mathcal{B}_{\text{wav}}$ be the unit ball in the Haar wavelet space, i.e., $\mathcal{B}_{\text{wav}} = \{f \in \mathcal{H}_{\text{wav}} : \|f\|_{\mathcal{H}_{\text{wav}}} \leq 1\}$. The quadrature error for a specific integrand and a specific quadrature rule is given by $\text{Err}(f; Q) = I(f) - Q(f; P, \{w_i\})$. Suppose that Q is random, i.e., the nodes, weights, and number of function evaluations are all chosen randomly. Specifically, let Q be chosen from some sample space, $\mathcal{Q}_n$, according to some probability distribution, μ, where the average number of function evaluations is n. (Deterministic quadrature rules are the case where $\mathcal{Q}_n$ has a single element.) The worst-case and random-case error criteria for the Haar wavelet space are:

$$\text{worst-case} \qquad e^{\text{w}}(\mathcal{H}_{\text{wav}}; \mathcal{Q}_n, \mu) := \operatorname*{rms}_{Q \in \mathcal{Q}_n} \sup_{f \in \mathcal{B}_{\text{wav}}} |\text{Err}(f; Q)|, \tag{3a}$$

$$\text{random-case:} \qquad e^{\text{r}}(\mathcal{H}_{\text{wav}}; \mathcal{Q}_n, \mu) := \sup_{f \in \mathcal{B}_{\text{wav}}} \operatorname*{rms}_{Q \in \mathcal{Q}_n} |\text{Err}(f; Q)|. \tag{3b}$$

The operator rms means root mean square. The worst-case error analysis corresponds to the case where your enemy chooses the worst possible integrand after you have chosen the particular quadrature rule. The random-case error analysis corresponds to the case where your enemy chooses the worst possible integrand after knowing your method for randomly choosing quadrature rules, but before you choose a particular one.

The optimal error criteria for the Haar wavelet space are defined as the infima of the above with respect to all possible quadrature rules:

$$e^{\text{w}}(\mathcal{H}_{\text{wav}}, n) := \inf_{\mathcal{Q}_n, \mu} e^{\text{w}}(\mathcal{H}_{\text{wav}}; \mathcal{Q}_n, \mu), \quad e^{\text{r}}(\mathcal{H}_{\text{wav}}, n) := \inf_{\mathcal{Q}_n, \mu} e^{\text{r}}(\mathcal{H}_{\text{wav}}; \mathcal{Q}_n, \mu).$$

A sequence of random quadrature rules $(\mathcal{Q}_{n_m}, \mu_m)$, $m = 0, 1, 2, \ldots$ is said to be optimal if it has the same asymptotic order as best possible quadrature rules. Specifically, one has worst-case and random-case optimality if there exists some nonzero constant C independent of n such that for all $n = 1, 2, \ldots$

$$\min_{n_m \leq n} e^{\mathrm{w}}(\mathcal{H}_{\mathrm{wav}}; \mathcal{Q}_{n_m}, \mu_m) \leq C e^{\mathrm{w}}(\mathcal{H}_{\mathrm{wav}}, n),$$
$$\min_{n_m \leq n} e^{\mathrm{r}}(\mathcal{H}_{\mathrm{wav}}; \mathcal{Q}_{n_m}, \mu_m) \leq C e^{\mathrm{r}}(\mathcal{H}_{\mathrm{wav}}, n).$$

It is possible for a sequence of quadrature rules to be optimal for one of the above criteria and not for the other.

5 Results

A key ingredient in the worst-case and random-case error analyses is the $\infty \times \infty$ matrix whose elements are the mean square errors of integrating the product of any two wavelet functions by a randomized quadrature. Define

$$\boldsymbol{\Lambda} := E_{Q \in \mathcal{Q}_n} \left\{ [\mathrm{Err}(\boldsymbol{\psi}_\omega; Q)][\mathrm{Err}(\boldsymbol{\psi}_\omega; Q)]^T \right\}.$$

Then the worst-case and random-case error analyses can be expressed as in the following theorem.

Theorem 2. *[2,5] Consider the case of random quadrature rules applied to multivariate Haar wavelet series. The error criteria defined in (3) are given by*

$$e^{w}(\mathcal{H}_{wav}; \mathcal{Q}_n) = \sqrt{\mathrm{trace}(\boldsymbol{\Lambda})}, \quad \textit{assuming } \alpha > 1/2,$$
$$e^{r}(\mathcal{H}_{wav}; \mathcal{Q}_n) = \sqrt{\rho(\boldsymbol{\Lambda})}, \quad \textit{assuming } \alpha \geq 0,$$

where trace *denotes the trace, and ρ denotes the spectral radius or largest eigenvalue.*

The assumption $\alpha \geq 0$ is required to insure that the Haar wavelet series are square integrable, so that the random-case error analysis is valid. The assumption $\alpha > 1/2$ is required to insure that the Haar wavelet series are absolutely summable, so that the worst-case error analysis is valid. From this theorem it can be seen that the worst-case error criterion is never smaller than the random-case error criterion because the trace of a matrix is never smaller than its spectral radius. The relationship in Theorem 2 in fact holds for all Hilbert spaces of functions [5].

The above formulas are difficult to evaluate precisely in general. However, for quasi-Monte Carlo quadrature rules based on scrambled digital nets one can derive a simple formula for $e^{\mathrm{w}}(\mathcal{H}_{\mathrm{wav}}; \mathcal{Q}_n)$. For any $\infty \times 1$ vector $\boldsymbol{\phi} = (\phi_1, \phi_2, \ldots)^T$, let $\xi(\boldsymbol{\phi})$ denote the number of zero elements in $\boldsymbol{\phi}$ preceding the first nonzero element:

$$\xi(\boldsymbol{\phi}) = \min\{k : \phi_{k+1} \neq 0\}.$$

In other words, the smallest interval of the form $[0, b^{-k})$, $k = 0, 1, \dots$ that contains z is $[0, b^{-\xi(\phi(z))})$. Next define the function $G(\xi; \alpha)$ as follows:

$$G(\xi;\alpha) = \begin{cases} -1, & \xi = 0, \\ (b^{2\alpha-1}-1)^{-1}[b^{2\alpha-1}(b-1-b^{1-(2\alpha-1)\xi}) + b^{-(2\alpha-1)\xi}], & 0 < \xi < \infty, \\ (b^{2\alpha-1}-1)^{-1}(b-1)b^{2\alpha-1}, & \xi = \infty. \end{cases}$$

The kernel function, $\mathcal{K}_{\mathrm{wav}}(\mathbf{x}, \mathbf{y})$ is defined in terms of G as

$$\mathcal{K}_{\mathrm{wav}}(\mathbf{x}, \mathbf{y}) = -1 + \prod_{r=1}^{s} [1 + \beta_r G(\xi(\phi(x_r) - \phi(y_r)); \alpha)].$$

This is, in fact, the reproducing kernel of the Hilbert space $\mathcal{H}_{\mathrm{wav}}$ [17].

Theorem 3. *[17] Let $\{\mathbf{x}_i\}$ be a basic, non-scrambled digital sequence in a prime power base b as defined in Definition 1. For quasi-Monte Carlo quadrature using any non-scrambled or randomly scrambled digital (λ, t, m, s)-net with $n = \lambda b^m$ points it follows that*

$$[e^{w}(\mathcal{H}_{wav}; \mathcal{Q}_n)]^2 = \frac{1}{n}\left[\sum_{\tilde{\imath}=0}^{b^m-1} \mathcal{K}_{wav}(\mathbf{x}_{\tilde{\imath}}, \mathbf{0}) + \sum_{\hat{\imath}=1}^{\lambda-1} \frac{2(\lambda-\hat{\imath})}{\lambda} \sum_{\tilde{\imath}=0}^{b^m-1} \mathcal{K}_{wav}(\mathbf{x}_{\hat{\imath}b^m+\tilde{\imath}}, \mathbf{0})\right].$$

Although analogous formulas for $e^{\mathrm{w}}(\mathcal{H}_{\mathrm{wav}}; \mathcal{Q}_n)$ exist for general reproducing kernel Hilbert spaces of integrands and general quadrature rules, they require $\mathrm{O}(n^2)$ operations to evaluate. Because of the good match between $\mathcal{H}_{\mathrm{wav}}$ and digital nets the above formula only requires $\mathrm{O}(n)$ operations to evaluate.

The asymptotic behaviour of $e^{\mathrm{w}}(\mathcal{H}_{\mathrm{wav}}; \mathcal{Q}_n)$ and $e^{\mathrm{r}}(\mathcal{H}_{\mathrm{wav}}; \mathcal{Q}_n)$ for scrambled net quadrature may be obtained by looking at the gain coefficients of nets as defined in [15] and analyzed in [6]. Lower bounds on the optimal convergence rates for quadrature rules may be obtained by constructing Haar wavelet series that fool any quadrature rule. Putting these results together leads to the following theorem.

Theorem 4. *[2] For quasi-Monte Carlo quadrature of Haar wavelet series using scrambled (λ, t, m, s)-nets in base b, the error criteria defined in (3) have the following asymptotic orders:*

$$\min_{\lambda b^m \le n} e^{w}(\mathcal{H}_{wav}; \mathcal{Q}_{sc,\lambda b^m}) \asymp e^{w}(\mathcal{H}_{wav}, n) \asymp n^{-\alpha}[\log n]^{(s-1)/2}, \qquad \alpha > 1/2,$$

$$\min_{\lambda b^m \le n} e^{r}(\mathcal{H}_{wav}; \mathcal{Q}_{sc,\lambda b^m}) \asymp e^{r}(\mathcal{H}_{wav}, n) \asymp n^{-\alpha-1/2}, \qquad \alpha \ge 0,$$

where $\asymp$ means "exactly the same asymptotic order".

6 Conclusion

The original reason for investigating the integration of multivariate Haar wavelet series arose from studies of quasi-Monte Carlo quadrature of arbitrary functions. If one uses scrambled nets as the sampling points then this has been shown to be equivalent to integrating Haar wavelet series [4,6]. Thus the results reported above have broader applicability.

However, no matter how smooth one assumes the integrand to be, the best convergence one can obtain using scrambled digital nets is $O(n^{-3/2+\epsilon})$ for the worst-case error. It appears that to handle smoother integrands well one must consider smoother wavelets and different quadrature rules. This is an open problem.

References

1. H. Faure, *Discrépance de suites associées à un système de numération (en dimension s)*, Acta Arith. **41** (1982), 337–351. 101
2. S. Heinrich, F. J. Hickernell, and R. X. Yue, *Optimal quadrature for Haar wavelet spaces*, 2001, submitted for publication to Math. Comp. 103, 104
3. P. Hellekalek and G. Larcher (eds.), *Random and quasi-random point sets*, Lecture Notes in Statistics, vol. 138, Springer-Verlag, New York, 1998. 105
4. F. J. Hickernell and H. S. Hong, *The asymptotic efficiency of randomized nets for quadrature*, Math. Comp. **68** (1999), 767–791. 105
5. F. J. Hickernell and H. Woźniakowski, *The price of pessimism for multidimensional quadrature*, J. Complexity **17** (2001), to appear. 102, 103
6. F. J. Hickernell and R. X. Yue, *The mean square discrepancy of scrambled (t,s)-sequences*, SIAM J. Numer. Anal. **38** (2001), 1089–1112. 104, 105
7. H. S. Hong and F. J. Hickernell, *Implementing scrambled digital nets*, 2001, submitted for publication to ACM TOMS. 101
8. G. Larcher, *Digital point sets: Analysis and applications*, In Hellekalek and Larcher [3], pp. 167–222. 99, 101
9. ———, *On the distribution of digital sequences*, Monte Carlo and quasi-Monte Carlo methods 1996 (H. Niederreiter, P. Hellekalek, G. Larcher, and P. Zinterhof, eds.), Lecture Notes in Statistics, vol. 127, Springer-Verlag, New York, 1998, pp. 109–123. 101
10. H. Niederreiter, *Low discrepancy and low dispersion sequences*, J. Number Theory **30** (1988), 51–70. 101
11. ———, *Random number generation and quasi-Monte Carlo methods*, CBMS-NSF Regional Conference Series in Applied Mathematics, SIAM, Philadelphia, 1992. 99, 101
12. H. Niederreiter and C. Xing, *Quasirandom points and global function fields*, Finite Fields and Applications (S. Cohen and H. Niederreiter, eds.), London Math. Society Lecture Note Series, no. 233, Cambridge University Press, 1996, pp. 269–296. 101
13. ———, *Nets, (t,s)-sequences and algebraic geometry*, In Hellekalek and Larcher [3], pp. 267–302. 101
14. A. B. Owen, *Randomly permuted (t,m,s)-nets and (t,s)-sequences*, Monte Carlo and Quasi-Monte Carlo Methods in Scientific Computing (H. Niederreiter and P. J.-S. Shiue, eds.), Lecture Notes in Statistics, vol. 106, Springer-Verlag, New York, 1995, pp. 299–317. 101

15. ———, *Monte Carlo variance of scrambled net quadrature*, SIAM J. Numer. Anal. **34** (1997), 1884–1910. 99, 100, 104
16. I. M. Sobol', *Multidimensional quadrature formulas and Haar functions (in Russian)*, Izdat. "Nauka", Moscow, 1969. 101
17. R. X. Yue and F. J. Hickernell, *The discrepancy of digital nets*, 2001, submitted to J. Complexity. 104

An Application of Continuous Wavelet Transform in Differential Equations

Qu Han-zhang[1], Xu Chen[2], and Zhao Ruizhen[3]

[1] Xi'an Post and Telecommunications Institute
Xi'an, P. R. China
[2] Xidian University
Xi'an, 710071, P. R.China
[3] Shenzhen University
518060, P.R.China

Abstract. The relation btween some differential equations and the integral equations is discussed;the differential equations can be transformed into the integral equations by using the continuous wavelet transform; the differential equations and the integral equations are equivalent not only in the weak topology but also in the strong topology; the discussion on the differential equations can be connected with the discussion on the integral equations.

1 Introduction

Wavelet theory includes the discret wavelet transform and continuous wavelet transform. On the discrete wavelet transform and its applications there are many papers. But on the continuous wavelet transform and its applications there are a few papers. Especially on the application of the continuous wavelet transform there are few papers. Therefore it is necessary to continue to discuss the wavelet transform and its applications.

On the continuous wavelet transform there are some results. These results mainly come from 'Ten Lecture on Wavelets' Wwritten by Ingrid Daubechies. Among those results there is the following result.

Lemma 1.1[1] $\psi(x) \in L^2(R), 0 < C_\psi < +\infty$, then for any $f(x) \in L^2(R)$

$$f(x) = (2\pi C_\psi)^{-1} \int_R \frac{da}{|a|^2} \int_R < f, \psi_{a,b} > \psi_{a,b} db$$

The above formula is true not only ine weak topology but also in the strong topology.

In this paper we connect some differential equatins with the integral equations by using the continuous wavelet transform, provide a method of the discussing the properties of the differential equations and enlarge the applications of continuous wavelet transform.

Y. Y. Tang et al. (Eds.): WAA 2001, LNCS 2251, pp. 107–116, 2001.

2 Some Differential Equations and the Integral Equations

We consider the following equation.

$$\sum_{k=0}^{n} a_k(x)y^{(k)} = b(x) \tag{1}$$

$\{a_k(x); k = 0, 1, \cdots, n\} \subset L^\infty(R)$, $\{y^{(k)}; k = 0, 1, \cdots, n\} \subset L^2(R)$, $b(x) \in L^2(R)$

Take $\{\psi^{(k)}(x); k = 0, 1, \cdots, n\} \subset L^2(R)$,$Supp(\psi) \subset [-L, L]$,

$$0 < C_\psi = \int_R \frac{|\widehat{\psi}(\eta)|^2}{|\eta|} d\eta = (2\pi)^{-1} < +\infty$$

According to lemma 1.1 there is the following.

$$y(x) = \int_R \frac{da}{|a|^2} \int_R < y, \psi_{a,b} > \psi_{a,b}(x) db \tag{2}$$

We differentiate formula (2). If we could commute the order between the differential and the integral, we should get

$$y^{(k)}(x) = \int_R \frac{da}{|a|^2} \int_R < y, \psi_{a,b} > a^{-k} \psi_{a,b}^{(k)}(x) db \tag{3}$$

$$\{k = 0, 1, \cdots, n\}$$

In formula (1) we should substitute the expressions of y and $y^{(k)}\{k = 0, 1, \ldots, n\}$ for y and $y^{(k)}\{k = 0, 1, \ldots, n\}$ respectively, and should get the following.

$$\int_R \frac{da}{|a|^2} \int_R < y, \psi_{a,b} > \sum_{k=0}^{n} a^{-k} a_k(x) \psi_{a,b}^{(k)}(x) db = b(x) \tag{4}$$

Because we don't know whether the order between the differential and the integral can be commuted, we don't know whether the integral operator in the left of formula (4) exists or not.

$$\sum_{k=0}^{n} a_k(x)y^{(k)} = \int_R \frac{da}{|a|^2} \int_R < y, \psi_{a,b} > \sum_{k=0}^{n} a^{-k} a_k(x) \psi_{a,b}^{(k)}(x) db \tag{5}$$

If formula (5) is true, we say that differential equation (1) is equivalent to integral equation (4). If formula (5) is true in the weak topology, we say that differential equation (1) is equivalent to integral equation (4) in the weak topology. If formula (5) is true in the strong topology, we say that differential equation (1) is equivalent to integral equation (4) in the strong topology.

Define

$$H = \{(f, f', \cdots, f^n); \{f, f', \cdots, f^n\} \subset L^2(R)\} \subset \overbrace{L^2(R) \times \cdots \times L^2(R)}^{n+1}$$

If formula (5) is true, the integral operator in the left of formula (5) is a bounded linear operator from H to $L^2(R)$.

From our assumption we can not know whether the order between the differential the integral can be commuted, so we can not know whether they are equivalent. In this paper we maily discuss whether formula (1) is equivalent to formula (4). Are they equivalent in the weak topology? Are they equivalent in the strong topology?

3 The Relation between Formula (1) and Formula (4) in the Weak Toplogy

In order to discuss whether they are equivalent, firstly we discuss whether they are equivalent in the weak topology.

Theorem 3.1 Formula (5) is true in the weak topology. That is, formula (1) is equivalent to formula (4) in the weak topology.

Proof: We only need to prove that the operator in the left of formula (1) is equivalent to the operator in the left of formula (4) in the weak topology. In order to do this we only need to prove that in the weak topology for $k = 0, 1, \cdots, n, a_k(x)y^{(k)}$ is equivalent to $\int_R \frac{da}{|a|^2} \int_R < y, \psi_{a,b} > a^{-k} a_k(x) \psi_{a,b}^{(k)}(x) db$. That is, for any $g(x) \in L^2(R)$, there is the following.

$$< a_k y^{(k)}, g >= \int_R \frac{da}{|a|^2} \int_R < y, \psi_{a,b} >< a^{-k} a_k \psi_{a,b}^{(k)}, g > db \tag{6}$$

Define

$$\begin{aligned} H_1 = \{(f, f', \cdots, f^n); \{f, f', \cdots, f^n\} \\ \subset L^2(R), k = 0, 1, \cdots, n, Supp(f^{(k)}) \textit{ is compact}\Big\} \end{aligned}$$

For any

$$X = (f, f', \cdots, f^n) \in H_1, \|x\|_{H_1} = \left\{ \sum_{k=0}^{n} \|f^{(k)}\|^2 \right\}^{\frac{1}{2}}.$$

Firstly we prove that formula is true for any $(y, y', \cdots, y^n) \in H_1$.

For any $(y, y', \cdots, y^n) \in H_1$, we only need to calculate the inner product in the left of formula (6).

$$\begin{aligned} &< a_k y^{(k)}, g >=< y^{(k)}, \bar{a}_k g >=< (y^{(k)}\widehat{)}, (\bar{a}_k \widehat{g)} > \\ &= 2\pi \int_R (y^{(k)}\widehat{)}(\eta)(\bar{a}_k \widehat{g)} d(\eta) \int_R \frac{|\widehat{\psi}(W)|^2}{|W|} dW \\ &= 2\pi \int_R (y^{(k)}\widehat{)}(\eta)(\bar{a}_k \widehat{g)} d(\eta) \int_R \frac{|\widehat{\psi}(a\eta)|^2}{|\eta|} d\eta \end{aligned}$$

$$(taking\ w = a\eta)$$

$$= 2\pi \int_R \frac{da}{|a|^2 \int_R (i\eta)^k \widehat{y}(\eta} \overline{\widehat{\psi}}(a\eta)\widehat{\psi}(a\eta)(\overline{\widehat{\overline{a}_k g}})d\eta$$

$$(according\ to\ Fubini's\ theorem^{[2]})$$

$$= \int_R \frac{da}{|a|^2} a^{-k} \int_R [\int_R \widehat{y}(\eta)e^{ib\eta}\overline{\widehat{\psi}}(a\eta)d\eta][\int_R e^{-ib\eta}(ia\eta)^k \widehat{\psi}(a\eta)(\overline{\widehat{\overline{a}_k g}})]d\eta$$

(according to the property that inverse Fourier transform preserves their inner products[3])

$$= \int_R \frac{da}{|a|^2} a^{-k} \int_R [\int_R \widehat{y}(\eta)e^{ib\eta}\overline{\widehat{\psi}}(a\eta)d\eta][\int_R e^{-ib\eta}(\psi^{(k)})\widehat{\ }(a\eta)(\overline{\widehat{\overline{a}_k g}})]d\eta$$

$$(according\ to\ the\ property\ that\ (\psi^{(k)})\widehat{\ }(a\eta) = (ia\eta)^k\widehat{\psi}(a\eta))$$

$$= \int_R \frac{da}{|a|^2} a^{-k} \int_R < \widehat{y}, (\psi_{a,b})\widehat{\ } >< (\psi^{(k)}_{a,b})\widehat{\ }, (\overline{a}_k g)\widehat{\ } > db$$

$$(substituting\ (\psi_{a,b})\widehat{\ }, (\psi^{(k)}_{a,b})\widehat{\ }\ for\ \widehat{\psi}, (\psi^{(k)})\widehat{\ }^{[4]})$$

$$= \int_R \frac{da}{|a|^2} a^{-k} \int_R < y, \psi_{a,b} >< \psi^{(k)}_{a,b}, \overline{a}_k g > db$$

(according to the property that Fourier transform preserves their inner products)

$$= \int_R \frac{da}{|a|^2} \int_R < y, \psi_{a,b} >< a^{-k} a_k \psi^{(k)}_{a,b}, g > db$$

That is, for any $(y, y', \cdots, y^{(n)}) \in H_1$ formula (6) is true. Thatis, the integral operator in the left of formula (6) is a bounded linear operator.

For any $(g_0, g_1, \cdots, g_n) \in \overline{H}_1$, there is a sequence $\left\{y_l, y'_l, \cdots, y_l^{(n)}\right\}$ of H_1 such that for any $m \geq l, \left\{\sum_{k=0}^n \|y_m^{(k)} - y_l^{(k)}\|^2\right\}^{\frac{1}{2}}$ $< 2^{-l}, \lim_{l\to 0} \left\{\sum_{k=0}^n \|g_{(k)} - y_l^{(k)}\|^2\right\}^{\frac{1}{2}} = 0$.

For $k = 0, 1, \cdots, n, g_k = y_1^{(k)} + \sum_{l=1}^{\infty}(y_{l+1}^{(k)} - y_l^{(k)})$. It is true in the strong topology.

According to the properties of the bounded linear functional we have the follwing.

$$< a_k g_k, g >=< g_k, \overline{a}_k g >=< y_1^{(k)} + \sum_{l=1}^{\infty}(y_{l+1}^{(k)} - y_l^{(k)}), \overline{a}_k g >$$

$$= \int_R \frac{da}{|a|^2} \int_R < y_1, \psi_{a,b} >< a^{-k}\psi^{(k)}_{a,b}, \overline{a}_k g > db$$

$$+\sum_{l=1}^{\infty}\int_R \frac{da}{|a|^2}\int_R < y_{l+1}-y_l, \psi_{a,b} >< a^{-k}\psi_{a,b}^{(k)}, \bar{a}_k g > db$$

$$= \int_R \frac{da}{|a|^2}\int_R < y_1+\sum_{l=1}^{\infty}(y_{l+1}-y_l), \psi_{a,b} >< a^{-k}\psi_{a,b}^{(k)}, \bar{a}_k g > db$$

$$= \int_R \frac{da}{|a|^2}\int_R < g_0, \psi_{a,b} >< a^{-k}\psi_{a,b}^{(k)}, \bar{a}_k g > db$$

That is, for any $(g_0, g_1, \cdots, g_n) \in \overline{H}_1$ formula (6) is true in the weak topology.

Because $H_1 \subset H \subset \overline{H}_1$, for any $(y, y', \cdots, y^{(n)}) \in H$ formula (6) is true in the weak topology.

We complete the proof.

4 The Relation between Formula (1) and Formula (4) in the Strong Topology

We prove that formula (1) is equivalent to formula (4) in the strong topology.

Theorem 4.1 Formula (1) is equivalent to formula (4) in the strong topology.

Proof: We prove that the following formula is true.

$$\lim_{A_1\to 0, A_2\to\infty, B\to\infty} \|\sum_{k=0}^{n} a_k(x)y^{(k)}(x) - \int_{A_1\le|a|\le A_2}\frac{da}{|a|^2}\int_{|b|\le B} < y, \psi_{a,b} > \sum_{k=0}^{n} a^{-k}a_k(x)\psi_{a,b}^{(k)}(x)db\| = 0$$

According to the triangle inequality and $a_k \in L^{\infty}(R)$ we only need to prove that for $k = 0, 1, \cdots, n$,

$$\lim_{A_1\to 0, A_2\to\infty, B\to\infty} \|y^{(k)}(x) - \int_{A_1\le|a|\le A_2}\frac{da}{|a|^2}\int_{|b|\le B} < y, \psi_{a,b} > a^{-k}\psi_{a,b}^{(k)}(x)db\| = 0 \tag{7}$$

Firstly we prove that formula (7) is true for any $(y, y', \cdots, y^{(n)}) \in H_1$.

According to Riez's lemma, we have

$$\|y^{(k)}(x) - \int_{A_1\le|a|\le A_2}\frac{da}{|a|^2}\int_{|b|\le B} < y, \psi_{a,b} > a^{-k}\psi_{a,b}^{(k)}(x)db\|$$

$$= \sup_{\|g\|=1} |< y^{(k)} - \int_{A_1\le|a|\le A_2}\frac{da}{|a|^2}\int_{|b|\le B} < y, \psi_{a,b} > a^{-k}\psi_{a,b}^{(k)}db, g > |$$

$$\le \sup_{\|g\|=1} |\int_{A_1\ge|a|}\frac{da}{|a|^2}\int_R < y, \psi_{a,b} >< a^{-k}\psi_{a,b}^{(k)}db, g > |$$

$$+ \sup_{\|g\|=1} |\int_{A_2\le|a|}\frac{da}{|a|^2}\int_R < y, \psi_{a,b} >< a^{-k}\psi_{a,b}^{(k)}db, g > |$$

$$+\sup_{\|g\|=1}\left|\int_{A_1\le|a|\le A_2}\frac{da}{|a|^2}\int_{|b|\ge B}<y,\psi_{a,b}><a^{-k}\psi_{a,b}^{(k)}db,g>\right| \tag{8}$$

$$(according\ to\ formula\ (6)\ and\ the\ triangle\ inequation)$$

According to the process of proving theorem 3.1 there is the following result.

$$\int_R <y,\psi_{a,b}><a^{-k}\psi_{a,b}^{(k)},g>db=\int_R <y^{(k)},\psi_{a,b}><\psi_{a,b},g>db \tag{9}$$

Firstly we prove that the first expression in the end of formula (8) converges to zero as $A_1\to 0,\ A_2,B\to\infty$.

$$\sup_{\|g\|=1}\left|\int_{A_1\ge|a|}\frac{da}{|a|^2}\int_R <y,\psi_{a,b}><a^{-k}\psi_{a,b}^{(k)}db,g>\right|$$

$$=\sup_{\|g\|=1}\left|\int_{A_1\ge|a|}\frac{da}{|a|^2}\int_R <y^{(k)},\psi_{a,b}><\psi_{a,b},g>db\right|$$

$$(according\ formula(9))$$

$$\le\sup_{\|g\|=1}\int_{A_1\ge|a|}\frac{da}{|a|^2}\int_R |<y^{(k)},\psi_{a,b}>||<\psi_{a,b},g>|db$$

$$\le\sup_{\|g\|=1}\left|\left\{\int_{A_1\ge|a|}\frac{da}{|a|^2}\int_R|<y^{(k)},\psi_{a,b}>|^2db\right\}^{\frac{1}{2}}\right.$$

$$\sup_{\|g\|=1}\left\{\int_{A_1\ge|a|}\frac{da}{|a|^2}\int_R|<\psi_{a,b},g>|^2db\right\}^{\frac{1}{2}}$$

$$\le\sup_{\|g\|=1}\left|\left\{\int_{A_1\ge|a|}\frac{da}{|a|^2}\int_R|<y^{(k)},\psi_{a,b}>|^2db\right\}^{\frac{1}{2}}\right.$$

$$(according\ to\ formula(2))$$

The integral converges to zero as $A_1\to 0$ because its infinite integral converges.

The second expression in formula(8) converges to zero as $A_2\to\infty$. Its proving is analogous to the first.

Finally we prove that the third expression in formula (8) converges to zero as $B\to\infty$.

Take

$$M=\sup_{\|g\|=1}\left|\int_{A_1\le|a|\le A_2}\frac{da}{|a|^2}\int_{B\le|b|}<y,\psi_{a,b}><a^K\psi_{a,b}^{(k)},g>db\right|$$

$$\le\sup_{\|g\|=1}\left|\int_{A_1\le|a|\le A_2\ and\ 1>|a|}\frac{da}{|a|^2}\int_{B\le|b|}<y,\psi_{a,b}><a^K\psi_{a,b}^{(k)},g>db\right|$$

$$+ \sup_{\|g\|=1} |\int_{A_1\le|a|\le A_2 \ and \ 1\le|a|} \frac{da}{|a|^2} \int_{B\le|b|} < y, \psi_{a,b} >< a^K \psi_{a,b}^{(k)}, g > db|$$

$$= M_1 + M_2$$

Because for $k = 0, 1, \cdots, n, Supp(y^{(k)})$ is compact, there is an $N_1 > 0$ such that for $k = 0, 1, \cdots, n, Supp(y^{(k)}) \subset [-N_1, N_1$.

If $|x| > n_1, y(x) = 0$. If we take $B > (L + N_1), |x| \le N_1$, and $|a| < 1$, then $|\frac{x-b}{a}| > L$. That is, $M_1 = 0$.

$$M_2 = \sup_{\|g\|=1} |\int_{A_1\le|a|\le A_2 \ and \ 1\le|a|} \frac{da}{|a|^2} \int_{B\le|b|} < y, \psi_{a,b} >< a^K \psi_{a,b}^{(k)}, g > db|$$

$$\le \sup_{\|g\|=1} \int_{A_1\le|a|\le A_2 \ and \ 1\le|a|} \frac{da}{|a|^2} \int_{B\le|b|} | < y, \psi_{a,b} > || < \psi_{a,b}^{(k)}, g > |db$$

$$\le \sup_{\|g\|=1} \left\{ \int_R \frac{da}{|a|^2} \int_{B\le|b|} | < y, \psi_{a,b} > |^2 db \right\}^{\frac{1}{2}}$$

$$\sup_{\|g\|=1} \left\{ \int_R \frac{da}{|a|^2} \int_R | < \psi_{a,b}^{(k)}, g > |^2 db \right\}^{\frac{1}{2}}$$

As $B \to \infty$ the first expression in the above formula converges to zero because its infinite integral converges.The second is bounded.

That is, the third expression in the formula (8) converges to zero as $B \to \infty$. Formula is true for any $(y, y', \cdots, y^{(n)}) \in H_1$.

In order to provr that formula (7) is true for any $(y, y', \cdots, y^{(n)}) \in \overline{H}_1$, we discuss the following bilinear forms.

$$T_{A_1,A_2,B}((f, f', \cdots, f^{(n)}), g) = \int_{A_1\le|a|\le A_2} \frac{da}{|a|^2} \int_{B\le|b|} < y, \psi_{a,b} >< a^K \psi_{a,b}^{(k)}, g > db|$$

$$T((f, f', \cdots, f^{(n)}), g) = \int_R \frac{da}{|a|^2} \int_R < y, \psi_{a,b} >< a^K \psi_{a,b}^{(k)}, g > db|$$

$$(f, f', \cdots, f^{(n)}) \in H_1, g \in L^2(R)$$

According to the definition of the integral we have

$$\lim_{A_1\to 0, \ A_2,B\to\infty} T_{A_1,A_2,B}((f, f', \cdots, f^{(n)}), g) = T((f, f', \cdots, f^{(n)}), g)$$

According to the properties of the bounded linear operators we can generalize $T_{A_1,A_2,B}, T$ from $H_1 \times L^2(R)$ to $\overline{H}_1 \times L^2(R)$.

We fix $g \in L^2(R)$. For any $x \in \overline{H}_1$, we have $\lim_{A_1\to 0, \ A_2,B\to\infty} T_{A_1,A_2,B}(x, g) = T(x, g)$, that is, $\{T_{A_1,A_2,B}(x, g); 0 < A_1 < A_2 < +\infty, 0 < B < +\infty\}$ is bounded.

$\overline{H}_1$ is a closed subspace of $\overbrace{L^2(R) \times \cdots \times L^2(R)}^{(n+1)}$. According to the uniform bounded principle we have that $\{\|T_{A_1,A_2,B}(g)\|; 0 < A_1 < A_2 < +\infty,\ 0 < B < +\infty\}$ is bounded. That is, there is a positive number $\sup\{\|T_{A_1,A_2,B}(g)\|; 0 < A_1 < A_2 < +\infty,\ 0 < B < +\infty\} = K(g)$ such that for any $0 < A_1 < A_2 < +\infty, 0 < B < +\infty, \|T_{A_1,A_2,B}(g)\| \leq K(g)$.

Because for any $g \in L^2(R), \{\|T_{A_1,A_2,B}(g)\|; 0 < A_1 < A_2 < +\infty,\ 0 < B < +\infty\}$ is bounded and $\|T_{A_1,A_2,B}(g_1 + g_2)\| \leq \|T_{A_1,A_2,B}(g_1)\| + \|T_{A_1,A_2,B}(g_2)\|$, $\|T_{A_1,A_2,B}(\alpha g)\| = |\alpha| \|T_{A_1,A_2,B}(g)\|$, $L^2(R)$ is a Hilbert space, according to the uniform bounded principle we have that $\{\|T_{A_1,A_2,B}\|; 0 < A_1 < A_2 < +\infty,\ 0 < B < +\infty\}$ is bounded, that is, there is a positive number K such that $\sup\{\|T_{A_1,A_2,B}\|; 0 < A_1 < A_2 < +\infty, 0 < B < +\infty\} \leq K$.

Secondly we prove that formula (7) is true for any $(g_0, g_1, \cdots, g_n) \in \overline{H}_1$.

For any $\epsilon > 0$, there is $(y, y', \cdots, Y^{(n)}) \in H_1$ such that

$$(\sum_{l=0}^{n} \|g_l - y^{(l)}\|) < \frac{\epsilon}{8}$$

Because for $k = 0, 1, \cdots, n$,

$$\lim_{A_1 \to 0,\ A_2, B \to \infty} \|y^{(k)} - \int_{A_1 \leq |a| \leq A_2} \frac{da}{|a|^2} \int_{B \leq |b|} < y, \psi_{a,b} >< a^{-K} \psi_{a,b}^{(k)} db\| = 0$$

we have $\delta_1 > 0, N_2 > 0, N_B > 0$ such that for $k = 0, 1, \cdots, n$, if $A_1 < \delta_1, A_2 > N_2, B > N_B$,

$$\|y^{(k)} - \int_{A_1 \leq |a| \leq A_2} \frac{da}{|a|^2} \int_{B \leq |b|} < y, \psi_{a,b} >< a^{-K} \psi_{a,b}^{(k)} db\| < \frac{\epsilon}{4}$$

If $A_1 < \delta_1, A_2 > N_2, B > N_B$, we have

$$\|g_{(k)} - \int_{A_1 \leq |a| \leq A_2} \frac{da}{|a|^2} \int_{B \leq |b|} < g_0, \psi_{a,b} >< a^{-K} \psi_{a,b}^{(k)} db\|$$

$$= \sup_{\|g\|=1} | < g_{(k)} - \int_{A_1 \leq |a| \leq A_2} \frac{da}{|a|^2} \int_{B \leq |b|} < g_0, \psi_{a,b} >< a^{-K} \psi_{a,b}^{(k)} db, g > |$$

$$\leq \sup_{\|g\|=1} | < g_k - y^{(k)}, g > |$$

$$+ \sup_{\|g\|=1} | < y^{(k)} - \int_{A_1 \leq |a| \leq A_2} \frac{da}{|a|^2} \int_{B \leq |b|} < y, \psi_{a,b} >< a^{-K} \psi_{a,b}^{(k)} db, g > |$$

$$+ \sup_{\|g\|=1} \int_{A_1 \leq |a| \leq A_2} \frac{da}{|a|^2} \int_{B \leq |b|} < y - g_0, \psi_{a,b} >< a^{-K} \psi_{a,b}^{(k)} db, g > |$$

$$\leq \|g_k - y^{(k)}\|$$

$$\| < y^{(k)} - \int_{A_1 \leq |a| \leq A_2} \frac{da}{|a|^2} \int_{B \leq |b|} < y, \psi_{a,b} >< a^{-K} \psi_{a,b}^{(k)} db\|$$

$$+K\sum_{l=0}^{n}\|y^{(l)}-g_l\| \leq \frac{\epsilon}{8}+\frac{\epsilon}{4}+\frac{\epsilon}{8}<\epsilon$$

Formula (7) is true for any $(g_0, g_1, \cdots, g_n) \in \overline{H}_1$.
Because $H \subset \overline{H}_1$, for any $(y, y', \cdots, y^{(n)}) \in H$,formula (7) is true.
We complete the proof.

5 Example and Conclusion

We introduce the following example.

Example 5.1 We consider the following differential equation.

$$\sum_{i=0}^{n} a_i(x)y^{(i)} = f(x) \qquad (10)$$

$$\{f(x), a_0(x), \cdots, a_n(x)\} \subset C[-\pi,\pi], \left\{y^{(i)}(x); i=0,1,\cdots,n\right\} \subset L^2(R) \qquad (11)$$

Require the solution of differential equation (10) that satisfies formula (11).

If $x \notin [-\pi,\pi]$, for any $i = 0, 1, \cdots, N$, we define $y^{(i)}(x) = 0, a_i(x) = 0$, $f(x) = 0$. Then $\{f(x), a_0(x), \cdots, a_n(x)\} \subset L^\infty(R)$, $\{f(x), a_0(x), \cdots, a_n(x),$ $y(x), \cdots, y^{(n)}(x)\} \subset L^2(R)$.

Take

$$\psi(x) = \begin{cases} cosx & x \in [-\pi,\pi] \\ 0 & x \notin [-\pi,\pi] \end{cases}$$

$$\widehat{\psi}(\eta) = \frac{1}{\sqrt{2\pi}}[\frac{sin(\eta+1)\pi}{\eta+1} + \frac{sin(\eta-1)\pi}{\eta-1}]$$

$$0 < C_\psi < +\infty$$

According to the above results there is the following.

$$(2\pi)^{-1}\int_{-\infty}^{+\infty}\frac{da}{|a|^2}\int_{-|a|\pi+x}^{|a|\pi+x}$$
$$\left\{[\int_{-|a|\pi+b}^{|a|\pi+b} y(z)cos\frac{b-z}{a}dz]\sum_{i=0}^{n} a_i(x)a^{-i}cos(\frac{b-x}{a}+\frac{i\pi}{2})\right\} db = f(x)$$

$$(x \in [-\pi,\pi])$$

In order to solve equation (10) we only need to solve the above integral equation.

We connect some differential equations with the integral equations by using the method of continuous wavelet transform. We obtain that formula (4) is equivalent to formula (4) not only in the weak topology but also in the strong toplogy.

References

1. Ingrid Daubechies. Ten Lectures on Wavelets. Philadelphia. Pennsyvania: Society for Industrial and Applied Mathematics. 1992
2. Zheng Wei-xing, Wang Sheng-wang. Outline of real function and functional analysis. Academical Education Press, China. 1991
3. Song Guo-xiang. Numerical Analysis and Introduction to Wavelet. Science and Technology Press of Henan,China. 1993
4. Charles K.Chui. An Introduction to Wavelets. Academic Press. Inc. 1992

Stability of Biorthogonal Wavelet Bases in $L_2(R)$

Paul F. Curran[1] and Gary McDarby[2]

[1] Department of Electronic and Electrical Engineering, University College Dublin
Belfield, Dublin 4, Ireland
paul.curran@ucd.ie
[2] Medialab Europe,
Crane St., Dublin 8, Ireland
gary@media.mit.edu

Abstract. For stability of biorthogonal wavelet bases associated with finite filter banks, two related Lawton matrices must have a simple eigenvalue at one and all remaining eigenvalues of modulus less than one. If the filters are perturbed these eigenvalues must be re-calculated to determine the stability of the new bases – a numerically intensive task. We present a simpler stability criterion. Starting with stable biorthogonal wavelet bases we perturb the associated filters while ensuring that the new Lawton matrices continue to have an eigenvalue at one. We show that stability of the new biorthogonal wavelet bases first breaks down, not just when a second eigenvalue attains a modulus of one, but rather when this second eigenvalue actually equals one. Stability is therefore established by counting eigenvalues at one of finite matrices. The new criterion, in conjunction with the lifting scheme, provides an algorithm for the custom design of stable filter banks.

1 Introduction

In 1988 Daubechies [1] discovered a class of compactly supported orthonormal bases for $L_2(R)$ which included the Haar basis as a special case. Mallat [2] established the relationship between wavelet transforms and multi-resolution analyses and showed that a discrete wavelet transform (relative to an orthonormal basis) can be implemented using orthogonal filter bank theory.

Whereas orthogonality of the basis is a useful property in the analysis and synthesis of signals, it is not indispensable. In 1992 Cohen, Daubechies and Feauveau [3] introduced the idea of biorthogonal wavelet bases. In this case two distinct bases are employed, one for analysis and one for synthesis. The two bases are not necessarily orthogonal in their own right but are orthogonal to one another. Biorthogonal bases offer increased flexibility in the design of the associated filter bank enabling, for example, the construction of filter banks from linear phase filters. Cohen, Daubechies and Feauveau [3], Cohen and Daubechies [4] and Strang [5] provide necessary and sufficient conditions for a pair of dual filters to generate biorthogonal compactly supported wavelet bases in $L_2(R)$.

Sweldens [6] introduced the lifting scheme for designing biorthogonal filter banks. This scheme formally maintains biorthogonality but does not guarantee

Y. Y. Tang et al. (Eds.): WAA 2001, LNCS 2251, pp. 117–128, 2001.

that the filter bank has associated compactly supported wavelet bases in $L_2(R)$. Whereas the lifting scheme in general contains many free parameters we reformulate it in terms of a single parameter. The principle contribution of the present work is the observation that for real, finite filters the single parameter dependent lifting scheme generates biorthogonal filter banks having associated wavelets in $L_2(R)$ provided the parameter lies in an open interval containing zero. We present an algorithm for finding the largest interval of this kind. In conjunction with the lifting scheme, this algorithm provides a method for the custom design of biorthogonal filter banks with associated wavelet bases in $L_2(R)$. While the method is cumbersome for large filters it has been found to be numerically tractable for filters having up to twenty taps. The resulting wavelet bases depend continuously upon the single parameter of the lifting scheme. In principle therefore it is possible, by employing a variety of different optimisation techniques, to select the value of this parameter such that the associated wavelet basis is optimal in some sense.

2 Lawton Matrices

Given $h = [h_{-m}, \ldots, h_{-1}, h_0, h_1, \ldots, h_m]$, a real filter of length $(2m+1)$, as usual we define the *z-transform* of the filter to be:

$$H(z) = \sum_{k=-m}^{m} h_k \, z^{-k} \, . \tag{1}$$

We say that filter h is *balanced* if $H(1) = 1$. We define also an associated real sequence η as $\eta_k = 2\sum_q h_{q+k} h_q$ where the filter coefficients with indices outside range $-m$ to m are defined to be zero. The real Lawton matrix [7] Λ associated with the filter is the $(4m+1) \times (4m+1)$ matrix:

$$\begin{bmatrix} \eta_{2m} & 0 & 0 & \cdots & 0 & 0 & 0 & \cdots & 0 & 0 & 0 \\ \eta_{2m-2} & \eta_{2m-1} & \eta_{2m} & \cdots & 0 & 0 & 0 & \cdots & 0 & 0 & 0 \\ \vdots & \vdots & \vdots & & \vdots & \vdots & \vdots & & \vdots & \vdots & \vdots \\ \eta_{2m} & \eta_{2m-1} & \eta_{2m-2} & \cdots & \eta_1 & \eta_0 & \eta_1 & \cdots & \eta_{2m-2} & \eta_{2m-1} & \eta_{2m} \\ 0 & 0 & \eta_{2m} & \cdots & \eta_3 & \eta_2 & \eta_1 & \cdots & \eta_{2m-4} & \eta_{2m-3} & \eta_{2m-2} \\ \vdots & \vdots & \vdots & & \vdots & \vdots & \vdots & & \vdots & \vdots & \vdots \\ 0 & 0 & 0 & \cdots & 0 & 0 & 0 & \cdots & \eta_{2m} & \eta_{2m-1} & \eta_{2m-2} \\ 0 & 0 & 0 & \cdots & 0 & 0 & 0 & \cdots & 0 & 0 & \eta_{2m} \end{bmatrix} . \tag{2}$$

We define a pair of *dual real finite filters* to be a set of two real balanced filters $(h, \tilde{h})$ of length $(2m+1)$ and $(2\tilde{m}+1)$ respectively such that

$$\tilde{H}\left(e^{i\theta}\right)\overline{H\left(e^{i\theta}\right)} + \tilde{H}\left(e^{i(\theta+\pi)}\right)\overline{H\left(e^{i(\theta+\pi)}\right)} = 1 \quad \text{for all } \theta. \tag{3}$$

Note that the overbar denotes complex conjugation. The following result is well known [4]:

Stability Condition: *A pair of dual real finite filters, $(h, \tilde{h})$, generate biorthogonal Riesz bases of compactly supported wavelets iff the Lawton matrix associated with each filter has a simple eigenvalue at one and all remaining eigenvalues have modulus less than one.*

Lemma 1. *A pair of dual real finite filters, $(h, \tilde{h})$, generate biorthogonal Riesz bases of compactly supported wavelets only if the sum of the elements in every column of the Lawton matrix associated with each of the filters is one.*

We call this necessary condition on the Lawton matrix associated with a balanced real filter the column sum condition. It transpires that the column sum condition corresponds to a simple condition on the filter itself [8]:

Lemma 2. *The Lawton matrix associated with a real balanced filter h of length $(2m+1)$ satisfies the column sum condition iff $H(-1) = 0$.*

3 Lawton Symmetry

Given a matrix $A \in C^{N \times M}$ with coefficients a_{ij} let matrix $A' \in C^{N \times M}$ be defined by $[A']_{ij} = \bar{a}_{N+1-i,M+1-j}$ for all i, j. Subject to this definition a matrix A is said to be *Lawton symmetric* if $A = A'$. It is not difficult to show that the Lawton matrix associated with a real filter of length $(2m+1)$ is real and Lawton symmetric. We observe also the following result:

Lemma 3. *A real, $(2M+1) \times (2M+1)$, Lawton symmetric matrix, L, has the following structure:*

$$L = \begin{bmatrix} A & a & B \\ b^T & c & (b')^T \\ B' & a' & A' \end{bmatrix} \tag{4}$$

where $A, B \in R^{M \times M}$, $a, b \in R^M$ and $c \in R$.

Employing this result and defining $w^T = [1, \ldots, 1] \in R^{1 \times M}$, $E \in R^{M \times M}$ such that $E_{ij} = \delta_{M+1-i,j}$ (where δ_{ij} denotes the Kronecker delta), we obtain the following:

Lemma 4. *The eigenvalues of a real, $(2M+1) \times (2M+1)$, Lawton symmetric matrix Λ satisfying the column sum condition may be classified as follows:*

1. *One of them equals 1.*
2. *A further M of them are the eigenvalues of the reduced order matrix $(A - BE)$ with one of these being $\frac{1}{2}$.*
3. *The remaining M are eigenvalues of the reduced order matrix $(A + BE - 2aw^T)$.*

We call the eigenvalue at 1 the *symmetric eigenvalue of type (1)* and that at $\frac{1}{2}$ the *skew-symmetric eigenvalue of type (1)*. The remaining $M-1$ eigenvalues of the second class we call the *skew-symmetric eigenvalues of type (2)* and the eigenvalues of the third class we call the *symmetric eigenvalues of type (2)*. The terminology is inspired by the readily established facts that symmetric eigenvalues have associated eigenvectors which are *symmetric*, i.e. have the following form:

$$\begin{bmatrix} x_1 \\ x_0 \\ Ex_1 \end{bmatrix} \quad \text{for some vector } x_1 \in C^{M\times 1} \text{ and } x_0 \in C \tag{5}$$

and similarly that skew-symmetric eigenvalues have associated eigenvectors which are *skew-symmetric*, i.e. which have the following form:

$$\begin{bmatrix} y_1 \\ 0 \\ -Ey_1 \end{bmatrix} \quad \text{for some vector } y_1 \in C^{M\times 1}. \tag{6}$$

4 Non-negativeness

Given any vector $v \in R^{2M+1}$, v will be said to be *non-negative*, denoted $v \geq 0$, if:

$$Re\left(V\left(e^{i\theta}\right)\right) + Im\left(V\left(e^{i\theta}\right)\right) \geq 0 \quad \forall\theta \in [0, 2\pi] \tag{7}$$

where $V(z)$ denotes the z-transform of v as above. All subsequent references to non-negative vectors are understood to be in this sense.

Kreîn and Rutman [9] define a convex cone in a finite dimensional, real vector space to be a subset, $\mathcal{C}$, of the vector space having the following properties:

1. If $x \in \mathcal{C}$ then $\alpha x \in \mathcal{C}$ for all scalars $\alpha \geq 0$.
2. If $x, y \in \mathcal{C}$ then $x + y \in \mathcal{C}$.
3. If $x, y \in \mathcal{C}$ then $x + y \neq 0$.
4. $\mathcal{C}$ is closed relative to the standard Euclidean norm-topology on the vector space.

Consider the set of all real, $(2M+1)\times 1$ non-negative vectors:

$$K = \left\{v \in R^{2M+1} | v \geq 0\right\} . \tag{8}$$

The set K has two properties that prove to be significant in the study of Lawton matrices:

Lemma 5. *K is a convex cone (in the sense of Kreîn and Rutman).*

Lemma 6. $K + (-K) = R^{2M+1}$.

The previous results permit a number of corollaries. Let $\mathcal{L} = \{v \in R^{2M+1} | v = v'\}$, i.e. $\mathcal{L}$ is the set of all real, Lawton symmetric, $(2M+1) \times 1$ vectors.

Corollary 1.

1. *$\mathcal{L}$ is a subspace of R^{2M+1}.*
2. *A real Lawton symmetric matrix maps $\mathcal{L}$ into itself.*
3. *$K \cap \mathcal{L}$ is a convex cone in $\mathcal{L}$.*
4. *$(K \cap \mathcal{L}) + (-K \cap \mathcal{L}) = \mathcal{L}$.*

Let $Z_0 = \{v \in R^{2M+1} \mid [1, \ldots, 1]v = 0,\ [M, \ldots, 1, 0, -1, \ldots, -M]v = 0\}$.

Corollary 2.

1. *Z_0 is a subspace of R^{2M+1}.*
2. *A real Lawton symmetric matrix that satisfies the column sum condition maps Z_0 into itself.*
3. *$(K \cap Z_0)$ is a convex cone in Z_0.*
4. *$(K \cap Z_0) + (-K \cap Z_0) = Z_0$.*

Corollary 3.

1. *$Z_0 \cap \mathcal{L}$ is a subspace of R^{2M+1}.*
2. *A real Lawton symmetric matrix that satisfies the column sum condition maps $Z_0 \cap \mathcal{L}$ into itself.*
3. *$K \cap Z_0 \cap \mathcal{L}$ is a convex cone in $Z_0 \cap \mathcal{L}$.*
4. *$(K \cap Z_0 \cap \mathcal{L}) + (-K \cap Z_0 \cap \mathcal{L}) = Z_0 \cap \mathcal{L}$.*

One further property of non-negative, real vectors that will be required in our subsequent discussion of Lawton matrices may be stated as follows:

Lemma 7. *There exists no non-zero, real, skew-symmetric, non-negative vector in R^{2M+1}.*

Corresponding to the definition of non-negative, real vectors given above we now propose a definition of non-negative, real matrices. Given any matrix $L \in R^{(2M+1)\times(2M+1)}$ we say that L is non-negative, denoted $L \geq 0$, if $Lv \geq 0$ for all $v \geq 0$ in R^{2M+1}.

A significant feature of Lawton matrices associated with real, finite filters is that they are non-negative. This observation is formally stated as follows:

Lemma 8. *The Lawton matrix, A, associated with a real filter of length $(2m+1)$ is non-negative.*

In terms of the cones introduced previously lemma 8 asserts that a Lawton matrix associated with a real, finite filter of length $(2m+1)$ defines a linear operator on real vector space R^{4m+1} which maps the convex cone K (with $M = 2m$) into itself. There exist some elementary, but important, corollaries to this result:

Corollary 4. *By restriction, a Lawton matrix associated with a real, finite filter defines linear operators on real vector spaces $\mathcal{L}$, Z_0, $Z_0 \cap \mathcal{L}$, which map the convex cones $(K \cap \mathcal{L})$, $(K \cap Z_0)$, $(K \cap Z_0 \cap \mathcal{L})$ (with $M = 2m$) respectively into themselves.*

5 Generalised Frobenius-Perron Theory

In their celebrated treatise, Kreĭn and Rutman [9] present a generalisation of the classical Frobenius-Perron theorem which we may paraphrase as follows:

Theorem 1. *Let $\mathcal{C}$ be a convex cone with non-null interior in a real, finite-dimensional vector space; if a linear mapping Q maps $\mathcal{C}$ into itself and is not nilpotent, then there is a real, positive eigenvalue $\lambda_{\mathcal{C}}$ of Q with an associated eigenvector lying in $\mathcal{C}$, having the property that no other eigenvalue of Q has modulus exceeding $\lambda_{\mathcal{C}}$.*

By employing the results of section 4 together with theorem 1, we may make a number of assertions concerning Lawton matrices associated with real, finite filters.

Lemma 9. *Let Λ be a Lawton matrix associated with a real, finite, balanced filter which satisfies the column sum condition, then there exists a real, positive eigenvalue, L, of Λ such that:*

(i) *all remaining eigenvalues of Λ have modulus less than or equal to L,*
(ii) *there exists a real, non-negative eigenvector, $v_{(L)}$, associated with L.*

Lemma 10. *Let Λ be a Lawton matrix associated with a real, finite, balanced filter which satisfies the column sum condition, then there exists a real, positive symmetric eigenvalue, S, of Λ such that:*

(i) *all remaining symmetric eigenvalues of Λ have modulus less than or equal to S,*
(ii) *there exists a real, non-negative eigenvector, $v_{(S)}$, associated with S.*

Lemma 11. *Let Λ be a Lawton matrix associated with a real, finite, balanced filter which satisfies the column sum condition, then there exists a real, positive eigenvalue, ρ, of Λ which is either symmetric of type (2) or skew-symmetric of type (2) such that*

(i) *all remaining symmetric and skew-symmetric eigenvalues of type (2) of* Λ *have modulus less than or equal to* ρ,
(ii) *there exists a real, non-negative eigenvector,* $v_{(\rho)}$, *associated with* ρ.

Lemma 12. *Let* Λ *be a Lawton matrix associated with a real, finite, balanced filter which satisfies the column sum condition, then there exists a real, positive eigenvalue,* σ, *of* Λ *which is symmetric of type (2) such that:*

(i) *all remaining symmetric eigenvalues of type (2) of* Λ *have modulus less than or equal to* σ,
(ii) *there exists a non-negative eigenvector,* $v_{(\sigma)}$, *associated with* σ.

Lemma 7 permits us to make a number of observations concerning the eigenvalues L, S, ρ and σ of lemmas 9-12. The proof of these observations is included to indicate the utility of lemma 7.

Lemma 13. *Eigenvalue* L *is symmetric and equals eigenvalue* S. *Eigenvalue* ρ *is symmetric of type (2) and equals eigenvalue* σ.

Proof. Lemma 9 assures that $v_{(L)}$ is real, non-zero and non-negative. By lemma 7 this vector cannot, therefore, be skew-symmetric. Hence eigenvalue L cannot be skew-symmetric and must, therefore, be symmetric. It is now trivial to show that L = S.

Lemma 11 assures that $v_{(\rho)}$ is real, non-zero and non-negative. As above, lemma 7 asserts that this vector cannot be skew-symmetric and, therefore, that eigenvalue ρ cannot be skew-symmetric. Hence ρ must be symmetric of type (2) and it is now trivial to show that $\rho = \sigma$.

Note that the eigenvalue σ, of lemma 12, is uniquely defined by the Lawton matrix (and hence by the real filter associated with it). We are finally in a position to state and prove the primary result of this investigation:

Theorem 2. *The Lawton matrix associated with a real, finite, balanced filter satisfying* $H(-1) = 0$ *has a simple eigenvalue at one and all remaining eigenvalues have modulus less than one iff the particular eigenvalue* σ *is less than 1.*

Proof. The conditions imposed imply that the associated Lawton matrix is real and satisfies the column sum condition. Hence, the division of eigenvalues into symmetric eigenvalues of types (1) and (2) and skew-symmetric eigenvalues of types (1) and (2) is valid.

If σ is greater than 1 then the Lawton matrix has a real, symmetric eigenvalue of type (2) greater than 1. It follows that the Lawton matrix does not satisfy the eigenvalue condition stated in the theorem.

If σ is equal to 1 then the Lawton matrix has a real, symmetric eigenvalue of type (2) equal to 1. Of course it also has a real, symmetric eigenvalue of type

(1) equal to 1. Hence the matrix has an eigenvalue at 1 of algebraic multiplicity greater than or equal to 2. It follows the Lawton matrix does not satisfy the eigenvalue condition.

If σ is less than 1 then, by lemma 12, all of the symmetric eigenvalues of type (2) of the Lawton matrix have modulus less than or equal to σ, i.e. less than 1. By lemma 13, $\rho = \sigma$, hence, by lemma 11, the skew-symmetric eigenvalues of type (2) of the associated Lawton matrix also have modulus less than 1. The skew-symmetric eigenvalue of type (1) equals $\frac{1}{2}$ and clearly has modulus less than 1. Of course the symmetric eigenvalue of type (1) equals 1. Hence the Lawton matrix satisfies the eigenvalue condition.

Note: the advantage of theorem 2 is that it permits us to test whether a Lawton matrix has a simple eigenvalue at one and all other eigenvalues of modulus less than one, not by checking all of the eigenvalues, but rather by testing a single eigenvalue σ which is known to be real, non-negative and symmetric of type (2). These known properties of σ significantly simplify the numerical task of finding this eigenvalue.

6 The Lifting Scheme

We outline a single parameter form of the lifting scheme as follows:

Theorem 3. *Take any initial set of real finite, balanced dual filters $\{h, \tilde{h}\}$, i.e. filters satisfying the biorthogonal constraint (3). Assume that these filters generate biorthogonal Riesz bases of compactly supported wavelets. Define companion filters g and $\tilde{g}$ as follows:*

$$G\left(e^{i\theta}\right) = e^{-i\theta}\overline{\tilde{H}\left(e^{i(\theta+\pi)}\right)} \quad , \quad \tilde{G}\left(e^{i\theta}\right) = e^{-i\theta}\overline{H\left(e^{i(\theta+\pi)}\right)} \tag{9}$$

then a new set of finite balanced filters $\{h, \tilde{h}^{new}\}$, together with their companion filters $\{g, \tilde{g}^{new}\}$, are generated as follows:

$$\begin{aligned} \tilde{H}^{new}\left(e^{i\theta}\right) &= \tilde{H}\left(e^{i\theta}\right) + \tau\tilde{G}\left(e^{i\theta}\right)\overline{S\left(e^{i2\theta}\right)} \\ G^{new}\left(e^{i\theta}\right) &= G\left(e^{i\theta}\right) - \tau H\left(e^{i\theta}\right)S\left(e^{i2\theta}\right) \end{aligned} \tag{10}$$

where $S\left(e^{i\theta}\right)$ is a real trigonometric polynomial and τ is a real parameter. These new filters also satisfy the biorthogonal constraint, i.e. are dual.

The question arises as to whether, for a given real trigonometric polynomial S and real parameter τ the dual filters $\{h, \tilde{h}^{new}\}$ generate biorthogonal Riesz bases of compactly supported wavelets. A simple necessary condition [8] is stated as follows:

Lemma 14. *The dual filters $\{h, \tilde{h}^{new}\}$ generate biorthogonal Riesz bases of compactly supported wavelets only if $S(1) = 0$.*

The principle contribution of the present work (theorem 2) leads directly to the following result:

Theorem 4. *Assuming $S(1) = 0$ the dual filters $\{h, \tilde{h}^{new}\}$ generate biorthogonal Riesz bases of compactly supported wavelets for all real τ in an open interval containing 0. Moreover, this interval is characterised by the facts that it is maximal and that at any boundary points, but at no interior points, the Lawton matrix associated with $\tilde{h}^{new}$ has a symmetric eigenvalue of type (2) equal to 1.*

By reference to lemmas 3 and 4 it is clear that the Lawton matrix associated with $\tilde{h}^{new}$ has a symmetric eigenvalue of type (2) equal to 1 iff $det(I - (A + BE - 2aw^T)) = 0$ where I is the identity matrix. It is elementary to show that the coefficients of the matrix $(I - (A + BE - 2aw^T)$ are quadratic polynomials in the variable τ. Consequently the evaluation of values of τ for which this determinant equals zero is a special case of the well-known quadratic eigenvalue problem [10]. By means of the standard method of linearisation [10] this problem may in general be converted to the problem of determining the eigenvalues of a matrix of twice the dimension. Specifically let

$$(I - (A + BE - 2aw^T) = I - C_0 - \tau C_1 - \tau^2 C_2 \tag{11}$$

for suitable constant, real matrices C_0 , C_1 , C_2 . Then, assuming $(I - C_0)$ is non-singular, $det(I - (A + BE - 2aw^T)) = 0$ for non-zero parameter value τ iff $(1/\tau)$ is an eigenvalue of the higher order matrix

$$\mathcal{Q} = \begin{bmatrix} 0 & I \\ C_2(I - C_0)^{-1} & C_1(I - C_0)^{-1} \end{bmatrix} . \tag{12}$$

Employing these observations yields a corollary to theorem 4 comprising a more readily tested stability condition.

Corollary 5. *If $S(1) = 0$ and if $(I - C_0)$ is non-singular, dual filters $\{h, \tilde{h}^{new}\}$ generate biorthogonal Riesz bases of compactly supported wavelets for all real τ in an open interval containing 0. Moreover, if they exist, the upper bound of this interval equals the reciprocal of the real, positive eigenvalue of $\mathcal{Q}$ of greatest modulus and the lower bound of this interval equals the reciprocal of the real, negative eigenvalue of $\mathcal{Q}$ of largest modulus.*

Although corollary 5 calls for inversion of matrix $(I - C_0)$ and determination of eigenvalues of the potentially large matrix $\mathcal{Q}$, numerical implemenation is facilitated by two observations: (i) $(I - C_0)$ is in general highly structured so that its inversion requires relatively little numerical effort, (ii) one does not seek all eigenvalues of matrix $\mathcal{Q}$, but rather the largest real positive and largest real negative eigenvalues only.

7 Example

To initialise the lifting scheme select the Haar filters $h = \left[0, \frac{1}{2}, \frac{1}{2}\right] = \tilde{h}$ and their companion filters $g = \left[0, -\frac{1}{2}, \frac{1}{2}\right] = \tilde{g}$. It is readily shown that filters h, $\tilde{h}$ satisfy the biorthogonal constraint (3). Note that filters h, $\tilde{h}$ are real, finite and balanced and that $H(-1) = \tilde{H}(-1) = 0$. They comprise a dual real finite pair of filters. The Lawton matrix associated with both filters is:

$$\begin{bmatrix} 0 & 0 & 0 & 0 & 0 \\ 1 & \frac{1}{2} & 0 & 0 & 0 \\ 0 & \frac{1}{2} & 1 & \frac{1}{2} & 0 \\ 0 & 0 & 0 & \frac{1}{2} & 1 \\ 0 & 0 & 0 & 0 & 0 \end{bmatrix} . \tag{13}$$

It satisfies the column sum condition and has eigenvalues $0, 0, 1, \frac{1}{2}, \frac{1}{2}$. One eigenvalue is 1. It is simple and strictly exceeds all other eigenvalues in modulus. It follows from [4] that filters $\{h, \tilde{h}\}$ generate biorthogonal Riesz bases of compactly supported wavelets. We apply the single parameter form of the lifting scheme using the fixed real trigonometric polynomial:

$$S\left(e^{i\theta}\right) = \left(-e^{i\theta} + e^{-i\theta}\right) \tag{14}$$

which clearly satisfies $S(1) = 0$. The new filters become:

$$h = \left[0, 0, 0, \frac{1}{2}, \frac{1}{2}, 0, 0\right] \quad , \quad \tilde{g} = \left[0, 0, 0, -\frac{1}{2}, \frac{1}{2}, 0, 0\right]$$

$$\tilde{h}^{new} = \left[0, -\frac{\tau}{2}, \frac{\tau}{2}, \frac{1}{2}, \frac{1}{2}, \frac{\tau}{2}, -\frac{\tau}{2}\right] \quad , \quad g^{new} = \left[0, \frac{\tau}{2}, \frac{\tau}{2}, -\frac{1}{2}, \frac{1}{2}, -\frac{\tau}{2}, -\frac{\tau}{2}\right] . \tag{15}$$

The Lawton matrix associated with filter $\tilde{h}^{new}$ is, of course, Lawton symmetric. As this matrix is 13×13 we elect not to write it out in full. However, by comparing with the canonical structure of lemma 3 we can identify the submatrices:

$$A = \begin{bmatrix} 0 & 0 & 0 & 0 & 0 & 0 \\ -\tau^2 & \frac{\tau^2}{2} & 0 & 0 & 0 & 0 \\ 0 & -\tau+\frac{\tau^2}{2} & -\tau^2 & \frac{\tau^2}{2} & 0 & 0 \\ 1+2\tau^2 & \frac{1}{2}+\tau-\tau^2 & 0 & -\tau+\frac{\tau^2}{2} & -\tau^2 & \frac{\tau^2}{2} \\ 0 & \frac{1}{2}+\tau-\tau^2 & 1+2\tau^2 & \frac{1}{2}+\tau-\tau^2 & 0 & -\tau+\frac{\tau^2}{2} \\ -\tau^2 & -\tau+\frac{\tau^2}{2} & 0 & \frac{1}{2}+\tau-\tau^2 & 1+2\tau^2 & \frac{1}{2}+\tau-\tau^2 \end{bmatrix} , \quad a = \begin{bmatrix} 0 \\ 0 \\ 0 \\ 0 \\ -\tau^2 \\ 0 \end{bmatrix} , \tag{16}$$

$$B = \begin{bmatrix} 0 & 0 & 0 & 0 & 0 & 0 \\ 0 & 0 & 0 & 0 & 0 & 0 \\ 0 & 0 & 0 & 0 & 0 & 0 \\ 0 & 0 & 0 & 0 & 0 & 0 \\ \frac{\tau^2}{2} & 0 & 0 & 0 & 0 & 0 \\ \left(-\tau+\frac{\tau^2}{2}\right) & -\tau^2 & \frac{\tau^2}{2} & 0 & 0 & 0 \end{bmatrix} , \quad b = \begin{bmatrix} 0 \\ \frac{\tau^2}{2} \\ -\tau^2 \\ -\tau+\frac{\tau^2}{2} \\ 0 \\ \frac{1}{2}+\tau-\frac{\tau^2}{2} \end{bmatrix} , \tag{17}$$

$$c = 1 + 2\tau^2 . \tag{18}$$

The symmetric eigenvalues of type (2) are the eigenvalues of the reduced order matrix $(A + BE - 2aw^T) =$

$$\begin{bmatrix} 0 & 0 & 0 & 0 & 0 & 0 \\ -\tau^2 & \frac{\tau^2}{2} & 0 & 0 & 0 & 0 \\ 0 & -\tau+\frac{\tau^2}{2} & -\tau^2 & \frac{\tau^2}{2} & 0 & 0 \\ 1+2\tau^2 & \frac{1}{2}+\tau-\tau^2 & 0 & -\tau+\frac{\tau^2}{2} & -\tau^2 & \frac{\tau^2}{2} \\ 2\tau^2 & \frac{1}{2}+\tau-\tau^2 & 1+4\tau^2 & \frac{1}{2}+\tau+\tau^2 & 2\tau^2 & -\tau+3\tau^2 \\ -\tau^2 & -\tau+\frac{\tau^2}{2} & 0 & \frac{1}{2}+\tau-\frac{\tau^2}{2} & 1+\tau^2 & \frac{1}{2}-\frac{\tau^2}{2} \end{bmatrix} . \tag{19}$$

The matrix has a symmetric eigenvalue of type (2) equal to 1 iff $det(I - (A + BE - 2aw^T)) = 0$ where I is the identity matrix. With reference to (11) we note that in the present case

$$I - C_0 = \begin{bmatrix} 1 & 0 & 0 & 0 & 0 & 0 \\ 0 & 1 & 0 & 0 & 0 & 0 \\ 0 & 0 & 1 & 0 & 0 & 0 \\ -1 & -\frac{1}{2} & 0 & 1 & 0 & 0 \\ 0 & -\frac{1}{2} & -1 & -\frac{1}{2} & 1 & 0 \\ 0 & 0 & 0 & -\frac{1}{2} & -1 & \frac{1}{2} \end{bmatrix} . \tag{20}$$

Not only is this matrix non-singular, it is also lower triangular and therefore readily invertible. In all cases examined by the authors the matrix $(I - C_0)$ turns out, not only to be non-singular, but also to be sparse and readily invertible. It is therefore feasible to construct matrix $\mathcal{Q}$ of (12) and to determine its eigenvalues.

In the present case, however, we do not actually need to employ linearisation. It is feasible to apply theorem 4 directly since $det(I - (A + BE - 2aw^T))$ is readily shown to equal the polynomial in τ:

$$\frac{1}{2}\left(1 - \frac{\tau^2}{2}\right)\left(1 + \tau^2\right)\left(1 + 2\tau - 8\tau^2\right)(1 + \tau) \tag{21}$$

whose roots are: $\pm\sqrt{2}, \pm i, -1, -\frac{1}{4}, \frac{1}{2}$. The maximal real open interval containing 0 with boundary points, but no interior points, in this set is given by $-\frac{1}{4} < \tau < \frac{1}{2}$. Hence, for any value of τ between $-\frac{1}{4}$ and $\frac{1}{2}$ the resulting filters $\{h, \tilde{h}^{new}\}$ generate biorthogonal Riesz bases of compactly supported wavelets.

8 Conclusions

We have formulated a single parameter form of the lifting scheme. We have shown that the scheme generates biorthogonal filter banks having associated wavelets in $L_2(R)$ provided the parameter lies in a certain open interval and have developed a method for finding the largest such interval. Numerically this method is equivalent to a special case of the quadratic eigenvalue problem. For low order filters the method of linearisation is in general appropriate. A single matrix inversion is required in the application of the linearisation method which reduces the problem to a standard eigenvalue problem (or rather to the problem of finding the largest and smallest non-zero, real eigenvalues of a matrix). It transpires that the matrix inversion often requires relatively little numerical effort. For high order filters more advanced techniques for solving the quadratic eigenvalue problem (e.g. the Jacobi-Davidson method) would be required. We note that the parameterised lifting scheme, in conjunction with this method, yields a class of biorthogonal filter banks with associated wavelet bases in $L_2(R)$ and that this class is itself parameterised. Clearly one may employ a stochastic algorithm to determine the filter bank in this parameterised class which is optimal with respect to some desirable property (such as maximum energy compaction, desired shape, etc.).

References

1. I. Daubechies, I.: Orthonormal Bases of Compactly Supported Wavelets. Comm. Pure Applied Math. **41** (1988) 909–996 117
2. Mallat, S.: A Theory for Multiresolution Signal Decomposition: The Wavelet Representation. IEEE Transaction on Pattern Analysis and Machine Intelligence **11** (1989) 674–693 117
3. Cohen, A., Daubechies, I., Feauveau, J. C.: Bi-orthogonal Bases of Compactly Supported Wavelets. Comm. Pure Applied Math. **45** (1992) 485–560 117
4. Cohen, A., Daubechies, I.: A Stability-Criterion for Biorthogonal Wavelet Bases and their Related Subband Coding Scheme. Duke Mathematical Journal **86** (1992) 313–335 117, 118, 126
5. Strang, G.: Eigenvalues of $(\downarrow 2)H$ and convergence of the cascade algorithm. IEEE transactions on signal processing **44** (1996) 233–238 117
6. Sweldens, W.: The Lifting Scheme: A Custom-Design Construction of Biorthogonal Wavelets. Appl. Comput. Harmon. Analysis **3** (1996) 186–200 117
7. Lawton, W. M.: Necessary and Sufficient Conditions for Constructing Orthonormal Wavelet Bases. Journal Math. Phys. **32** (1991) 57–61 118
8. McDarby, G., Curran, P., Heneghan, C., Celler, B.: Necessary Conditions on the Lifting Scheme for Existence of Wavelets in $L_2(R)$. ICASSP, Istanbul, (2000) 119, 124
9. Kreîn, M. G., Rutman, M. A.: Linear Operators Leaving Invariant a Cone in a Banach Space. Functional Analysis and Measure Theory. American Mathematical Society, Providence R. I., 10 Translation Series 1 (1962) 199–325 120, 122
10. Gohberg, I., Lancaster, P., Rodman, L.: Matrix Polynomials. Academic Press, New York, (1982) 125

Characterization of Dirac Edge with New Wavelet Transform

Lihua Yang[1], Xinge You[2], Robert M. Haralick[3], Ihsin T. Phillips[4], and Yuan Y. Tang[2]

[1] Department of Mathematics, Zhongshan University Guangzhou 510275, P. R. China
[2] Department of Computer Science, Hong Kong Baptist University Kowloon Tong, Hong Kong
{yytang,xyou}@comp.hkbu.edu.hk
[3] Department of Computer Science, Graduate Center, City University of New York 365 Fifth Ave., New York, NY 10016, USA
haralick@gc.cuny.edu
[4] Department of Computer Science, Queens College, City University of New York 65-30 Kissena Blvd., Flushing, NY 11367 USA
yun@image.cs.qc.edu

Abstract. This paper aims at studying the characterization of Dirac-structure edges with a novel wavelet transform, and selecting the suitable wavelet functions to detect them. Three significant characteristics of the local maximum modulus of the wavelet transform with respect to the Dirac-structure edges are presented. By utilizing a novel continuous wavelet, it is proven that the local maxima modulus of such continuous wavelet transform of a Dirac-structure edge forms two new curves which are located symmetrically at the two sides of the original one and have the same direction with it and the distance between the two curves is estimated. An algorithm to detect curves in an image by utilizing the above invariants is developed. Several experiments are conducted, and positive results are obtained.

1 Introduction

In our previous paper [7], we presented a novel method based on the quadratic spline wavelet, to identify different structures of edges, and thereafter, to extract the Dirac-structure ones. Furthermore, a very important characterization of the Dirac-structure edges by wavelet transform was provided. Three significant characteristics of the local maximum modulus of the wavelet transform with respect to the Dirac-structure edges were presented, namely: (1) slope invariant: the local maximum modulus of the wavelet transform of a Dirac-structure edge is independent on the slope of the edge. (2) grey-level invariant: the local maximum modulus of the wavelet transform with respect to a Dirac-structure edge takes place at the same points when the images with different grey-levels are to be processed. (3) width light-dependent: for various widths of the Dirac-structure

Y. Y. Tang et al. (Eds.): WAA 2001, LNCS 2251, pp. 129–138, 2001.

edge images, the location of maximum modulus of the wavelet transform varies slightly when the scale s of the wavelet transform is larger than the width d of the Dirac-structure edge images. Based on the characteristics, a novel algorithm to detect the Dirac-structure edges from an image has been developed. Some examples of applying this algorithm to detect the Dirac-structure edge can be found in [7]. However, there are some weaknesses in the method. This paper proposes a great improvement of that work.

Noted the foregoing third property in [7], it says "width light-dependent", does not say "width invariant". This means that for various widths of the Dirac-structure edge images, the location of maximum modulus of the wavelet transform may change. These changes are small. What we want is that the location of maximum modulus does not change, i.e. the location of maximum modulus has the property of width invariant. Let us look at Fig. 1. The first row of Fig. 1 has three circles. The left image is the original one which contains a circle with various width. The middle one is the location of maximum modulus of the wavelet transform with scale $s = 6$, which depends on the width of the circle in some way. Finally, by utilizing the algorithm proposed in [7], the central line of the circle is extracted and displayed on the right of Fig. 1. We can find that the central line of the circle is broken. The second row of Fig. 1 has trees, where the sizes of the branches vary, some are thick and some are thin. The left image is the original one, and the right of Fig. 1 is the central line extracted utilizing the algorithm proposed in [7]. It is easy to see that some branches of the tree are lost. To overcome such a defect, In this paper, a novel wavelet is utilized, so

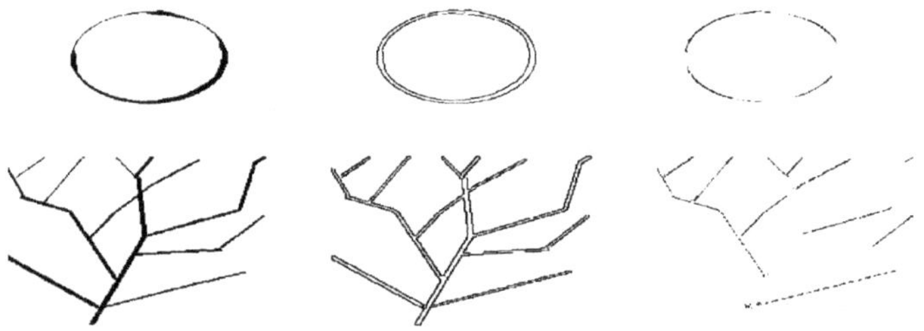

Fig. 1. Left: the original images; Middle: the location of maximum modulus of the wavelet transform with $s = 6$; Right: the central line images extracted by the algorithm of [7]

that the above "width light-dependent" properties can be improved to "width invariant" without losing the "slope invariant" and "grey-level invariant". Due to this improvement, the detection of curves is more accurate.

This paper is organized as follows: Section 2 will be a brief review of the scale wavelet transform followed by the construction of a special wavelet. In Section 3, a characterization of the Dirac-structure edges in an image by wavelet transform will be developed. Then, In Section 4, an algorithm to extract the central line of a curve will be presented, and several experiments will be illustrated. At last, some conclusions will be provided in Section 5.

2 Continuous Wavelet Transform with New Wavelet Function

Let $L^2(R^2)$ be the Hilbert space of all the square-integrable 2-D functions on plane R^2, $\psi \in L^2(R^2)$ is called a wavelet function, if

$$\int_R \int_R \psi(x,y)dxdy = 0, \tag{1}$$

For $f \in L^2(R^2)$ and scale $s > 0$, the scale wavelet transform of $f(x,y)$ is defined by

$$\begin{aligned} W_s f(x,y) &:= (f * \psi_s)(x,y) \\ &= \int_R \int_R f(u,v) \frac{1}{s^2} \psi(\frac{x-u}{s}, \frac{y-v}{s}) dudv, \end{aligned} \tag{2}$$

where * denotes the convolution operator, and $\psi_s(u,v) := \frac{1}{s^2}\psi(\frac{u}{s}, \frac{v}{s})$. Obviously, the scale wavelet transform described in Eq. (2) is a filter, and since $\psi \in L^2(R^2)$, its Fourier transform can be defined by $\hat{\psi}(\xi,\eta) := \int_R \int_R \psi(x,y) e^{-i(\xi x + \eta y)} dxdy$ which satisfies the condition of $\hat{\psi} \in L^2(R^2)$. Thus, both functions ψ and $\hat{\psi}$ decrease at infinity. For a general theory of the scale wavelet transform, it can be found in [2,3].

However, the wavelet transform differs from Fourier transform. There is only one basic function in the latter, while there exist many different wavelet functions in the former. Therefore, it is very important to select the one that is as "good" as possible according to its particular applications.

Theoretically, Eq. (1), i.e. $\hat{\psi}(0,0) = 0$, implies that $\psi(x,y)$ is a band-pass filter, but a high-pass one because of the decrease of its Fourier transform at infinity. It is easy to see that the partial derivatives of a low-pass function can become the candidates of the wavelet functions. In this paper, we consider such kind of wavelets, i.e., $\psi^1(x,y) := \frac{\partial}{\partial x}\theta(x,y)$, $\psi^2(x,y) := \frac{\partial}{\partial y}\theta(x,y)$ where $\theta(u,v)$ denotes a real function satisfying: 1)$\theta(u,v)$ fast decreases at infinity; 2)$\theta(u,v)$ is an even function on both u and v; 3)$\hat{\theta}(0,0) = 1$.

For wavelet $\psi^1(x,y)$ defined above, its scale wavelet transform is

$$W_s^1 f(x,y) = s\frac{\partial}{\partial x}(f * \theta_s)(x,y)$$

where $\theta_s(x,y) := \frac{1}{s^2}\theta(\frac{x}{s}, \frac{y}{s})$.

This formula is equivalent to the classical multi-scale edge detection [1,5], if $\theta(x,y)$ is set to be a Gaussian. A similar explanation for wavelet $\psi^2(x,y)$ defined above can be made. However, the partial derivative is along the vertical direction instead of the horizontal one.

Gaussian function has been employing in image processing. It possesses some excellent properties, such as, the locality in both the time domain and frequency domain, the same widths in both the time-window and frequency-window, and so on. All these properties make it applied extensively and deeply in the area of the filtering, and it already almost becomes the best candidate of low-pass filter in practice. Unfortunately, Gaussian function is not always the best one for all applications. In fact, we have shown that it is not the best candidate for characterizing a Dirac-structure edge [7]. Even the quadratic spline wavelet is better than it, although, the quadratic spline wavelet is still not a perfect one for such applications. In [7] it has been proved that the location of maximum modulus of the wavelet transform with respect to a Dirac-structure edge is not width invariant. It still depends on the width of the edge even though it depends lightly. To avoid such dissatisfaction, a novel wavelet is constructed and used in this paper, and its definition is described below.

Let

$$\begin{cases} \psi_1(x) = -\frac{2}{\pi}(-8x \ln \frac{1+\sqrt{1-16x^2}}{4x} + \frac{1}{2x}\sqrt{1-16x^2}) \\ \psi_2(x) = -\frac{2}{\pi}(8x \ln \frac{3+\sqrt{9-16x^2}}{4x} - \frac{3}{2x}\sqrt{9-16x^2}) \\ \psi_3(x) = -\frac{2}{\pi}(-4x \ln \frac{1+\sqrt{1-x^2}}{x} + \frac{4}{x}\sqrt{1-x^2}) \end{cases}$$

Then, the 1-D wavelet $\psi(x)$ is an odd function defined on $(0, \infty)$ by

$$\psi(x) := \begin{cases} \psi_1(x) + \psi_2(x) + \psi_3(x) & x \in (0, \frac{1}{4}) \\ \psi_2(x) + \psi_3(x) & x \in [\frac{1}{4}, \frac{3}{4}) \\ \psi_3(x) & x \in [\frac{3}{4}, 1) \\ 0 & x \in [1, \infty) \end{cases} \tag{3}$$

Let $\phi(x) := \int_0^x \psi(x)dx$. Then $\phi(x)$ is an even function, compactly supported on [-1, 1], and $\phi'(x) = \psi(x)$. The smoothness function $\theta(x, y)$ is then defined by $\theta(x, y) := \phi(\sqrt{x^2 + y^2})$, and the 2-D wavelets are defined by

$$\begin{cases} \psi^1(x, y) := \frac{\partial}{\partial x}\theta(x, y) = \phi'(\sqrt{x^2 + y^2})\frac{x}{\sqrt{x^2+y^2}} \\ \psi^2(x, y) := \frac{\partial}{\partial y}\theta(x, y) = \phi'(\sqrt{x^2 + y^2})\frac{y}{\sqrt{x^2+y^2}}. \end{cases} \tag{4}$$

and are illustrated in Fig. 2.

The gradient direction and the amplitude of the wavelet transform are denoted respectively by

$$\nabla W_s f(x, y) := \begin{pmatrix} W_s^1 f(x, y) \\ W_s^2 f(x, y) \end{pmatrix}, \tag{5}$$

and

$$|\nabla W_s f(x, y)| := \sqrt{|W_s^1 f(x, y)|^2 + |W_s^2 f(x, y)|^2}. \tag{6}$$

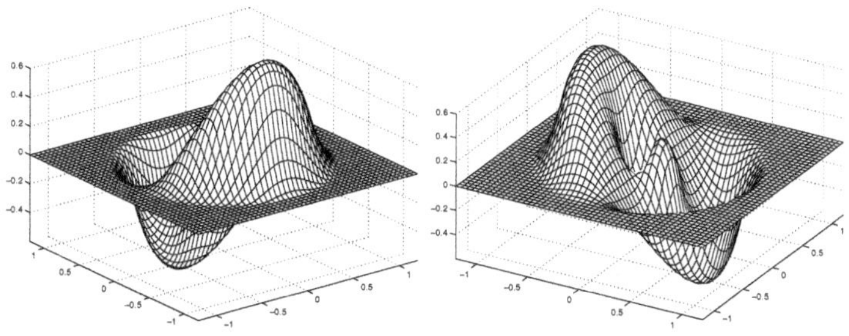

Fig. 2. The graphical descriptions of 2-D wavelet functions: left - function $\psi^1(x, y)$; right - function $\psi^2(x, y)$

By locating the local maxima of $|\nabla W_s f(x, y)|$, we can detect the edges of the images.

3 Characterization of Curves through New Wavelet Transform

In this section, three significant characteristics of the local maximum modulus of the wavelet transform with respect to the Dirac-structure edges in images will be presented, namely:

- Grey-level invariant: the local maximum modulus of the wavelet transform with respect to a Dirac-structure edge takes place at the same points when the images with different grey-levels are to be processed.
- Slope invariant: the local maximum modulus of the wavelet transform of a Dirac-structure edge is independent on the slope of the edge.
- Width invariant: for various widths of the Dirac-structure edges in an image, the location of maximum modulus of the wavelet transform does not vary under certain circumstance.

The proof of the above characteristics may be obtained similarly to our previous work [7].

However, it concluded mathematically that the amplitude of of the wavelet transform $|\nabla W_s f(x, y)|$ reaches the local maximum if and only if the scale $s \geq d$. Namely, the local maxima of $|\nabla W_s f_{l_d}(x_\rho, y_\rho)|$ arrive at both sides of the central line l of l_d and the distance from l is $\frac{s}{2}$, which is independent on the width d. In summary, The above three invariance properties can be rewritten as the following theorem:

Theorem 1. *Let l_d be a Dirac-structure edge with width d and l be its central line. The local maxima modulus of the wavelet transform corresponding to the wavelets of Eq. (4) forms two new lines which are located symmetrically on both sides of the central line, and have the same direction with it. If scale $s \geq d$, then the distance between the two new ones equals to s.*

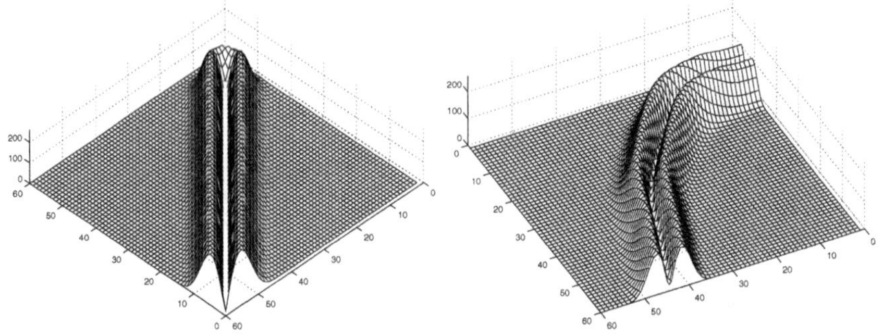

Fig. 3. Modulus of wavelet transforms corresponding a segment of straight line and a curve

This theorem describes the property of width-invariance, which is important. It improves our former results in [7]. Namely, for each scale s, the local maximum moduli of the wavelet transforms with respect to the curves of different widths are located at the same positions. A couple of graphical examples are shown in Fig. 3.

4 Algorithm and Experiments

In this section, the algorithm for extracting the Dirac-structure edges will be presented. Several experiments will also be conducted.

4.1 Algorithm

In practice, the wavelet transform should be calculated discretely. We have the following formula:

$$\begin{aligned} W_s^i f(n,m) &= \int\int f(u,v)\psi_s^i(n-u,m-v)dudv \\ &= \sum_{k,l} f(k,l)\int_k^{k+1}\int_l^{l+1} \psi_s^i(n-u,m-v)dudv \\ &= \sum_{k,l} f(n-k-1,m-l-1)\psi_{k,l}^{s,i}\,, \quad (i=1,2), \end{aligned}$$

where $\psi_{k,l}^{s,i} = \int_k^{k+1}\int_l^{l+1}\psi_s^i(u,v)dudv = \int_{k/s}^{(k+1)/s}\int_{l/s}^{(l+1)/s}\psi^i(u,v)dudv$, $(i = 1,2)$. Next, we give the calculating formulae of the coefficients $\{\psi_{k,l}^{s,i}\}$. The details can be found in [4,6,7]. It is deduced easily that

$$\psi_{k,l}^{s,1} = \psi_{k,l}^{s,2}, \quad \psi_{-k,l}^{s,1} = -\psi_{k-1,l}^{s,1} \quad \psi_{k,-l}^{s,1} = \psi_{k,l-1}^{s,1}, \quad \psi_{-k,-l}^{s,1} = -\psi_{k-1,l-1}^{s,1}.$$

Through further calculating, we have $\psi_{k,l}^{s,1} = \phi_{l,k+1}^{s} + \phi_{l+1,k}^{s} - \phi_{l+1,k+1}^{s} - \phi_{l,k}^{s}$, for all non-negative integers k and l, $\phi_{k,l}^{s} = \int_{\frac{k}{s}\sqrt{k^2+l^2}}^{1} \left[\frac{l}{s} - \sqrt{v^2 - (l/s)^2}\right] \psi(v)dv$. On the other hand, it is easy to see that $\phi_{k,l}^{s} = 0$ for all integers k, l satisfying $k^2 + l^2 \geq s^2$ due to the compact support $[-1, 1]$ of $\phi(x)$. we can calculate

Table 1. The nonzero coefficients $\{\phi_{k,l}^{s}\}$ for $s = 2$

$l \backslash k$	$l = 0$	$l = 1$
$k = 0$	0.2500	0.1250
$k = 1$	0.0497	0.0111

all the coefficients $\phi_{k,l}^{s}$ numerically for non-negative integers k, l. The possible nonzero items of $\phi_{k,l}^{s}$ for $s = 2, 4, 6, 8$ are listed in Tables 1 - 4. Based on the

Table 2. The nonzero coefficients $\{\phi_{k,l}^{s}\}$ for $s = 4$

$k \backslash l$	$l = 0$	$l = 1$	$l = 2$	$l = 3$
$k = 0$	0.2500	0.2292	0.1250	0.0208
$k = 1$	0.1468	0.1206	0.0552	0.0060
$k = 2$	0.0497	0.0366	0.0111	0.0003
$k = 3$	0.0047	0.0026	0.0002	0

characterization of a straight line in an image developed in Section 3, an algorithm to detect straight lines in an image can be designed. The result is also valid for general curves since a short segment of a curve can be regarded as a straight line approximately. In fact, wavelet transforms are essentially local analysis. Therefore the result of Theorem 1 can be applied to the general curves in an image. Our algorithm to detect curves in an image is designed as follows.

Table 3. The nonzero coefficients $\{\phi_{k,l}^{s}\}$ for $s = 6$

$k \backslash l$	$l = 0$	$l = 1$	$l = 2$	$l = 3$	$l = 4$	$l = 5$
$k = 0$	0.2500	0.2438	0.2022	0.1250	0.0478	0.0062
$k = 1$	0.1831	0.1718	0.1333	0.0767	0.0257	0.0025
$k = 2$	0.1106	0.1003	0.0723	0.0367	0.0094	0.0005
$k = 3$	0.0497	0.0436	0.0281	0.0111	0.0017	0.0000
$k = 4$	0.0133	0.0109	0.0056	0.0014	0.0000	0
$k = 5$	0.0011	0.0008	0.0002	0.0000	0	0

Table 4. The nonzero coefficients $\{\phi^s_{k,l}\}$ for $s = 8$

$k\backslash l$	$l=0$	$l=1$	$l=2$	$l=3$	$l=4$	$l=5$	$l=6$	$l=7$
$k=0$	0.2500	0.2474	0.2292	0.1849	0.1250	0.0651	0.0208	0.0026
$k=1$	0.2006	0.1950	0.1741	0.1358	0.0884	0.0433	0.0126	0.0013
$k=2$	0.1468	0.1403	0.1206	0.0902	0.0552	0.0244	0.0060	0.0004
$k=3$	0.0935	0.0882	0.0733	0.0517	0.0287	0.0107	0.0020	0.0000
$k=4$	0.0497	0.0462	0.0366	0.0236	0.0111	0.0032	0.0003	0
$k=5$	0.0199	0.0180	0.0132	0.0072	0.0026	0.0004	0.0000	0
$k=6$	0.0047	0.0041	0.0026	0.0011	0.0002	0.0000	0	0
$k=7$	0.0004	0.0003	0.0001	0.0000	0	0	0	0

Algorithm 1 Let $f(x, y)$ be an image containing curves. For a scale $s > 0$,

Step 1 Calculate all the wavelet transforms $\{W^1_s f(x, y),\ W^2_s f(x, y)\}$ with respect to the wavelets defined by Eq.(4).

Step 2 Calculate the local maxima f_{locmax} of $|\nabla W_s f(x, y)|$ and the gradient direction $f_{gradient}$.

Step 3 For each point (x, y) with local maximum, search the point whose distance along the gradient direction from (x, y) is s. If it is a point of local maxima, the center point is detected.

Step 4 The curves formed by all the points detected in Step 3 are what we need.

4.2 Experiments

Let us turn back to Section 1, and look at Fig. 1. The particular task is that we are required to extract the central line of the circle with various widths. Unfortunately, as we have shown in Section 1, the algorithm based on the spline wavelet in [7] can not work well due to the width dependence of the detection. Fortunately, as described in detail in Section 3, the new method developed in this paper possesses the width invariant, grey-level invariant as well as slope invariant According to these properties, the central line of the circle and tree in Fig. 1 can be extracted. After applying Steps 1 and 2 of the above algorithm to the original image as displayed on the left column of Fig. 1, the local maximum modulus of the wavelet transform with respect to them can be computed and presented on the middle column in Fig. 1. At last, the central lines are extracted using Steps 3 and 4 of the above algorithm, and presented on the right column in Fig. 1. Next, some interesting examples are shown. In Fig. 5, the left image consist of a face with various widths. By carrying out the algorithm of this paper, the central line is extracted, which is shown graphically on the right in Fig. 5. For the Chinese character "peace", the original image, the maximum modulus image of the wavelet transform corresponding to $s = 2$ and the central line extracted by the proposed algorithm are shown respectively from the left to right in Fig. 6.

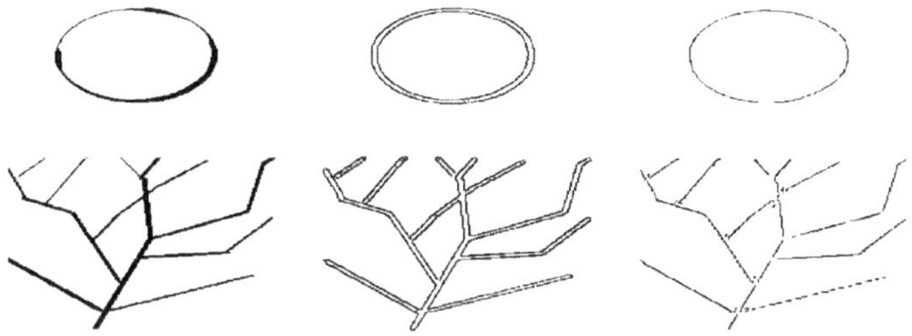

Fig. 4. Left: the original image; Middle: the location of maximum modulus of the wavelet transform corresponding to $s = 6$; Right: the central line extracted by the algorithm in this paper

Fig. 5. Left: the original image; Middle: the maximum modulus image of the wavelet transform corresponding to $s = 6$; Right: the central line extracted by the proposed algorithm

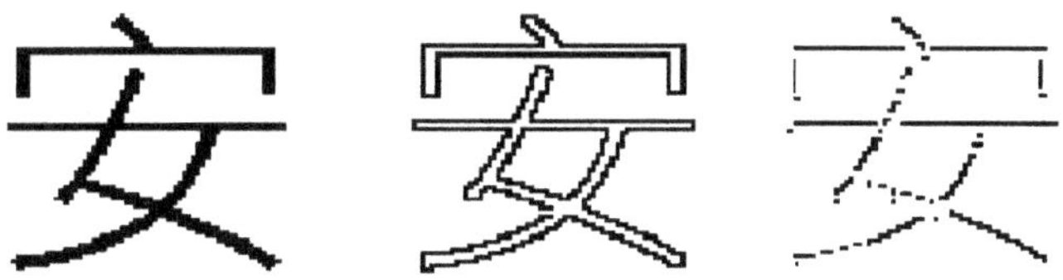

Fig. 6. Left: the original image; Middle: the location of maximum modulus of the wavelet transform corresponding to $s = 2$; Right: the central line extracted by the proposed algorithm

5 Conclusions

We have improved our previous work [7] in this paper. By utilizing a novel wavelet, we have shown three significant characteristics of the local maximum modulus of the wavelet transform with respect to the the Dirac-structure edges, namely:

- Slope invariant; the local maximum modulus of the wavelet transform of a Dirac-structure edge is independent on the slope of the edge.
- Grey-level invariant: the local maximum modulus of the wavelet transform with respect to a Dirac-structure edge takes place at the same points when the images with different grey-levels are to be processed.
- Width invariant. for various widths of the Dirac-structure edge images, the location of maximum modulus of the wavelet transform does not vary when the scale s of the wavelet transform is not less than the width d of the curve.

Based on the invariance of the wavelet transform, an algorithm to extract the Dirac-structure edge by wavelet transform has been developed. Then several experiments have been conducted, and positive results have been obtained in this paper.

References

1. J. Canny. "A Computational Approach to Edge Detection". *IEEE Trans. on Pattern Analysis and Machine Intelligence*, 8:679–698, 1986. 131
2. C. K. Chui. *An Introduction to Wavelets.* Academic Press, Boston, 1992. 131
3. I. Daubechies. *Ten Lectures on Wavelets.* Society for Industrial and Applied Mathemathics, Philadelphia, 1992. 131
4. S. Mallat and W. L. Hwang. "Singularity Detection and Processing with Wavelets". *IEEE Trans. Information Theory*, 38(2):617–643, March 1992. 134
5. D. Marr and E. C. Hildreth. "Theory of Edge Detection". In *Proc. Roy. Soc.*, pages 187–217, London B 207, 1980. 131
6. Y. Y. Tang, Qi Sun, Lihua Yang, and Li Feng. "Two-Dimensional Overlap-Save Method in Handwriting Recognition". In *6th International Workshop on Frontiers in Handwriting Recognition(IWFHR'98)*, pages 627–633, Taejon, Korea, August 12-14 1998. 134
7. Y. Y. Tang, Lihua Yang, and Jiming Liu. "Characterization of Dirac-Structure Edges with Wavelet Transform". *IEEE Trans. Systems, Man, Cybernetics (B)*, 30(1):93–109, 2000. 129, 130, 132, 133, 134, 136, 138

Wavelet Algorithm for the Numerical Solution of Plane Elasticity Problem

Youjian Shen[1] * and Wei Lin[2] **

[1] Department of Mathematics, Zhongshan University and Hainan Normal University
Haikou. 571158, P. R. China
syjian@hainnu.edu.cn

[2] Department of Mathematics, Zhongshan University
Guangzhou, 510275, P. R. China
stslw@zsu.edu.cn

Abstract. In this paper, we apply Shannon wavelet and Galerkin method to deal with the numerical solution of the natural boundary integral equation of plane elasticity probem in the upper half-plane. The fast algorithm is given and only $3K$ entries need to be computed for one $4K \times 4K$ stiffness matrix.

Keyword: plane elasticity problem, natural integral equation, Shannon wavelet, Galerkin-wavelet method.

1 Introduction

The plane elasticity problem arises from the plane strain problem and the plane stress problem which are widely applied in engineering. For the plane elasticity problem in a disc we have obtained the fast algorithm for the numerical solution by the wavelet method [1]. Now we consider the problem in the upper half-plane which has been considersd by Yu in [2], but he did not give the algorithm of the numerical solution. In this paper, as in [1], to reduce the problem into the integral equation we use the natural boundary element method which first introduced by Kang Feng and Dehao Yu [3]. In the last decade, the natural boundary element method has been efficiently used to solve some elliptic problems [1,2,4]. One of the advantages of the natural boundary integral element method is that the energy functional of the original partial differential equation preserves unchanged which results in the unique existense and stability of the solution of the natural boundary integral equation. The natural boundary integral equation possesses the kernel with hypersingularity in Hadamard finite part sense. Nowadays many methods have been developed to deal with the hypersingular integrals [1,2,5]. In this paper, we utilize Galerkin-wavelet method and the Fourier Transform of the singular kernel in the distribution sense to tackle the difficulty of hypersingularity. It is a potential numerical technique for using wavelet to solve partial

* Supported in part by NSF of Hainan normal university
** Supported in part by NSF of Guangdong

Y. Y. Tang et al. (Eds.): WAA 2001, LNCS 2251, pp. 139–144, 2001.

differential equations and integral equations in recent years([1],[4], [6]-[8]). The wavelet we use is Shannon wavelet that is a important wavelet in signal process which has excellent localization property in frequency [9]. We find that our Galerkin-wavelet method is very suitable to solve the natural boundary equation of the plane elasticity problem in the upper half-plane. As a result, the computational formulae of the stiffness matrices are simple and only $3K$ entries need to be computed for a $4K \times 4K$stiffness matrix. So that our fast algorithm requires less computational cost and the solution error is small in practical computation.

We organize this paper as follows. In Section 2, we introduce the Poisson integral formula and the natural integral equation of the plane elasticity in the upper half-plane. In Section 3, we use the Galerkin-wavelet method to solve the natural integral equation and give the computational formulae of the stiffness matrices, and in Section 4, we consider the convergence of the numerical solution. Lastly, the results of numerical experiments are presented in Section 5.

2 Plane Elasticity Problem

We consider the second boundary problem of the plane elasticity equation in the upper half-plane

$$\begin{cases} L\boldsymbol{u} = 0 & in\ \Omega := \{(x,y)|y>0\} \\ \beta\boldsymbol{u} = \boldsymbol{g} & on\ R. \end{cases} \tag{2.1}$$

where $\boldsymbol{u} = (u_1, u_2)$

$$L\boldsymbol{u} = \begin{pmatrix} a\frac{\partial^2}{\partial x^2} + b\frac{\partial^2}{\partial y^2} & (a-b)\frac{\partial^2}{\partial x \partial y} \\ (a-b)\frac{\partial^2}{\partial x \partial y} & b\frac{\partial^2}{\partial x^2} + a\frac{\partial^2}{\partial y^2} \end{pmatrix} \begin{pmatrix} u_1 \\ u_2 \end{pmatrix}.$$

$$\beta\boldsymbol{u} = \left[\begin{pmatrix} -b\frac{\partial}{\partial y} & -b\frac{\partial}{\partial x} \\ (2b-a)\frac{\partial}{\partial x} & -a\frac{\partial}{\partial y} \end{pmatrix} \begin{pmatrix} u_1 \\ u_2 \end{pmatrix} \right]_{y=0}$$

with $a = \lambda + 2\mu, b = \mu$ (λ, μ are Lamè constants), and $\boldsymbol{g} = (g_1, g_2)$ is a given vector function on R and satisfies the following compatible conditions:

$$\int_{-\infty}^{\infty} g_i(x)dx = 0, \qquad i = 1, 2. \tag{2.2}$$

Set

$$W_0^1(\Omega) = \{u| \frac{u}{\sqrt{1+x^2+y^2}\ln(2+x^2+y^2)}, \frac{\partial u}{\partial x}, \frac{\partial u}{\partial y} \in L^2(\Omega)\}$$

From Green formula, it is not difficult to show that for any $\boldsymbol{u}, \boldsymbol{v} \in W_0^1(\Omega)^2$

$$\iint_{\Omega} (\boldsymbol{v} \cdot L\boldsymbol{u} - \boldsymbol{u} \cdot L\boldsymbol{v}) dxdy = \int_R (\boldsymbol{v} \cdot \beta\boldsymbol{u} - \boldsymbol{u} \cdot \beta\boldsymbol{v}) dx \tag{2.3}$$

From this and the Green function of the equation (2.1)([2], chapter IV) we can get the Poisson formula

$$\boldsymbol{u} = P * \boldsymbol{u}_0, \qquad y > 0 \tag{2.4}$$

where $\boldsymbol{u}_0 = \boldsymbol{u}|_{y=0}$,

$$P = \begin{pmatrix} \frac{y}{\pi(x^2+y^2)} + \frac{(a-b)y(x^2-y^2)}{\pi(a+b)(x^2+y^2)^2} & \frac{2(a-b)xy^2}{\pi(a+b)(x^2+y^2)^2} \\ \frac{2(a-b)xy^2}{\pi(a+b)(x^2+y^2)^2} & \frac{y}{\pi(x^2+y^2)} - \frac{(a-b)y(x^2-y^2)}{\pi(a+b)(x^2+y^2)^2} \end{pmatrix}.$$

Substituting (2.4) into the boundary value condition $\beta\boldsymbol{u} = \boldsymbol{g}$, we obtain the following natural boundary integral equation of the problem (2.1)

$$\mathcal{K}\boldsymbol{u}_0 = \boldsymbol{g} \tag{2.5}$$

where

$$\mathcal{K}\boldsymbol{u}_0 = \begin{pmatrix} -\frac{2ab}{\pi(a+b)x^2} & -\frac{2b^2}{a+b}\delta'(x) \\ \frac{2b^2}{a+b}\delta'(x) & -\frac{2ab}{\pi(a+b)x^2} \end{pmatrix} * \boldsymbol{u}_0.$$

where $\delta(x)$ is the Dirac function. It is obvious that the kernel of the natural integral operator $\mathcal{K}$ possesses 2nd-order singularity. On the other hand, if boundary load $\boldsymbol{g} \in H^{-1/2}(R)^2$ satisfies the compatible condition (2.2) , then the natural boundary integral equation (2.5) has a unique solution in $H^{1/2}(R)^2$ [2].

Introduce the bilinear form

$$\hat{D}(\boldsymbol{u}_0, \boldsymbol{v}_0) = \int_{-\infty}^{\infty} \boldsymbol{v}_0 \cdot \mathcal{K}\boldsymbol{u}_0 dx,$$

and the linear functional

$$\hat{F}(\boldsymbol{v}_0) = \int_{-\infty}^{\infty} \boldsymbol{g} \cdot \boldsymbol{v}_0 dx$$

then the natural boundary integral eqution (2.5) is equivalent to the following variational problem:

$$\begin{cases} find \quad \boldsymbol{u}_0 \in H^{1/2}(R)^2, & s.t. \\ \hat{D}(\boldsymbol{u}_0, \boldsymbol{v}_0) = \hat{F}(\boldsymbol{v}_0), & \forall \boldsymbol{v}_0 \in H^{1/2}(R)^2 \end{cases} \tag{2.6}$$

3 Galerkin-Wavelet Methods

Set

$$\hat{\phi}(\xi) = \chi_{[-\pi,\pi]}(\xi) \tag{3.1}$$

then

$$\phi(x) = \frac{1}{2\pi} \int_{-\infty}^{\infty} \hat{\phi}(\xi) e^{i\xi x} d\xi = \frac{\sin \pi x}{\pi x} \tag{3.2}$$

This is well known as the scaling function of Shannon wavelet. Now, we use Galerkin method to solve the variational problem (2.6). For $K \in N, j \in Z$ let

$$V_j^K = \overline{Span\{\phi_{j,k}(x) | \phi_{j,k}(x) = 2^{j/2}\phi(2^j x - k), k = -K, -K+1, \cdots, K-1\}},$$

Substituting $(V_j^K)^2$ for $H^{1/2}(R)^2$ in (2.6) leads to the following approximate variational problem:

$$\begin{cases} find \quad \boldsymbol{u}_0^{j,K} \in (V_j^K)^2 \quad s.t. \\ \hat{D}(\boldsymbol{u}_0^{j,K}, \boldsymbol{v}_0^{j,K}) = \hat{F}(\boldsymbol{v}_0^{j,K}), \forall \boldsymbol{v}_0^{j,K} \in (V_j^K)^2 \end{cases} \tag{3.3}$$

We express $\boldsymbol{u}_{01}^{j,K}, \boldsymbol{u}_{02}^{j,K}$ as

$$\begin{cases} \boldsymbol{u}_{01}^{j,K} = \sum\limits_{k=-K}^{K-1} \alpha_{j,k}^1 \phi_{j,k}(x) \\ \boldsymbol{u}_{02}^{j,K} = \sum\limits_{k=-K}^{K-1} \alpha_{j,k}^2 \phi_{j,k}(x) \end{cases}$$

select $\boldsymbol{v}_0^{j,K} = (\phi_{j,m}(x), 0)$ and $\boldsymbol{v}_0^{j,K} = (0, \phi_{j,m}(x))(m = -K, -K+1, \cdots K-1)$ respectively, then we get the following linear algebraic system:

$$\begin{pmatrix} Q_{11} & Q_{12} \\ Q_{21} & Q_{22} \end{pmatrix} \begin{pmatrix} \alpha^1 \\ \alpha^2 \end{pmatrix} = \begin{pmatrix} f_1 \\ f_2 \end{pmatrix}. \tag{3.4}$$

where

$$\begin{aligned} \alpha^i &= (\alpha_{j,-K}^i, \alpha_{j,-K+1}^i, \cdots, \alpha_{j,K-1}^i)^T, & i = 1,2. \\ Q_{ps} &= (q_{mn}^{ps})_{m,n=-K,-K+1,\cdots,K-1}, & p,s = 1,2 \\ q_{mn}^{ps} &= \hat{D}((\delta_{1,s}, \delta_{2,s})\phi_{j,n}(x), (\delta_{1,p}, \delta_{2,p})\phi_{j,m}(x)) & \\ f_i &= (b_{-K}^i, b_{-K+1}^i, \cdots, b_{K-1}^i)^T, & i = 1,2 \\ b_m^i &= \int_{-\infty}^{\infty} \boldsymbol{g}(x) \cdot (\delta_{1,i}, \delta_{2,i})\phi_{j,m}(x) dx & \end{aligned}$$

and δ_{ij} is the Kronecker's symbol.

Theorem 1. *The entries of the stiffness matrix of the linear algebraic system can be expressed as*

$$q_{mn}^{11} = q_{mn}^{22} = \begin{cases} \frac{2^j ab\pi}{a+b}, & r = 0 \\ \frac{2^{j+1} ab}{\pi(a+b)r^2}((-1)^r - 1), & r \neq 0 \end{cases} \tag{3.5}$$

$$q_{mn}^{21} = q_{mn}^{12} = \begin{cases} 0, & r = 0 \\ \frac{2^{j+1} b^2}{(a+b)r}(-1)^r, & r \neq 0 \end{cases} \tag{3.6}$$

where $r = m - n$.

By the Theorem (3.1) only $3K$ entries need to be computed for one $4K \times 4K$ stiffiness matrix.

4 Convergence of Numerical Solution

For $j \in Z, K \in N$, we define $L_j^K : L^2(R) \to V_j^K$ as

$$L_j^K f = \sum_{k=-K}^{K-1} \langle f, \phi_{j,k} \rangle \phi_{j,k}$$

Lemma 1. *For all $f \in H^1(R)$, we have*

$$\lim_{j \to \infty} \lim_{K \to \infty} \|L_j^K f - f\|_{H^1(R)} = 0 \tag{4.1}$$

Theorem 2. *If $\boldsymbol{u}_0 \in H^1(R)^2$, then*

$$\lim_{j \to \infty} \lim_{K \to \infty} \|\boldsymbol{u}_0 - \boldsymbol{u}_0^{j,K}\|_{\hat{D}} = 0 \tag{4.2}$$

where $\|\cdot\|_{\hat{D}}$ is energy norm induced by bilinear form $\hat{D}(\cdot,\cdot)$, i.e.$\|\cdot\|_{\hat{D}} = \hat{D}(\cdot,\cdot)^{1/2}$

5 Numerical Results

In this section, we present the numerical results of a test example to illustrate our algorithm for the natural boundary integral equation (2.5) discussed in section 3.

Example Consider the problem

$$\mathcal{K}\boldsymbol{u}_0 = \left(\frac{3x^3 - x}{10(x^2+1)^3}, \frac{3x^2 - 1}{10(x^2+1)^3} \right) \qquad \textit{on } R.$$

Selecting Lamè constants $\lambda = 1, \mu = 0.5$. then the exact solution is

$$\boldsymbol{u}_0 = \left(-\frac{x(4x^2+1)}{30(x^2+1)^2}, \frac{x^2 - 5}{60(x^2+1)^2} \right).$$

Table 1. Numerical results ($K = 2^j$)

j	1	2	3
$\|\boldsymbol{u}_0^{j,K}(x) - \boldsymbol{u}_0(x)\|_{L^2(R)}$	0.14351	0.14349	0.14345
$\max_{-5 \le m \le 5} \lvert\boldsymbol{u}_0^{j,K}(m) - \boldsymbol{u}_0(m)\rvert$	$1.28227e^{-17}$	$2.666464e^{-17}$	2.63968^{-17}

j	4	5	6
$\|\boldsymbol{u}_0^{j,K}(x) - \boldsymbol{u}_0(x)\|_{L^2(R)}$	0.14337	0.14318	0.14265
$\max_{-5 \le m \le 5} \lvert\boldsymbol{u}_0^{j,K}(m) - \boldsymbol{u}_0(m)\rvert$	$4.46704e^{-17}$	$4.65589e^{-17}$	$5.11821e^{-17}$

The computational results of above examples show that our algorithm privides high accuracy with low computing cost.

References

1. W. Lin, Y. J. Shen, Wavelet solutions to the natural integral equations of the plane elasticity problem, Proceedings of the second ISAAC Congress, Vol. 2, 1471-1480. (2000), Kluwer Academic Publishers. 139, 140
2. Dehao Yu, Mathematical theory of natural boundary element methods, Science press (in chinese), Beijing (1993). 139, 141
3. K. Feng and D. Yu, Canonical integral equations of elliptic boundary value problems and their numerical solutions, Proc. of China-France Symp. on FEM, Science Press, Beijing (1983), 211–252. 139
4. Wensheng Chen and Wei Lin, Hadamard singular integral equations and its Hermite wavelet, Proc. of the fifth international colloquium on finite or infinite dimensional complex analysis , (Z. Li, S. Wu and L. Yang. Eds.) Beijing, China (1997), 13–22. 139, 140
5. C.-Y. Hui, D. Shia, Evaluations of hypersingular integrals using Gaussian quadrature, Int. J. for Numer. Meth. in Engng. 44, 205-214 (1999). 139
6. R. P. Gilbert and Wei Lin, Wavelet solutions for time harmonic acoutic waves in a finite ocean, Journal of Computional Acoustic Vol. 1, No. 1 (1993) 31–60. 140
7. C. A. Micchelli, Y. Xu and Y. Zhao, Wavelet Galerkin methods for second-kind integral equations. J. Comp. Appl. Math. 86 (1997), 251-270.
8. Tobias Von Petersdorff, Christoph Schwab, Wavelet approximations for first kind boundary integral equations on polygons, Numer, Math, 74 (1996), 479-519. 140
9. I. Daubechies, Ten lectures on wavelets, Capital City Press, Montpelier, Vermont, 1992. 140

Three Novel Models of Threshold Estimator for Wavelet Coefficients

Song Guoxiang and Zhao Ruizhen

School of Science, Xidian University
Xi'an, 710071, P. R. China

Abstract. The soft-thresholding and the hard-thresholding method to estimate wavelet coefficients in wavelet threshold denoising are firstly discussed. To avoid the discontinuity in the hard-thresholding and biased estimation in the soft-thresholding, three novel models of threshold estimator are presented, which are polynomial interpolating thresholding method, compromising method of hard- and soft-thresholding and modulus square thresholding method respectively. They all overcome the disadvantages of the hard- and soft-thresholding method. Finally, an example is given and the experimental results show that the improved techniques presented in this paper are efficient.

1 Introduction

Wavelet theory has recently become a popular mathematical tool in many research fileds. It throws a new light on such applications as image and signal processing. In this paper, we concentrate on the problem of signal denoising. Generally, there are three approaches which are used to distinguish noise from regular wavelet coefficients. The first one is based on the principle of modulus maximum with wavelet transform presented by Mallat[1][2]. The second approach is grounded on the different correlation properties between the wavelet coefficients of the noise and the regular signal. And the third approach is the wavelet thresholding technique presented by Donoho[3][4]. For the third case, the idea of the hard-thresholding or the soft-thresholding method is to replace the small coefficients by zero and keep or shrink the large coefficients. However, the Estimated Wavelet Coefficients (EWC) obtained in the hard-thresholding are not continuous at the threshold, so it may induce the oscillation of the reconstructed signal. In the soft-thresholding case, EWC are mathematically tractable due to the good continuity, but when the wavelet coefficients become larger, there are deviations in EWC. Thus the error will certainly bring to the reconstructed signal. The methods discussed in this paper belong to the third case, namely, noise reduction based on thresholding. Combining with the hard-thresholding and the soft-thresholding, three improved techniques are presented in this paper to avoid the disadvantages. They are polynomial interpolating thresholding method, compromising method of the hard- and soft-thresholding and modulus squared thresholding method respectively. The wavelet coefficients estimated

Y. Y. Tang et al. (Eds.): WAA 2001, LNCS 2251, pp. 145–150, 2001.

through two of the methods given in this paper are continuous at the threshold and nearly unbiased when the original coefficients become larger. And all the three methods obtain good results. The correspondence is organized as follows. Section 2 briefly introduces some basic notations of wavelet transform and the hard-thresholding and the soft-thresholding method. And three novel models of threshold estimator for wavelet coefficients are presented in Section 3. Finally, Section 4 gives some experimental results, and a brief conclusion is stated in Section 5.

2 Wavelet Transform and the Thresholding Method

Suppose there is an observed signal

$$f(t) = s(t) + n(t) \tag{1}$$

where $s(t)$ is the original signal,$n(t)$ is Gaussian white noise with mean 0 and variance σ^2. If $f(t)$ is sampled as N point discrete signal $f(k)$, then the wavelet fast algorithm is

$$Sf(j+1,k) = Sf(j,k) * h(j,k) \tag{2}$$

$$Wf(j+1,k) = Sf(j,k) * g(j,k) \tag{3}$$

where $Sf(0,k)$ is original signal $f(k)$, $Sf(j,k)$ are approximated coefficients and $Wf(j,k)$ wavelet coefficients. h and g are respectively low-pass and high-pass filters. For convenience, we abbreviate $Wf(j,k)$ to $w_{j,k}$. Accordingly, wavelet reconstruction formula is

$$Sf(j-1,k) = Sf(j,k) * \bar{h}(j,k) + Wf(j,k) * \bar{g}(j,k) \tag{4}$$

Due to the linear property of wavelet transform, the wavelet coefficients $w_{j,k}$ of observed data $f(k) = s(k) + n(k)$ consist of two parts. One is $Ws(j,k)$ (abbreviated to $u_{j,k}$) corresponding to $s(k)$ and the other is $Wn(j,k)$ (abbreviated to $v_{j,k}$) corresponding to $n(k)$.

The idea of the wavelet threshold denoising is

1. Getting the wavelet coefficients $w_{j,k}$ from the noisy signal $f(k)$ by using (2) and (3);
2. Determining the estimated wavelet coefficients $\hat{w}_{j,k}$ from $w_{j,k}$ by the thresholding method such that $||\hat{w}_{j,k} - u_{j,k}||$ are as small as possible;
3. Reconstructing the denoised signal $\hat{f}(k)$ from $\hat{w}_{j,k}$ by (4).

Donoho has presented a very concise method to estimate the wavelet coefficients $w_{j,k}$. A proper threshold λ should be firstly chosen. Then the coefficients with absolute values smaller than λ are replaced by zero and those larger than λ are kept in the hard-thresholding case and shrunk in the soft-thresholding case.

The threshold of Donoho and Johnstone[4] is $\lambda = \sigma\sqrt{2\log N}$. Define

$$\hat{w}_{j,k} = \begin{cases} w_{j,k}, & |w_{j,k}| \geq \lambda \\ 0, & |w_{j,k}| < \lambda \end{cases} \tag{5}$$

It is called the hard-thresholding estimator. The soft-thresholding estimator is defined as

$$\hat{w}_{j,k} = \begin{cases} sign(w_{j,k})(|w_{j,k}| - \lambda), & |w_{j,k}| \geq \lambda \\ 0, & |w_{j,k}| < \lambda \end{cases} \tag{6}$$

Although these methods are widely used in applications, they have some underlying disadvantages. For instance, the estimated wavelet coefficients $\hat{w}_{j,k}$ by the hard-thresholding method are not continuous at the threshold λ, which may lead to the oscillation of the reconstructed signal. In the soft-thresholding case, when $|w_{j,k}| > \lambda$, there are deviations between $\hat{w}_{j,k}$ and $w_{j,k}$, which directly influence the accuracy of the reconstructed signal. To overcome the above disadvantages of the hard-thresholding and the soft thresholding method, we present some improved schemes in Section 3.

3 Three Novel Models of Threshold Estimators

3.1 The Polynomial Interpolating Thresholding Method

Because the hard-threshold estimation is not continuous and the soft-threshold estimation has some deviations, the applications of these methods are somewhat limited. So we have a chance to improve them. A natural approach is to design an estimator from which the estimated wavelet coefficients $w_{j,k}$ will be continuous at the threshold λ and with the $w_{j,k}$ increase, little deviation exists in EWC. For example, we can design an estimator such that for $|w_{j,k}| > t$, $(t > \lambda)$, $\hat{w}_{j,k}$ and $w_{j,k}$ are completely the same. Such an assumption can be realized through polynomial interpolating. The model is as follows:

$$\hat{w}_{j,k} = \begin{cases} w_{j,k}, & |w_{j,k}| \leq t \\ sign(w_{j,k})P(|w_{j,k}|), & \lambda \leq |w_{j,k}|\, t \\ 0, & |w_{j,k}| < \lambda \end{cases} \tag{7}$$

where $P(|w_{j,k}|)$ is an interpolating polynomial. Generally $P(|w_{j,k}|)$ can be quadratic or cubic polynomial. The corresponding interpolating conditions are

$$\begin{cases} P(\lambda) = 0 \\ P(t) = t \\ P'(t) = 1 \end{cases} and \begin{cases} P(\lambda) = 0 \\ P'(\lambda) = 0 \\ P(t) = t \\ P'(t) = 1 \end{cases} \tag{8}$$

respectively. Very simple derivations can lead to the quadratic polynomial

$$P(x) = -\frac{1}{(t-\lambda)^2}[\lambda x^2 - (\lambda^2 + t^2)x + \lambda t^2], (\lambda \leq x \leq t) \tag{9}$$

and the cubic polynomial

$$P(x) = -\frac{1}{(t-\lambda)^3}[(t+\lambda)x^3 - 2(t^2 + t\lambda + \lambda^2)x^2 + \lambda(4t^2 + t\lambda + \lambda^2)x - 2t^2\lambda^2], (\lambda \leq x \leq t) \tag{10}$$

The estimated wavelet coefficients $\hat{w}_{j,k}$ obtained from the above method are continuous everywhere. Moreover, if $P(x)$ is a cubic polynomial, then $\hat{w}_{j,k}$ are derivative in the whole domain as well. For $|w_{j,k}| > t$, $\hat{w}_{j,k}$ are unbiased estimated, which makes up for the shortage of the soft-thresholding.

3.2 The Compromising Method of the Hard- and Soft-Thresholding

Define

$$\hat{w}_{j,k} = \begin{cases} sign(w_{j,k})(|w_{j,k}| - \alpha\lambda), & |w_{j,k}| \geq \lambda \\ 0, & |w_{j,k}| < \lambda \end{cases}, (0 \leq \alpha \leq 1) \tag{11}$$

This model of estimator for wavelet coefficients is called the compromising method of the hard- and soft-thresholding. Particularly, (10) will turn to the hard-thresholding (5) if α equals 0 and the soft-thresholding (6) if α is 1. For $0 < \alpha < 1$, it is clear that the data $\hat{w}_{j,k}$ by (10) lie between those by (5) and (6). So it is called the compromising method of the hard- and soft-thresholding.

This method is quite efficient in noise reduction although it is simple and straight forward. It is no wonder if we pay a little attention to the thresholding method itself. For the soft-threshlding case, the absolute value of the estimated coefficient $\hat{w}_{j,k}$ is always smaller than that of $w_{j,k}$ by λ (when $w_{j,k} \leq \lambda$). Therefore, the deviation should be cut as small as possible. However, the deviation being zero (corresponding to the hard-thresholding) is not the best case as well in that $|w_{j,k}|$ is always larger than $|u_{j,k}|$ in most cases because $w_{j,k}$ consists of $u_{j,k}$ and $v_{j,k}$. While our aim is to find proper $\hat{w}_{j,k}$ such that $||\hat{w}_{j,k} - u_{j,k}||$ are minimum. Therefore, the value of $\hat{w}_{j,k}$ should lie between $|w_{j,k}| - \lambda$ and $|w_{j,k}|$, which will make $\hat{w}_{j,k}$ be closer to $u_{j,k}$. Based on this idea, we add a factor α in the soft-thresholding estimator (6) to improve the performance. α is any real number between 0 and 1. An appropriate α may better the denoising result. In this correspondence, we choose $\alpha = 0.5$.

3.3 The Modulus Squared Thresholding Method

We firstly consider the case $w_{j,k} > 0$, then generalize the result to $w_{j,k} < 0$. In the soft-thresholding method, (6) is equivalent to

$$\hat{w}_{j,k} = \begin{cases} \lambda(w_{j,k}/\lambda - 1), & w_{j,k}/\lambda \geq 1 \\ 0, & w_{j,k}/\lambda < 1 \end{cases} \tag{12}$$

when $w_{j,k} > 0$. If we see $w_{j,k}/\lambda$ as a whole, then (12) means that when $w_{j,k}/\lambda \geq 1$, $w_{j,k}$ can be thought as the coefficients of the signal and hence are kept, otherwise $w_{j,k}$ should be removed since they are considered as the coefficients of the noise. Although it is equivalent to (6), (12) is easier to be extended. We can modify (12) as the following model

$$\hat{w}_{j,k} = \begin{cases} \lambda\sqrt{(w_{j,k}/\lambda)^2 - 1}, & w_{j,k}/\lambda \geq 1 \\ 0, & w_{j,k}/\lambda < 1 \end{cases} \tag{13}$$

The difference between (13) and (12) is that in (13) $w_{j,k}/\lambda$ is in its square form. The advantage of this modification is that if $w_{j,k}/\lambda$ is above 1, then the square of $w_{j,k}/\lambda$ will become larger; if $w_{j,k}/\lambda$ is below 1, then the square of $w_{j,k}/\lambda$ will become smaller. Such a procession will speed the separation of noise from signal.

(13) is true only if $w_{j,k} > 0$. For general case we have

$$\hat{w}_{j,k} = \begin{cases} sign(w_{j,k})\sqrt{(w_{j,k})^2 - \lambda^2}, & |w_{j,k}| \geq \lambda \\ 0, & |w_{j,k}| < \lambda \end{cases} \tag{14}$$

It is easy to prove that when $|w_{j,k}| \geq \lambda$,

$$|w_{j,k}| - \lambda \leq \sqrt{(w_{j,k})^2 - \lambda^2} \leq |w_{j,k}| \tag{15}$$

holds. From (15) we can know that the value of $\hat{w}_{j,k}$ estimated by (14) still lies between those by (5) and (6). When $|w_{j,k}| \geq \lambda$, $\hat{w}_{j,k}$ is a nonlinear function. And $\hat{w}_{j,k}$ becomes closer and closer to $w_{j,k}$ with $|w_{j,k}|$ increasing.

4 Experimental Results

A comparison is made in signal denoising with the above threshold methods presented in this paper. Instead of the fixed threshold $\lambda = \sigma\sqrt{2\log(N)}$ presented by Donoho, we take the different thresholds $\lambda_j = \sigma\sqrt{2\log(N)}/\log(j+1)$ at different scales. A noisy signal is processed by the above five methods. Before denoising, the Signal to Noise Ratio (SNR) is 8.226270. Table 1 shows the comparison of the SNR and relatively mean square error (RMSE) of the reconstructed signal with the above methods. From Table 1, we can see that the compromising method of the hard- and soft-thresholding and the modulus square thresholding method are obviously superior to the hard-thresholding and the soft-thresholding method. The polynomial interpolating thresholding method is only superior to the hard-thresholding method and is equivalent to the soft-thresholding method.

Table 1. Comparison of estimators for wavelet coefficients by SNR and RMSE

Estimator	SNR	RMSE
Soft-thresholding	15.276322	0.172260
Hard-thresholding	14.331342	0.192058
Square Interpolating	15.152992	0.174723
Cubic Interpolating	15.288729	0.172014
Compromising of Hard and Soft	15.582417	0.166295
Modulus Squared Method	15.367344	0.170464

5 Conclusion

We will indicate that λ_j in this paper are not the best. If λ_j are properly selected, the superiority of our methods will be more remarkable. In addition, for different λ_j, the experimental results may be slightly different. However, from a mass of experiments the authors have made, a conclusion can be drawn that the pure hard- or soft-thresholding method has poor stability and is strongly dependent on λ_j. Moreover, at least one of the two methods can not reach a satisfactory result. By comparison, the modulus square thresholding method and the polynomial interpolating thresholding method are more stable and can obtain nearly the same results as the better one of the hard- or soft-thresholding. Finally, whatever λ_j is, the compromising method of the hard- and soft-thresholding is obviously superior to the hard- or soft-thresholding method.

In addition, we only make some improvements on the threshoding method itself. There are some other problems in wavelet threshold denoising such as the selection of the threshold λ and nonstationary noise, say, Poisson noise case and so on. Some achievements have been made in [5], [6], [7].

References

1. Mallat S. and Zhong S. Characterization of signals from multiscale edges. IEEE Trans. on PAMI, 1992, 14(7): 710-732
2. Mallat S. and Hwang W. L. Singularity detection and processing with wavelets. IEEE Trans. on IT, 1992, 38(2): 617-643
3. Donoho D. L. De-noising by soft-thresholding. IEEE Trans. on IT., 1995, 41(3):613-627
4. Donoho D. L. and Johnstone I. M. Ideal spatial adaption via wavelet shrinkage. Biometrika, 1994, 81:425-455
5. Jansen M. and Bultheel A. Multiple wavelet threshold estimation by generalized cross validation for Images with correlated noise. IEEE Trans. on IP., 1999, 8(7):947-953
6. Nowak R. D. and Baraniuk R. G. Wavelet-domain filtering for photon imaging systems. IEEE Trans. on IP., 1999, 8(5):666-678
7. Ching P. C., So H. C. and Wu S. Q. On wavelet denoising and its applications to time delay estimation. IEEE Trans. on SP., 1999, 47(10):2879-288

The PSD of the Wavelet-Packet Modulation

Mingqi Li[1], Qicong Peng[2], and Shouming Zhong[1]

[1] Applied Mathematics Department of the University of Electronic Science and Technology of China
Chengdu, 610054, P. R. China
lmqi2001@yahoo.com.cn

[2] Institute of Communication & Information Engineering, University of Electronic Science and Technology of China,
Chengdu, 610054, P. R. China

Abstract. On the wavelet-packet modulation scheme, wavelet packets are used as carriers, where information to be transmitted is encoded, via an inverse wavelet packet transform, as the coefficient of wavelet packets. The power spectrum density (PSD) of the modulated signals describes the property of modulation in frequency domain, which is discussed by many researchers by simulation. In this paper, the formula of the PSD of the modulated signals is derived. The characteristics of the modulated signals, such as spread spectrum and spectral flatness, is shown from the formula.

1 Introduction

Wavelet and wavelet-packet transform with its desirable characteristics, such as location in time and frequency, and orthogonality across scale and translation, has brought out many useful properties. Wavelet-packet modulation is one of its important application, where wavelet packets are used as the waveform for information transmission. This new kind of modulation scheme generalizes the traditional baseband modulation scheme. In fact, we find now, both rectangular and sinc pulse are scaling function corresponding Haar and Meyer wavelet respectively. On the other hand, wavelet packets modulation is seen as a new kind of multiplexing for its time and frequency overlapping, where TDM and FDM are taken as the special cases according to [1] .

In the wavelet-packet modulation scheme, each user is assigned a set of waveform in a group of wavelet packets. The information of each user is impressed on the corresponding waveform via the coefficients. At the receiver, the desired signal is recovered by cross-correlation with a known reference signal in the wavelet packet basis.

Spread spectrum and spectral flatness are very important in a communication system especially for channel fading and security. The simulation of wavelet-packet modulation, discussed by many references, gives many advantages in communication, including covert and featureless waveforms (see [2],[3],[4],[5]). We find to catch the PSD of the modulated signals is very important in its application. In this paper, the formula of the PSD of the wavelet-packet modulated

Y. Y. Tang et al. (Eds.): WAA 2001, LNCS 2251, pp. 151–156, 2001.

signals is arrived (theorem 1). As a special case of wavelet-packet modulation, the PSD of wavelet modulation (theorem 2) is achieved, too.

Now, we briefly summarize the concepts of wavelet and wavelet packets. A multiresolution analysis consists of a collection of embedded subspace sequence in the space of finite energy signal $L^2(R)$. That is

$$... \subset V_{-2} \subset V_{-1} \subset V_0 \subset V_1 \subset V_2 ...$$

Each subspace V_j has an orthogonormal basis $\{\varphi_{j,k} : \varphi_{j,k} = 2^{\frac{j}{2}}\varphi(2^j t-k), k \in Z\}$. $\varphi(t)$ is called the scaling function satisfying $\varphi(t) = \sqrt{2}\sum_{k=-\infty}^{\infty} h[k]\varphi(2t-k)$. If functions $\{\varphi_{j,k} : k \in Z\}$ forms an orthonormal basis of the space V_j, the following orthonormal constraint on $h[n]$ must be satisfied:

$$\begin{aligned} \sum_{k=-\infty}^{\infty} h[k-2n]h[k-2m] &= \delta_{m,n} \\ \sum_{k=-\infty}^{\infty} h[k] &= \sqrt{2} \end{aligned} \tag{1}$$

Then we get a quadrature mirror filter (QMF) $h[n]$, $g[n] := (-1)^n h[1-n]$. The function $\psi(t)$, defined by $\psi(t) = \sqrt{2}\sum_{k=-\infty}^{\infty} g[k]\varphi(2t-k)$, is called a wavelet function induced by $\varphi(t)$.

We define the recursive function sequence $\{P_n(t)\}$ as Coifman,

$$\begin{aligned} p_{2n}(t) &= \sqrt{2}\sum_{k=-\infty}^{\infty} h[k]p_n(2t-k) \\ p_{2n+1}(t) &= \sqrt{2}\sum_{k=-\infty}^{\infty} g[k]p_n(2t-k) \end{aligned} \tag{2}$$

where $p_0(t) = \varphi(t), p_1(t) = \psi(t)$.

We need also the following notation:

$$\phi_{l,m}(t) := 2^{\frac{l}{2}} p_m(2^l t) \tag{3}$$

$$U_l^m := Clos\{2^{\frac{N-l}{2}} p_m(2^{N-l}t-k) : k \in Z, m \in Z_+\}. \tag{4}$$

Then we get $U_l^m = U_{l-1}^{2m} \bigoplus U_{l-1}^{2m+1}, V_N = U_N^0$ and $W_N = U_N^1$. In order to decompose the subspace $V_N = U_N^0$, the set of subspace U_l^m may be organized as a binary tree, where U_N^0 is on the top and U_{N-l}^m is on the $(m+1)$th node of level l. There are 2^l nodes in the same level. We can grow or prune the tree in any desired fashion, and the different fashion provides a different set of basis functions.

2 The Modulation and Its PSD

Firstly, we consider a TDM system in which there are $K_{l,m}$ independent binary message signals interlaced with each other. Between two consecutive binary symbols of the same message, there are $K_{l,m}-1$ other binary symbol: one from each of the other message signals. The combined sequence forms a composite sequence of binary symbols $\sigma_{l,m}[n]$, where $\sigma_{l,m}[n] = \pm 1$. The system we propose here seeks the representation of the binary symbols 1 and -1 by $\phi_{l,m}(t)$ and $-\phi_{l,m}(t)$, respectively. Then the modulated signal of the TDM sequence $\{\sigma_{l,m}[n]\}$, encoded by $\phi_{l,m}(t-2^{l-N}n)$, can be given by

$$s_{l,m}(t) = \sum_{k=-\infty}^{\infty} \sigma_{l,m}[k]\phi_{l,m}(t-2^{l-N}k). \tag{5}$$

Let the set M of (l,m) satisfy $\bigoplus_{(l,m)\in M} U_l^m = V_N$. Since all the constituent terminal functions in a given tree structure M are orthogonal to each other, we may employ all of these functions to carry binary data from deferent TDM groups of users. So the total number of users is $\sum_{(l,m)\in M} K_{l,m}$. Let $\sigma_{l,m}^k[n]$ represent the information sequence of the kth user while its assigned waveform is $\phi_{l,m}(t-2^{l-N}(nK_{l,m}+k))$ in U_{N-l}^m. We get the modulated signal $s(t)$ satisfying

$$s(t) = \sum_{(l,m)\in M}\sum_{k=1}^{K_{l,m}}\sum_{n=-\infty}^{\infty} \sigma_{l,m}^k[n]\phi_{l,m}(t-2^{l-N}(nK_{l,m}+k)) \tag{6}$$

We make the following reasonable assumptions to simplify the calculation of PSD:
1. Information sequence $\{\sigma_{l,m}^k[n]\}$ of user (l,m) is stationary process;
2. Different user $\{\sigma_{l,m}^k[n]\}$ with different (l,m) are statistically independent and $E(\sigma_{l,m}^k[n]) = 0$.

We denote the correlation coefficient of $\{\sigma_{l,m}^k[n] : n \in Z\}$ as

$$R_{l,m}^k[n] := E(\sigma_{l,m}^k[n]\sigma_{l,m}^k[n+h]). \tag{7}$$

Then we have

$$E(s(t+\tau)s^*(t)) = E((\sum_{(l,m)\in M}\sum_{k=1}^{K_{l,m}}\sum_{n=-\infty}^{\infty} \sigma_{l,m}^k[n]\phi_{l,m}^*(t-2^{l-N}(nK_{l,m}+$$

$$k)))(\sum_{(a,b)\in M}\sum_{c=1}^{K_{a,b}}\sum_{d=-\infty}^{\infty} \sigma_{a,b}^k[n]\phi_{l,m}(t+\tau-2^{a-N}(dK_{a,b}+k))))$$

so,we get

$$E(s(t+\tau)s^*(t)) == \sum_{(l,m)\in M}\sum_{k=1}^{K_{l,m}}\sum_{n=-\infty}^{\infty}\sum_{h=-\infty}^{\infty} R_{l,m}^k[h]\phi_{l,m}^*(t-2^{l-N}(nK_{l,m}+$$

$$k))\phi_{l,m}(t+\tau-2^{l-N}((n+h)K_{l,m}+k))$$

Define function $g(t)$:

$$g(t) := \sum_{(l,m)\in M} \sum_{k=1}^{K_{l,m}} \sum_{n=-\infty}^{\infty} \sum_{h=-\infty}^{\infty} R_{l,m}^{k}[h]\phi_{l,m}^{*}(t - 2^{l-N}(nK_{l,m} + k))\phi_{l,m}(t + \tau - 2^{l-N}((n+h)K_{l,m} + k)). \tag{8}$$

When $K_{l,m}$ is a constant K indepent l, m, each of waveforms has the same number of users. Then $g(t)$ is a periodic function. That is $g(t) = g(t+K)$. We know from the assumptions above that stochastic process $s(t)$ is generalized cyclostationary with period $T = K_{l,m} = K$. So the correlation function of $s(t)$ can be defined

$$R(\tau) := \frac{1}{K}\int_{0}^{K} g(t)dt \tag{9}$$

Then the $R(\tau)$ of $s(t)$ is

$$R(\tau) = \frac{1}{K} \sum_{(l,m)\in M} \sum_{k=1}^{K} \sum_{h=-\infty}^{\infty} R_{l,m}^{k}[h] \sum_{n=-\infty}^{\infty} \int_{-2^{l-N}(nK+k)}^{K-2^{l-N}(nK+k)} \phi_{l,m}^{*}(u)\phi_{l,m}(u + \tau - 2^{l-N}hK)du$$

$$= \frac{1}{K} \sum_{(l,m)\in M} \sum_{k=1}^{K} \sum_{h=-\infty}^{\infty} R_{l,m}^{k}[h]2^{N-l} \int_{-\infty}^{\infty} \phi_{l,m}^{*}(u)\phi_{l,m}(u + \tau - 2^{l-N}hK)du$$

So we get the PSD of $s(t)$. That is

$$\hat{R}(\omega) = \frac{1}{K} \sum_{(l,m)\in M} \sum_{k=1}^{K} \sum_{h=-\infty}^{\infty} R_{l,m}^{k}[h]2^{N-l}|\hat{\phi}_{l,m}(\omega)|^2 e^{-j\omega h 2^{l-N}K}$$

$$= \frac{1}{K} \sum_{(l,m)\in M} \sum_{k=1}^{K} |\hat{\phi}_{l,m}(\omega)|^2 \sum_{h=-\infty}^{\infty} R_{l,m}^{k}[h] e^{-j\omega h 2^{l-N}K}$$

So $\hat{R}(\omega)$ is arrived,

$$\hat{R}(\omega) = \frac{1}{K} \sum_{(l,m)\in M} \sum_{k=1}^{K} 2^{N-l}|\hat{\phi}_{l,m}(\omega)|^2 \hat{R}_{l,m}^{k}(2^{l-N}\omega) \tag{10}$$

where $\hat{R}_{l,m}^{k}(\omega) := \sum_{h=-\infty}^{\infty} R_{l,m}^{k}[h]e^{-j\omega hK}$.

Summarizing the description above, we get the following theorem:

Theorem 1. *If assumption (1) and (2) are satisfied and $K_{l,m}$ is a constant K independent on l, m, the PSD of $s(t)$ is*

$$\hat{R}(\omega) = \frac{1}{K} \sum_{(l,m)\in M} \sum_{k=1}^{K} 2^{N-l}|\hat{\phi}_{l,m}(\omega)|^2 \hat{R}_{l,m}^{k}(2^{l-N}\omega) \tag{11}$$

where $\hat{R}_{l,m}^{k}(\omega) = \sum_{h=-\infty}^{\infty} R_{l,m}^{k}[h]e^{-j\omega hK}$.

When we select the waveforms of nodes $(l, 1)$, the wavelet-packet modulation turns into wavelet modulation. We can get easily a similar result.

Now, we discuss a wavelet modulation with m_0 users and the waveforms

$$\psi_{l,n}(t) = 2^{\frac{l}{2}}\psi(2^l t - n), m_1 \le l \le m_0 + m_1, m_1 \in \mathrm{Z}.$$

We get the modulated signal $s(t)$. That is

$$s(t) = \sum_{l=m_1}^{m_0+m_1} \sum_{n=-\infty}^{\infty} \sigma_l[n]\psi_{l,n}(t) \tag{12}$$

where $\sigma_l[n] = \pm 1$ is the nth data of lth user. We denote

$$\begin{aligned} R_l(k) := E(s_l(n+k)s_l^*(n)), \hat{R}_l(\omega) := \sum_{k=-\infty}^{\infty} R_l[k]e^{-j\omega K} \\ R(t+\tau, t) := E(s(t+\tau)s^*(t)) \end{aligned} \tag{13}$$

Theorem 2. *If $\{\sigma_l[n] : n \in Z\}$ is a stationary process and $\{\sigma_l[n] : n \in Z\}$ are statistically independent with $E(\sigma_l[n]) = 0$, the PSD of $s(t)$ is*

$$\hat{R}(\omega) = \sum_{l=m_1}^{m_0+m_1} |\hat{\psi}(2^{-l}\omega)|^2 \hat{R}_l(2^{-l}\omega) \tag{14}$$

Proof. The correlation function of $s(t)$ is

$$R(\tau) = \int_0^1 R(t+\tau, t)dt \tag{15}$$

Then we have

$$R(\tau) = \sum_{l=m_1}^{m_0+m_1} \sum_{k=-\infty}^{\infty} R_l[k]2^l \int_{-\infty}^{\infty} \psi^*(u)\psi(u + 2^l\tau - k)du. \tag{16}$$

So, the PSD of $s(t)$ is

$$\hat{R}(\omega) = \sum_{l=m_1}^{m_0+m_1} \sum_{k=-\infty}^{\infty} R_l[k]|\hat{\psi}(2^{-l}\omega)|^2 e^{-j\omega 2^{-l}K} = \sum_{l=m_1}^{m_0+m_1} |\hat{\psi}(2^{-l}\omega)|^2 \hat{R}_l(2^{-l}\omega).$$

From the theorems above, we know clearly that wavelet-packet (wavelet) modulation spreads the spectrum of the original signals. We will get wider spectrum of the modulated signals with greater N in theorem 1 and m_1 in theorem 2. Because wavelet (especially wavelet packets) has wonderful time-frequency localization property, We can select N, l and m_1 so that the spectrum of the modulated signals is in the domain we need. That's very important for a communication system in the freqency-selective channel. We find also, from the formula, that the PSD of the modulated signals will vary slowly with frequency f. Further more, the PSD of the modulated signals will be flatter with lager N and m_1. The featureless waveform would be helpful for the covert in a communication system.

3 Conclusion

The PSD formula of the wavelet-packet and wavelet modulated signals is derived in the paper. It gives the properties of the modulated signals in frequency domain. It will be helpful in application, especially in the design of a communication system based on wavelet-packet.

References

1. K. M. Wong, Jiangfeng Wu, et. al.: Performance of wavelet packet- division multiplexing in impulsive and gaussian noise, IEEE transactions on comm., Vol. 48, No. 7, pp.1083-1086, July. 2000. 151
2. A. R. Lindsey, J. C. Dill Proc.: A digital transceiver for wavelet-packet modulation, SPIE, Vol. 3391/255-264. 151
3. R. S. Orr, C. Pike, M. J. Lyall: Wavelet transform domain communication systems, Proc. SPIE, Vol. 2491/271-282. 151
4. Prashant P. Gandhi, Sathyanarayan S. Rao, et.al: Wavelets for Waveform Coding of Digital System, IEEE transactions on signal processing, Vol. 45, No. 9, pp.2387-2390, Sep. 1997. 151
5. R. E. Learned, et al: Wavelet-packet-based multiple access communication, Proc. SPIE, Vol. 2303/246-264. 151

Orthogonal Multiwavelets with Dilation Factor a

Shouzhi Yang, Zhengxing Cheng, and Hongyong Wang

Department of Mathematics, Xi'an Jiaotong University
Xi'an, 710049, P.R.China
yangshouzhi@china.com

Abstract. There are perfect construction formulas for the orthonormal uniwavelet. However, it seems that there is not such a good formula with similar structure for multiwavelets. Especially, construction of multiwavelets with dilation factor $a(a \geq 2, a \in Z)$ lacks effective methods. In this paper, a procedure for constructing compactly supported orthonormal multiscale functions is first given, and then based on the constructed multiscale functions, we propose a method of constructing multiwavelets, which is similar to that of uniwavelet. Finally, we give a specific example illustrating how to use our method to construct multiwavelets.

1 Introduction

Since Geronimo, Hardin and Massopust [1] presented the first example of multiwavelets by using fractal interpolation functions, the study of multiwavelets has drawn many researcher's attention(e.g. See [2],[3] and [4]). Later, more examples were provided in [5] and [6]. As we know, Daubechies [7] obtained perfect constructing formulas for the uniwavelet. Since multiwavelets is a vector-vauled function, the construction of multiwavelets is more difficult than the that of uniwavelet. Multiwavelets can possess simultaneously many desirable properties, such as continuty, compact and short supportedness, orthonormality, interpolating, and very important symmetry or antisymmetry. However, for the uniwavelet, some of these properties are impossible or incompatible. From this respect, applications of multiwavelets are more extensive than those of uniwavelet. Therefore, finding the approaches of construction for the multiwavelets is very significant both in theory and in applications. Donovan, Geronimo, and Hardin [8] discussed the above problem by using fractal interpolation functions, but their construction procedure is very complicated. The main objective of this paper is to give a way of constructing compactly supported multiscale functions and the associated multiwavelets.

2 Multiresolution Analysis

Let $\mathbf{\Phi}(x) = (\phi_1, \phi_2, \cdots, \phi_r)^{\mathrm{T}}, \phi_1, \phi_2, \cdots, \phi_r \in L^2(R)$, satisfy the following two-scale matrix equation:

$$\mathbf{\Phi}(x) = \sum_{k=0}^{M} P_k \mathbf{\Phi}(ax - k), \tag{1}$$

Y. Y. Tang et al. (Eds.): WAA 2001, LNCS 2251, pp. 157–163, 2001.

where some $r \times r$ matrices $\{P_k\}$ are called the two-scale matrix sequence. $\mathbf{\Phi}(x)$ is termed multiscale functions with dilation $a(a \geq 2, a \in Z)$ and multiplicity r.

Applying Fourier transformation to (1), we obtain

$$\hat{\mathbf{\Phi}}(w) = P(z)\hat{\mathbf{\Phi}}(\frac{w}{a}), \quad P(z) = \frac{1}{a}\sum_{k=0}^{M} P_k z^k, z = e^{-\frac{iw}{a}}. \tag{2}$$

$P(z)$ is called the two-scale matrix symbol of matrix sequence $\{P_k\}$ of $\mathbf{\Phi}$.

Define subspace $\mathbf{V_j} = \text{clos}_{L^2(R)}\langle \phi_{\ell:j,k} : 1 \leq \ell \leq r, k \in Z\rangle, j \in Z$, here and afterwards, for $f_\ell \in L^2$, we will use the notation $f_{\ell:j,k} = a^{\frac{j}{2}} f_\ell(a^j x - k)$.

As usual, $\mathbf{\Phi}(x)$ in (1) generates a multiresolution analysis $\{\mathbf{V_j}\}_{j\in Z}$ of $L^2(R)$, if $\{\mathbf{V_j}\}_{j\in Z}$ satisfy the nestedness, $\cdots \subset \mathbf{V_0} \subset \mathbf{V_1} \subset \mathbf{V_2} \cdots$. Let $\mathbf{W_j}, j \in Z$, denote the orthogonal complementary subspace of $\mathbf{V_j}$ in $\mathbf{V_{j+1}}$, and vector-valued function $\mathbf{\Psi}(x) = (\psi_1, \psi_2, \cdots, \psi_{(a-1)r})^{\mathrm{T}}, \psi_\ell \in L^2, \ell = 1, 2, \cdots, (a-1)r$, constitutes a Riesz basis for $\mathbf{W_j}$, i.e., $\mathbf{W_j} = \text{clos}_{L^2(R)}\langle \psi_{\ell:j,k} : 1 \leq \ell \leq (a-1)r, k \in Z\rangle, j \in Z$. It is clear that $\psi_1(x), \psi_2(x), \cdots, \psi_{(a-1)r}(x)$ are in $\mathbf{W_0} \subset \mathbf{V_1}$, Hence there exists a sequence of matrices $\{Q_k\}_{k\in Z}$ such that

$$\mathbf{\Psi}(x) = \sum_{k=0}^{M} Q_k \mathbf{\Phi}(ax - k). \tag{3}$$

From the two-scale relation (3), we obtain

$$\hat{\mathbf{\Psi}}(w) = Q(z)\hat{\mathbf{\Phi}}(\frac{w}{a}), \quad Q(z) = \frac{1}{a}\sum_{k=0}^{M} Q_k z^k. \tag{4}$$

For column vector functions Λ and Γ with elements in $L^2(R)$, define $\langle \Lambda, \Gamma\rangle = \int_R \Lambda(x)\Gamma(x)^{\mathrm{T}} dx$. We call $\mathbf{\Phi}(x) = (\phi_1, \phi_2, \cdots, \phi_r)^{\mathrm{T}}$ orthogonal multiscaling function, if $\langle \mathbf{\Phi}(\cdot), \mathbf{\Phi}(\cdot - n)\rangle = \delta_{0,n} I_r$, $n \in Z$. $\mathbf{\Psi}(x) = (\psi_1, \psi_2, \cdots, \psi_{(a-1)r})^{\mathrm{T}}$ will be said to be orthogonal multiwavelets associated with multiscaling functions $\mathbf{\Phi}$, if $\mathbf{\Psi}(x)$ satisfy the following equations $\langle \mathbf{\Phi}(\cdot), \mathbf{\Psi}(\cdot - n)\rangle = \langle \mathbf{\Psi}(\cdot), \mathbf{\Phi}(\cdot - n)\rangle = O_{r\times(a-1)r}$ and $\langle \mathbf{\Psi}(\cdot), \mathbf{\Psi}(\cdot - n)\rangle = \delta_{0,n} I_{(a-1)r}$, $n \in Z$, where $O_{r\times(a-1)r}$ and $I_{(a-1)r}$ denote the zero matrix and unit matrix, respectively.

Lemma 1 Let $\eta = (\eta_1, \eta_2, \cdots, \eta_r)^{\mathrm{T}}$, where $\eta_1, \eta_2, \cdots, \eta_r \in L^2$, then $\{\eta_\ell(x-k) : 1 \leq \ell \leq r,\ k \in Z\}$ is a family of orthogonal functions if and only if $\sum_{k\in Z} \hat{\eta}(\omega + 2k\pi)\hat{\eta}(\omega + 2k\pi)^* = I_r$, $|z| = 1$, here and throughout, the asterisk denotes complex conjugation of transpose.

Lemma 2 let $\mathbf{\Phi}(x)$ be a multiscale function satisfying (1), $P(z)$ be two-scale matrix symbol, then (i) $\mathbf{\Phi}(x)$ is compactly supported, with supp $\mathbf{\Phi}(x) \subset [0, \frac{M}{a-1}]$; (ii) P(1) has eigenvalue 1, and $[P(1)]^n$ converges as $n \to \infty$; (iii) the vector $u = \hat{\mathbf{\Phi}}(0)$ is an eigenvector corresponding to the eigenvalue 1 of P(1);

Similar to the case of $a = 2$(See [9]), (i) can be proved analogously. (ii) and (iii) also can be deduced by using the similar method in [2]

Lemma 3 Let $\mathbf{\Phi}$ be a multiscale function satisfying (1). If both P_0, P_M are not nilpotent, then Supp $\Phi = [0, \frac{M}{a-1}]$.

The Lemma 3 can be proved by using the Similar method in [9].

3 Construction of Orthonormal Multiwavelets

Theorem 1 Let $\mathbf{\Phi}(x)$ be the orthogonal multiscaling functions defined in (1), $P(z)$ be the two -scale matrix symbol, $\omega_j,\ j = 1, 2, \cdots, a$ be a roots of equation $z^a - 1 = 0$,then $\sum\limits_{j=1}^{M} P(\omega_j z)P(\omega_j z)^* = I_r,\ |z| = 1$. i.e.,

$$\sum_{i=0}^{M} P_i P_{i+ak}^* = a\delta_{k,0} I_r, \quad |z| = 1. \tag{5}$$

Further, suppose $\mathbf{\Psi} = (\psi_1, \psi_2, \cdots, \psi_{(a-1)r})^{\mathrm{T}}$ is an orthogonal multiwavelets associated with $\mathbf{\Phi}$, $Q(z)$ is two-scale matrix symbol, then

$$\sum_{j=1}^{M} P(\omega_j z)Q(\omega_j z)^* = O, \quad \sum_{j=1}^{M} Q(\omega_j z)Q(\omega_j z)^* = I_{(a-1)r}. \tag{6}$$

Eqs. (6) are equivalent to the following Eqs.(7), respectively,

$$\sum_{i=0}^{M} P_i Q_{i+ak}^{\mathrm{T}} = O, \quad \sum_{i=0}^{M} Q_i Q_{i+ak}^{\mathrm{T}} = a\delta_{0,k} I_{(a-1)r}. \tag{7}$$

By using Lemma 1, we can easily prove Theorem 1

Analogous to Hermite cardinal spline interpolation, $\mathbf{\Phi}(x) = (\phi_1, \phi_2, \cdots, \phi_r)^{\mathrm{T}}$ with common support is said to be interpolatory, if it satisfies the following condition:

$$\begin{cases} \mathbf{\Phi}^{(j-1)}(k+k_0) = \phi_j^{(j-1)}(k_0)\delta_{k,0}\mathbf{e}_j \\ \phi_j^{(j-1)}(k_0) \neq 0 \end{cases}, \mathbf{e}_1 = (1, 0, \cdots, 0)^{\mathrm{T}}, \cdots, \mathbf{e}_r = (0, \cdots, 0, 1)^{\mathrm{T}} \tag{8}$$

Theorem 2 Let $\mathbf{\Phi}(x)$ be a multiscale function with dilation a and multiplicity r as in (1) and satisfy (8) for some positive integer $k_0 (1 \leq k_0 \leq [\frac{M}{a-1}])$,then we have

$$P_{ak+k_0} = \delta_{k,0} P_{k_0}, \quad P_{k_0} = diag(1, \frac{1}{a}, \cdots, \frac{1}{a^{r-1}}), k \in Z \tag{9}$$

proof Taking $j-1$ derivatives to (1) and applying the interpolation condition (8), we have $P_{ak+k_0} e_j = \frac{1}{a^{j-1}} \delta_{k,0} e_j, 1 \leq j \leq r$, which implies (9).

Theorem 3 Let $\mathbf{\Phi}(x) = (\phi_1, \phi_2, \cdots, \phi_r)^{\mathrm{T}}$ be a multiscale function with dilation a as in (1) , $P(z)$ be two-scale matrix symbol, if supp$\phi_i = [h_i, g_i],\ 1 \leq i \leq r$, then

(i) ϕ_{2i-1} are symmetric and ϕ_{2i} antisymmetric for all j in the following sense $\phi_i(x) = (-1)^{i-1}\phi_i(h_i + g_i - x), 1 \le i \le r$ if and only if the entries $P_{i,j}$ of the matrix $P(z)$ satisfy

$$P_{i,j}(z) = (-1)^{i+j} z^{a(h_i+g_i)-(h_j+g_j)} P_{i,j}(\overline{z}), 1 \le i, j \le r \tag{10}$$

(ii) $\phi_1, \phi_2, \cdots, \phi_{r_1}$ are symmetric, the remainder $\phi_{r_1+1}, \cdots, \phi_r$ are antisymmetric in the sense $\phi_i(x) = \phi_i(h_i + g_i - x), i = 1, 2, \cdots, r_1$, and $\phi_i(x) = -\phi_i(h_i + g_i - x), i = r_1, r_1 + 1, \cdots, r$ if and only if the entries $P_{i,j}$ of the matrix $P(z)$ satisfy

$$P_{i,j} = \begin{cases} z^{a(h_i+g_i)-(h_j+g_j)} P_{i,j}(\overline{z}), 1 \le i, j \le r_1 \text{ or } r_1 + 1 \le i, j \le r \\ -z^{a(h_i+g_i)-(h_j+g_j)} P_{i,j}(\overline{z}), 1 \le i \le r_1 \text{ and } r_1 + 1 \le j \le r \\ \qquad \text{or } r_1 + 1 \le i \le r \text{ and } 1 \le j \le r_1 \end{cases} \tag{11}$$

(iii) If $a(h_i + g_i) - (h_j + g_j)(1 \le i, j \le r)$ strictly is less than zero or isn't an integer, then $P_{i,j} = 0$

Proof If $\phi_1, \phi_2, \cdots, \phi_r$ satify $\phi_i(x) = (-1)^{i-1}\phi_i(h_i + g_i - x), 1 \le i \le r$, let $S_r = \text{diag}(1, -1, \cdots, (-1)^r)$, then $\mathbf{\Phi}(x) = (\phi_1(x), \phi_2(x), \cdots, \phi_r(x))^{\mathrm{T}} = S_r(\phi_1(h_1 + g_1 - x), \phi_2(h_2 + g_2 - x), \cdots, \phi_r(h_r + g_r - x))^{\mathrm{T}}$, hence,

$$\begin{cases} \hat{\mathbf{\Phi}}(\omega) = S_r D_r(z^a)\overline{\hat{\mathbf{\Phi}}(\omega)} \\ D_r(z) = diag(z^{h_1+g_1}, z^{h_2+g_2}, \cdots, z^{h_r+g_r}) \end{cases}$$

Successively using (2), we obtain $P(z)\hat{\mathbf{\Phi}}(\frac{\omega}{a}) = S_r D_r(z^a)\overline{P(z)} D_r(\overline{z}) S_r \hat{\mathbf{\Phi}}(\frac{\omega}{a})$. Since $\{\phi_\ell(x-k) : 1 \le \ell \le r, k \in Z\}$ is a Riesz basis of V_0, so $P(z) = S_r D_r(z^a)\overline{P(z)}\, D_r(\overline{z}) S_r$. Or equivalently, $S_r P(z) S_r = D_r(z^a)\overline{P(z)} D_r(\overline{z})$, which implies (10) holds. This completes the proof of Theorem 3

Corollary 1 If $\text{supp}\phi_1 = \text{supp}\phi_2 = \cdots = \text{supp}\phi_r = [0, \frac{M}{a-1}]$, then ϕ_{2i-1} are symmetric and ϕ_{2i} antisymmetric for all j if and only if $P_k = S_r P_{M-k} S_r$.

In fact, since $\text{supp}\phi_i = [0, \frac{M}{a-1}]$, $a(h_i + g_i) - (h_j + g_j) \equiv M$, we obtain $P(z) = z^M S_r P(\overline{z}) S_r$ by (10). Hence, Corollary 1 holds.

As we know, for a multiscale function $\mathbf{\Phi}(x)$, if $\text{supp}\mathbf{\Phi}(x) = [0, M]$, then $\text{supp}\mathbf{\Phi}'(x) = [0, \lceil\frac{M}{a}\rceil]$, where $\mathbf{\Phi}'(x) = [\mathbf{\Phi}^{\mathrm{T}}(ax), \mathbf{\Phi}^{\mathrm{T}}(ax-1), \cdots, \mathbf{\Phi}^{\mathrm{T}}(ax - a + 1)]^{\mathrm{T}}$. Hence, without loss of generality, we only investigate the construction of multiwavelets with $a+1$-coefficient. i.e., $\mathbf{\Phi}^{\mathrm{T}}(x)$ satisfies the following equation

$$\mathbf{\Phi}(x) = \sum_{k=0}^{a} P_k \mathbf{\Phi}(ax - k) \tag{12}$$

In the applications of multiwavelets, certain special properties is desirable , such as interpolating and symmetry. In the two-scale matrix sequence $\{P_k\}$, associated with those multiwavelets with these properties , there must exists some $P_i, 0 \le i \le a$ such that the matrix $(aI - P_i P_i^{\mathrm{T}})^{-1} P_i P_i^{\mathrm{T}}$ is a positive definite matrix

Lemma 4 Let $\mathbf{\Phi}(x)$ be the orthogonal compactly supported multiscale function with dilation a and multiplicity r satisfing (12), Assume that there exists an $P_i, 0 \le i \le a$ such that the matrix H defined in following equation is a positive definite matrix

$$H^2 = (aI_r - P_i P_i^{\mathrm{T}})^{-1} P_i P_i^{\mathrm{T}}, \tag{13}$$

Let $H_s (s = 1, 2, \cdots, a-1)$ be $(a-1)$ essentialy different symmetric matrices satisfing (13), define $q_j^{(s)} = H_s P_j (j \ne i)$, and $q_j^{(s)} = -H_s^{-1} P_j (j = i)$, here$j = 0, 1, \cdots, a; s = 1, 2, \cdots, a-1$. then

$$P_0 (q_a^{(s)})^{\mathrm{T}} = O, \tag{14}$$

$$P_0 (q_0^{(s)})^{\mathrm{T}} + P_1 (q_1^{(s)})^{\mathrm{T}} + \cdots + P_a (q_a^{(s)})^{\mathrm{T}} = O \tag{15}$$

$$(q_0^{(\ell)})(q_a^{(s)})^{\mathrm{T}} = O, \ \ell, s = 1, 2, \cdots, a-1 \tag{16}$$

$$(q_0^{(s)})(q_0^{(s)})^{\mathrm{T}} + (q_1^{(s)})(q_1^{(s)})^{\mathrm{T}} + \cdots + (q_a^{(s)})(q_a^{(s)})^{\mathrm{T}} = aI_r. \tag{17}$$

Proof For convenience, let i=1. (14) and (16) can be proved easily by using (6). For (15) and (17), we have from (6) that

$$\begin{aligned}
&\sum_{\ell=0}^{a} P_\ell (q_\ell^{(s)})^{\mathrm{T}} = P_0 P_0^{\mathrm{T}} H_s - P_1 P_1^{\mathrm{T}} (H_s^{-1}) + \cdots + P_a P_a^{\mathrm{T}} H_s \\
&= [P_0 P_0^{\mathrm{T}} + P_2 P_2^{\mathrm{T}} + \cdots + P_a P_a^{\mathrm{T}}] H_s - P_1 P_1^{\mathrm{T}} (H_s)^{-1} \\
&= [aI_r - P_1 P_1^{\mathrm{T}}] H_s - P_1 P_1^{\mathrm{T}} (H_s)^{-1} \\
&= [(aI_r - P_1 P_1^{\mathrm{T}})(H_s)^2 - P_1 P_1^{\mathrm{T}}](H_s)^{-1} = O \\
&\sum_{\ell=0}^{a} q_\ell^{(s)} (q_\ell^{(s)})^{\mathrm{T}} = H_s P_0 P_0^{\mathrm{T}} H_s + (H_s)^{-1} P_1 P_1^{\mathrm{T}} (H_s)^{-1} + \cdots + H_s P_a P_a^{\mathrm{T}} H_s \\
&= H_s [P_0 P_0^{\mathrm{T}} + P_2 P_2^{\mathrm{T}} + \cdots + P_a P_a^{\mathrm{T}}] H_s + (H_s)^{-1} P_1 P_1^{\mathrm{T}} (H_s)^{-1} \\
&= H_s [aI_r - P_1 P_1^{\mathrm{T}}] H_s + (H_s)^{-1} P_1 P_1^{\mathrm{T}} (H_s)^{-1} \\
&= (H_s)^{-1} [(H_s)^2 (aI_r - P_1 P_1^{\mathrm{T}})(H_s)^2 - P_1 P_1^{\mathrm{T}}](H_s)^{-1} \\
&= (H_s)^{-1} [(H_s)^2 P_1 P_1^{\mathrm{T}} + P_1 P_1^{\mathrm{T}}](H_s)^{-1} = (H_s)^{-1} [(H_s)^2 + I_r] P_1 P_1^{\mathrm{T}} (H_s)^{-1} \\
&= H_s [P_1 P_1^{\mathrm{T}} + (H_s)^{-2} P_1 P_1^{\mathrm{T}}](H_s)^{-1} = H_s a I_r (H_s)^{-1} = aI_r
\end{aligned}$$

This completes the proof of Lemma 4.

In the setting of Lemma 4, we can generate $a-1$ sequences $\{q_k^{(s)}\}, s = 1, 2, \cdots, a-1$. We construct the following functions in terms of these sequences,

$$\psi_s(x) = \sum_{k=0}^{a} q_k^{(s)} \mathbf{\Phi}(ax - k), \ \ s = 1, 2, \cdots, a-1. \tag{18}$$

Appling *Schmidt* orthonormalizing to a functions $\mathbf{\Phi}(x), \psi_s(x), s = 1, 2, \cdots, a-1$, and generating a functions $\mathbf{\Phi}(x), \Psi_s(x), s = 1, 2, \cdots, a-1$, we can conclude that there must exist $a-1$ sequences $\{Q_k^{(s)}\}, s = 1, 2, \cdots, a-1$, such that

$$\mathbf{\Psi}_s(x) = \sum_{k=0}^{a} Q_k^{(s)} \mathbf{\Phi}(ax-k), s = 1, 2, \cdots, a-1. \tag{19}$$

Hence, we have the following theorem:

Theorem 4 In the setting of Lemma 4, let $\mathbf{\Psi}_s(x), s = 1, 2, \cdots, a-1$ be defined as in (19). Define $\mathbf{\Psi}(x) = [\mathbf{\Psi}_1(x)^{\mathrm{T}}, \mathbf{\Psi}_2(x)^{\mathrm{T}}, \cdots, \mathbf{\Psi}_{a-1}(x)^{\mathrm{T}}]^{\mathrm{T}}$, then $\mathbf{\Psi}(x)$ is compactly supported orthogonal multiwavelets with dilation a associated with $\mathbf{\Phi}(x)$, and satisfies the following two-scale matrix equation

$$\mathbf{\Psi}(x) = \sum_{k=0}^{a} [(Q_k^{(1)})^{\mathrm{T}}, (Q_k^{(2)})^{\mathrm{T}}, \cdots, (Q_k^{(a-1)})^{\mathrm{T}}]^{\mathrm{T}} \mathbf{\Phi}(ax-k) \tag{20}$$

Corollary 2 In the setting of Lemma 4, (i) If dilation factor $a = 2$, then $\psi_1(x)$ defined in (18) is compactly supported orthogonal multiwavelets with dilation 2 associated with $\mathbf{\Phi}(x)$; (ii) If dilation factor $a = 3$, and $\mathbf{\Psi}_s(x) = \sum_{k=0}^{3} Q_k^{(s)} \mathbf{\Phi}(ax-k), s = 1, 2$. Let $\mathbf{\Psi}(x) = [\mathbf{\Psi}_1(x)^{\mathrm{T}}, \mathbf{\Psi}_2(x)^{\mathrm{T}}]^{\mathrm{T}}$, then $\mathbf{\Psi}(x)$ is compactly supported orthogonal multiwavelets with dilation 3 associated with $\mathbf{\Phi}(x)$, and satisfies (20) in which ,

$$\begin{cases} Q_k^{(1)} = q_k^{(1)} \\ Q_k^{(2)} = \frac{1}{2}[q_k^{(2)} - \sum\limits_{h=0}^{3} q_h^{(2)} (Q_h^{(1)})^{\mathrm{T}} Q_k^{(1)}] \end{cases}, k = 0, 1, 2, 3. \tag{21}$$

4 Example

We will illustrate by a specific example how to construct orthogonal multiwavelets based on our method.

Example (Construction of orthogonal multiwavelets with dilation 3 and multiplicity 3)

Let $\mathbf{\Phi}(x) = (\phi_1, \phi_2, \phi_3)^{\mathrm{T}}$, satisfy $\mathbf{\Phi}(x) = P_0\mathbf{\Phi}(3x) + P_1\mathbf{\Phi}(3x-1) + P_2\mathbf{\Phi}(3x-2)$. By Lemma 2 ,supp$\mathbf{\Phi}(x) \subset [0, 1]$. Suppose both ϕ_1 and ϕ_3 are symmetric and ϕ_2 is antisymmetric, $\mathbf{\Phi}(x)$ satisfies the interpolatory condition (8) with $k_0 = 1$, then in view of (9), taking $i = 1$ and using Theorem 4, we obtain

$$q_0^{(1)} = \begin{bmatrix} \frac{1}{2} & \frac{1}{2} & 0 \\ -\frac{\sqrt{26}}{52} & \frac{\sqrt{182}}{156} & \frac{\sqrt{26}}{26} \\ 0 & 0 & \frac{\sqrt{2}}{18} \end{bmatrix}, \quad q_1^{(1)} = \begin{bmatrix} -\sqrt{2} & 0 & 0 \\ 0 & -\frac{\sqrt{26}}{3} & 0 \\ 0 & 0 & -\frac{11\sqrt{2}}{9} \end{bmatrix},$$

$$q_2^{(1)} = \begin{bmatrix} \frac{1}{2} & -\frac{1}{2} & 0 \\ \frac{\sqrt{26}}{52} & \frac{\sqrt{182}}{156} & -\frac{\sqrt{26}}{26} \\ 0 & 0 & \frac{\sqrt{2}}{18} \end{bmatrix}, \quad q_0^{(2)} = \begin{bmatrix} \frac{1}{2} & \frac{1}{2} & 0 \\ -\frac{\sqrt{26}}{52} & \frac{\sqrt{182}}{156} & \frac{\sqrt{26}}{26} \\ 0 & 0 & -\frac{\sqrt{2}}{18} \end{bmatrix},$$

$$q_1^{(2)} = \begin{bmatrix} -\sqrt{2} & 0 & 0 \\ 0 & -\frac{\sqrt{26}}{3} & 0 \\ 0 & 0 & \frac{11\sqrt{2}}{9} \end{bmatrix}, \quad q_2^{(2)} = \begin{bmatrix} \frac{1}{2} & -\frac{1}{2} & 0 \\ \frac{\sqrt{26}}{52} & \frac{\sqrt{182}}{156} & -\frac{\sqrt{26}}{26} \\ 0 & 0 & -\frac{\sqrt{2}}{18} \end{bmatrix}.$$

Finally, we obtain orthogonal multiwavelets by (20) and (21).

References

1. Geronimo,J., Hardin,D. P., Massopust,P.: Fractal Functions and Wavelet Expansions Based on Several Scaling Functions. J. Approx. Theory. 78(1998) 373-401
2. Chui,C. K., Lian,J.: A Study on Orthonormal Multiwavelets. J. Appl. Numer. Math., 20(1996) 273-298
3. Lian,J.: Orthogonal Criteria for Multiscaling Functions. Appl. Comp. Harm. Anal. 5(1998) 277-311
4. Hardin,D. P., Marasovich,J. A.: Biorthogonal Multiwavelets on [-1,1]. Appl. Comp. Harm. Anal. 7(1999) 34-53
5. Goh,S. S., Yap,V. B.: Matrix Extension and Biorthogonal Multiwavelets Construction. Linear Algebra and Applictions. 269(1998) 139-157
6. Marasovich,J.:Biorthogonal Multiwavelets, Dissertation, Vanderbilt University, Nashville, TN, (1996)
7. Daubechies,I.: Ten lectures on wavelets,SIAM, Philadelphia, PA, (1992)
8. Donovan,G. C., Geronimo,J., Hardin,D. P.: Construction of Orthogonal Wavelets Using Fractal Interpolution Functions. SIAM J. Math. Anal. 27(1996) 1158-1192
9. Wang So, Jianzhang Wang, Estimating the Support of a Scaling Vector. SIAM J. Matrix Anal. Appl. 1(1997) 66-73

A Wavelet-Based Image Indexing, Clustering, and Retrieval Technique Based on Edge Feature

Masaaki Kubo[1], Zaher Aghbari[1], Kun Seok Oh[2], and Akifumi Makinouchi[1]

[1] Graduate School of Information Science and Electrical Engineering, Department of Intelligent Systems, Kyushu University
6-10-1 Hakozaki, Higashi-ku, Fukuoka-shi 812-8581, Japan
{kubo,zaher,akifumi}@db.is.kyushu-u.ac.jp

[2] Division of Computer Engineering College of Engineering Chosun University
375 Susuk-dong Dong-gu Kwangju 501-759 Korea
okseak38@hotmail.com

Abstract. This paper proposes a technique for indexing, clustering and retrieving images based on their edge features. In this technique, images are decomposed into several frequency bands using the Haar wavelet transform. From the one-level decomposition sub-bands an edge image is formed. Next, the higher order auto-correlation function is applied on the edge image to extract the edge features. These higher order autocorrelation features are normalized to generate a compact feature vector, which is invariant to shift, image size and gray level. Then, these feature vectors are clustered by a self-organizing map (SOM) based on their edge feature similarity. The performed experiments show the high precision of this technique in clustering and retrieving images in a large image database environment.

1 Introduction

In the past decade, the number of digital images has increased tremendously due to the steady growth of computer power, decline of storage cost, and rapid increase in access to the Internet. Therefore, fast and effective methods to organize and search images in large image database environments are essential. In particular, images need to be effectively clustered and then fast content-based mechanism is required to retrieve desired images.

Currently, two main indexing approaches exist: (1) indexing images based on features from raw image data [1][2], such as pixel intensity, histogram, etc. (2) indexing images based on coefficients in the transform domain [3][4][5][6], such as total energy of wavelet coefficients. These extracted features are represented by means of a *feature vector*, which is a compact representation of the image content. These feature vectors are then organized by a spatial access method (SAM), such as B-tree, R-tree, etc., or a clustering method, such as self-organizing map (SOM)[7]. When a query Q, such as *"Find images similar to Q"*, is issued, Q is compared with the database of feature vectors that represent the image database. As a result, the K most similar images to Q are returned to the user.

Y. Y. Tang et al. (Eds.): WAA 2001, LNCS 2251, pp. 164–176, 2001.

This paper presents a Haar wavelet-based technique that extracts the edge features from an edge image, which is generated from the one-level decomposition sub-bands of an image, by means of the higher order autocorrelation method. Since image in large databases are found in different sizes and gray levels, it is essential to adopt the extracted features to tolerate such differences, an important property lacked in previous work [3][4]. Thus, in this work, the extracted higher order autocorrelation features are normalized; as a result, they become invariant to shift, image size and gray level. The normalized features of an image are combined into a compact feature vector (25 feature values). Then, the feature vectors of all images are clustered by a SOM method. The system supports *query-by-example* access to the images.

The rest of this paper is organized as follows: the related work is surveyed in Sect. 2. In Sect. 3, we present the system architectur. The indexing and clustering technique of our system is discussed in Sect. 4. Then, the querying method and experimental results are discussed in Sect. 5. Finally, we conclude the paper in Sect. 6.

2 Related Work

An example of indexing based on raw image data is the (QBIC) system [1] of IBM that indexes images on multiple features, such as color histograms, texture, shapes, etc. Although such multiple features provides an effective representation of an image, they are computationally expensive during both the index computation phase and the query processing phase. Another example of indexing based on raw image data is the VisualSEEK system [2] which indexes each image in the database by its salient color regions.

For indexing based on the transformed-domain coefficients, Wang et al. [5] have proposed a wavelet-based image indexing and searching (WBIIS) algorithm. In the WBIIS project, Daubechies' wavelet transform are employed to produce color feature vectors that provide better frequency localization than other traditional color layout coding algorithms, as argued by the authors. Another example of indexing based on the transformed-domain is proposed by Jacobs et al. [8] in which an image searching algorithm that makes use of the multiresolution Haar wavelet decompositions of images is presented.

In large image databases, it is essential to organize and/or classify feature vectors into different clusters to speed up the search. This organization and/or clustering of images is based on the similarity of feature vectors of images. Here we introduce some examples that utilize such algorithms. Albuz et al. [3] have proposed an algorithm to cluster the feature vectors, which represent images, in a modified k order B-tree data structure, where k is the maximum number of clusters. This approach have utilized the multiresolution property of the wavelet transform to compute the feature vectors. The problem with this approach is that the number of clusters have to be decided by the user before inserting keys into the B-tree. Oja et al. [9] have introduced the PicSOM system to cluster images based on a Tree Structured Self-Organizing Maps (TS-SOMs). The TS-

SOM is a tree-structured vector quantization algorithm that uses SOMs [7] at each of its hierarchal levels. However, since the SOM algorithm is not scalable to new classes, if a new class of images is to be inserted into the database the TS-SOM, which is a hierarchy of SOMs, has to undergo a computationally expensive process of retraining at each hierarchy.

3 System Architecture

The basic architecture of our system is shown in Fig. 1. The solid arrows show the sequence of processes of indexing and clustering images and the dashed arrows follow the sequence of processes of querying. As shown in Fig. 1, both the images to be indexed and the query image go through the same sequence of processes. However, in case of indexing and clustering, after the *SOM-Based Clustering* process the feature vector of an image is added to the corresponding cluster in the database. In case of querying, the images associated with the best matching node (cluster) to the query image are returned to the user. A detailed discussion of these processes is in Sect. 4.

4 Indexing and Clustering

As shown in Fig. 1, the system applies several processes on an image to index and cluster it. In this Sect., we discuss these processes.

4.1 Haar Wavelet Transform

The wavelet transform describes the image in terms of a coarse overall shape, plus some details that range from broad to narrow. The Haar wavelet transform

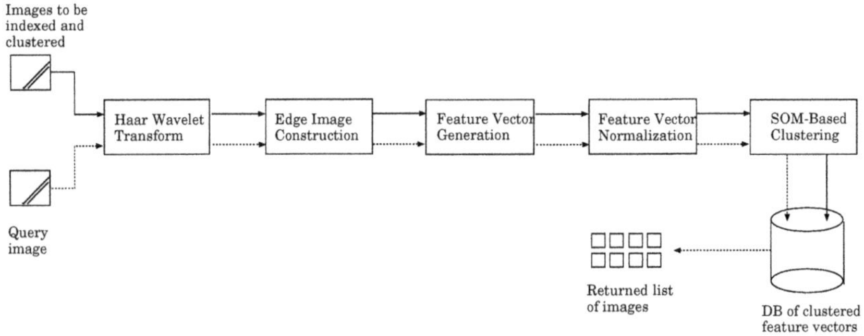

Fig. 1. Basic architecture of a system: The solid arrows show the path of image indexing and clustering. The dashed arrows show the path of image querying

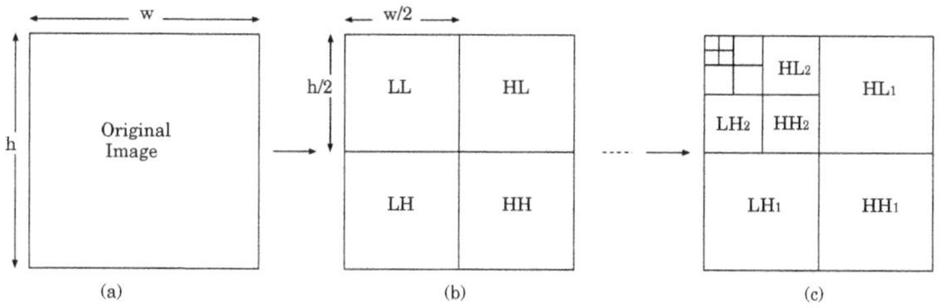

Fig. 2. Wavelet multiresolution property of an image: (a) represents original image, (b) a one-level decomposition produces 4 sub-bands, namely LL, LH, HL and HH, (c) a four-level decomposition produces 13 sub-bands

is applied iteratively on an image to generate multi-level decomposition (see Fig. 2). At level l decomposition, $3l + 1$ sub-bands are produced. In a large image database environment, it is essential to represent images by a method that supports the following requirements on a feature vector:

(1) Compact,
(2) Fast to compute, and
(3) Supports similarity retrieval.

Therefore, we are using a Haar wavelet transform to decompose images into several frequency bands and then compute a feature vector from these bands. The above requirements are satisfied as follows:

1. Compact: By making use of the wavelet multiresolution property, we can decompose an image and then use only a few coefficients to represent the image content sufficiently. As shown in Fig. 3, the Haar wavelet transform decomposed the original image (see Fig. 3.a) into four sub-bands: LL, LH, HL and HH (see Fig. 3.b). The Haar wavelet coefficients, *Haar basis* and *coefficient details*, are computed by Equations 1 and 2, respectively.

Fig. 3. An Example of wavelet decomposition: (a) original image, (b) one-level decomposition

$$F_0(x(n)) = \frac{1}{\sqrt{2}}(x(n) + x(n+1)) \tag{1}$$

$$F_1(x(n)) = \frac{1}{\sqrt{2}}(x(n) - x(n+1)) \tag{2}$$

Where, $x(n)$ and $x(n+1)$ are the current and next values of an image, respectively. The LH and HL sub-bands are used to generate an edge image (see Subsection 4.2). Then, we use the higher order autocorrelation function to extract the edge features from the edge image (see Subsection 4.3). Only a few (25 coefficients) of the extracted higher order autocorrelation coefficients are used to produce a feature vector. Thus, the feature vector of an image is compact.

2. Fast to Compute: The Haar wavelet basis is the simplest wavelet basis, in terms of implementation, and the fastest to compute [6][8]. From Equation 1, we notice that the Haar wavelet transform is mathematically equivalent to the averaging of color blocks [5]. Because Haar wavelets are fast to compute, they become a key to several applications such as data compression, data transmission, denoising, and edge detection. However, one drawback of Haar basis for lossy compression is that it tends to produce blocky image artifacts for high compression rates [8]. However, in our application, the result of compression is never viewed; therefore, these artifacts do not affect our indexing and querying processes.

3. Supports Similarity Retrieval: Similarity retrieval is preferred in image databases because users can simply select an image that is similar to the wanted image and then issue a query '*Find images that are similar to this query image*'. Or, a user can simply make a rough sketch, such as the dominant edges, of a wanted image and issue a query '*Find images that are similar to this sketch*'. To achieve this goal, we use only 25 normalized higher order autocorrelation coefficients to represent the image. These 25 coefficients sufficiently approximate the image and provide some margin for similarity retrieval.

4.2 Edge Image Construction

From a signal processing point of view, the wavelet transform is basically a convolution operation, which is equivalent to passing an image through low-pass and high-pass filters. Let the original image be $I(w, h)$, then the LH sub-band represents the vertical edges and HL sub-band represents the horizontal edges of $I(w, h)$. Using these properties of the LH and HL sub-bands, we construct an edge image. If an element in the LH sub-band is $v_{m,n}$ and an element in the HL sub-band is $h_{m,n}$, then the corresponding element $e_{m,n}$ in the edge image is given by Equation 3. Let w and h be the width and height, respectively, of the LH and HL sub-bands.

$$e_{m,n} = \sqrt{(v_{m,n})^2 + (h_{m,n})^2} \tag{3}$$

Where, $1 \leq m \leq w$ and $1 \leq n \leq h$. Fig. 4 shows the constructed edge image from the LH and HL sub-bands of Fig. 3.b using Equation 3. In our system, we use the LH and HL sub-bands of the one-level decomposition because they have more detailed information of the dominant edges in the original image.

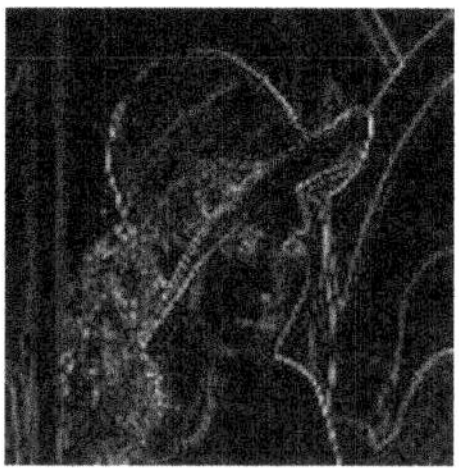

Fig. 4. Constructed edge image from the LH and HL sub-bands of Fig. 3.b

4.3 Feature Vector Generation

The higher order autocorrelation features are the primitive edge features that we use to index and retrieve images. Such features are shift-invariant (irrelevant to where the objects are located in the image), which is a useful property in image querying.

As defined in [10] and [11], let an image plane be P and a function $I(r)$ represents an image intensity function on the retinal plane P such that $r \in P$. That is, r is the image coordinate vector. A shift (translation) of $I(r)$ within P is represented by $I(r+a_i)$, where a_i is the displacement vector. Therefore, the Nth-order autocorrelation functions with N displacements $a_1, ..., a_N$ are defined by

$$R^N(a_1, ..., a_N) = \sum_P I(r)I(r + a_1)...I(r + a_N) \qquad (4)$$

It is obvious from Equation 4 that the number of autocorrelation functions obtained by the possible combinations of the displacements over the image plane is large. Therefore, it is essential to reduce this large number for practical applications. Here, we limit the order N up to 2 ($N = 0, 1, 2$). Also, the range of displacements is limited to within a local 3×3 window, of which the center is the reference local point.

The local mask pattern for extracting higher order autocorrelation features is shown in Fig. 5. The 0th-order autocorrelation function corresponds to the average gray level of the image $I(r)$. By eliminating the displacements that are equivalent by shift, the number of unique patterns is reduced to 25 as shown in Fig. 5. Using these mask patterns, the feature vector f^v that contains the higher order autocorrelation functions is defined as follows:

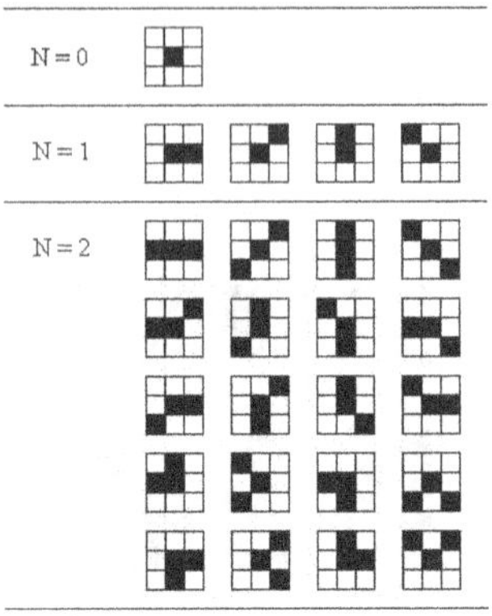

Fig. 5. The 25 Local mask patterns for extracting higher order autocorrelation features, where the order N is limited to 2

$$f^v = f_1, ..., f_{25} \tag{5}$$

Let the position of the mask pattern in the 3 × 3 window be denoted by x and y coordinates, such that $I_{x,y}$ denotes the mask pattern of the 0th-order autocorrelation function f_1. Also, let the width and height of the edge image I be w and h. Thus, each f_i is defined as:

$$f_1 = \sum_x \sum_y (I_{x,y})$$

$$f_2 = \sum_x \sum_y (I_{x,y})(I_{x,y+1})$$

$$\vdots$$

$$f_5 = \sum_x \sum_y (I_{x,y})(I_{x-1,y-1})$$

$$f_6 = \sum_x \sum_y (I_{x,y})(I_{x-1,y})(I_{x+1,y})$$

$$\vdots$$

$$f_{25} = \sum_x \sum_y (I_{x,y})(I_{x-1,y-1})(I_{x+1,y-1})$$

4.4 Feature Vector Normalization

As mentioned in Sect. 4.3, the extracted higher order autocorrelation features (see Equation 5) are invariant to shift (location of objects in the image). However, in large collections of images, such as images in the Internet, digital libraries, image archives, etc, images exist in different sizes and different intensities (gray levels). Therefore, it is important to design our features so that the search result should include the wanted image even if the selected image query (during the query by example process) is shifted, different in size, or different in gray level as compared with the wanted image.

In addition to being invariant to shift, we consider the following essential requirements on features for practical image search:

1. Features should be invariant to the size of an image.
2. Features should be invariant to the gray level of an image.

invariant to image size: For the first requirement, we divide the higher order autocorrelation functions by the width w and height h of the original image. As a result, the Feature values will not be proportional to the size of the original image, hence reducing the effect of the size difference between the query image and wanted image (see Equation 6).

invariant to gray level: we notice from Equation 5 that the extracted values of the higher order autocorrelation functions are proportional to the order of autocorrelation. For example, say that the sum of values of gray level of the original image equal to S, then when the order $N = 0$ the gray level value of $f_1 = S^1$, and when $N = 2$ the gray level value of, say $f_5 \approx S^3$. Therefore, we normalize the gray level values of the extracted higher order autocorrelation features by raising them to the power $1/N$, where N is the order of autocorrelation (see Equation 6).

$$\frac{1}{wh} \sum_{P} \sqrt[N]{I(r)I(r+a_1)\cdots I(r+a_N)} \tag{6}$$

For example, $f_2 = \frac{1}{wh} \sum_x \sum_y \sqrt[2]{(I_{x,y})(I_{x,y+1})}$. The effect of normalizing the feature vectors are shown in Fig. 6 and Fig. 7. The two images in Fig. 6 are different in size and gray scale. Figures 7.a and 7.b show the feature vectors (higher order autocorrelation features) of the two images before and after the normalizing process, respectively. Hence, the normalization process of feature vectors brings similar images, but different in their sizes and gray levels, closer together, which is a useful property in similarity-based retrieval. After being normalized, the feature vectors are inserted into a SOM to be clustered. The next section introduces briefly the SOM-based clustering process.

4.5 SOM-Based Clustering

The self-organizing Map (SOM) [7] is unsupervised neural network that maps high-dimensional input data $\Re^n$ (in our case, normalized higher order autocorrelation features of an image) onto a usually two-dimensional output space while preserving the topological relations (similarities) between the data items. The SOM consists of nodes (neurons) arranged in a two-dimensional rectangular or hexagonal grid. In our system, we simply arranged the SOM nodes in a

Fig. 6. An example to show the effect of normalizing feature vectors: the two images are different in size and gray scale

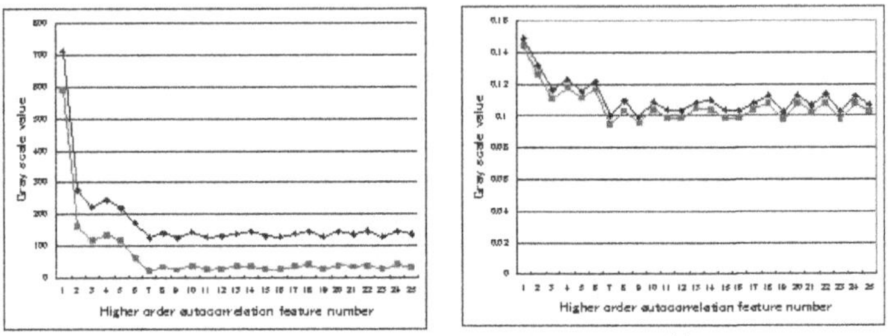

(a) (b)

Fig. 7. Effect of normalizing the feature vectors: (a)feature vectors of the two images in Fig. 6 before the normalizing process, (b) after the normalizing process

2-dimensional rectangular grid. With every node i, a weight vector $\mathbf{m}_i \in \Re^n$ is associated. An input vector $\mathbf{x} \in \Re^n$ is compared with $\mathbf{m}_i$, and the best-match-node (BMN), which has the smallest angle θ_{BMN} (see Equation 7), is determined. The input is thus mapped onto the location of the determined BMN.

$$\theta_{BMN} = \arccos\left\{\frac{\mathbf{x}\cdot\mathbf{m}_i}{\|\mathbf{x}\|\,\|\mathbf{m}_i\|}\right\} \tag{7}$$

The reason we used angle θ between vectors as a measure of distance rather than a simple Euclidean distance is illustrated in Fig. 8. The distances $d(\mathbf{a},\mathbf{c})$ and $d(\mathbf{b},\mathbf{c})$ between vectors are correctly expressed by the angles θ_{ac} and θ_{bc} rather than the Euclidean distances $\|\mathbf{a}-\mathbf{c}\|_2$ and $\|\mathbf{b}-\mathbf{c}\|_2$ between the vectors, respectively.

The weight vector $\mathbf{m}_c$ of the BMN is adopted to match the input vector. That is done by moving $\mathbf{m}_c$ towards $\mathbf{x}$ by a certain fraction of the angle θ_{BMN}. Moreover, the weight vectors of nodes in the neighborhood of the BMN are moved towards $\mathbf{x}$, but to a lesser extent than the BMN. This learning process finally leads to a topologically-ordered mapping of the input vectors. That is the cluster structure within the data and the inter-cluster similarity is represented

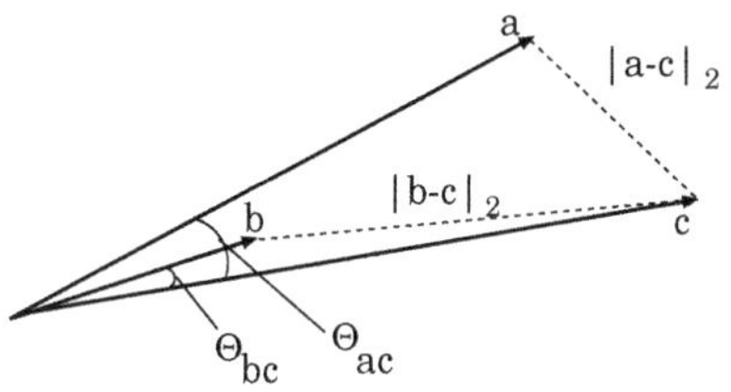

Fig. 8. An example to illustrate the effectiveness of using the angle θ between vectors instead of the 2-norm (Euclidean distance) as a measure of dissimilarity (distance)

clearly in the map. The map is called a *topological feature map* and the weight vector held by a node is called a *codebook vector.*

5 Querying and Results

The programs of our system are written in C++. The database (images, codebook vector and clusters of images) is managed by the Jasmine-supported object database management system developed by FUJITSU and Computer Associates. Hence, we used Jasmine's Weblink to build a user interface that supports *query-by-example* type of queries and create the HTML retrieval templates to display the results on a web browser. The experiments are performed on 620 images. The system runs on a sparc Ultra-5_10, 270 MHz, 128 MBytes Sun Workstation.

5.1 Querying

Currently, our system supports *query-by-example* in which a user selects a query image Q that is most similar to the wanted image(s) from a set of displayed images. Then, Q undergoes the same sequence of processes described in Sect. 4. Briefly, Q is decomposed, an edge image is generated, the higher order autocorrelation feature vector is extracted, normalized, and compared with the codebook vectors of all nodes of the SOM. Again, the BMN that is most similar (has the smallest θ_{BMN}, see Equation 7) to Q is determined. Finally, the images associated with the BMN are returned to the user for further manual browsing.

5.2 Results

To provide numerical results, we tested 7 sample queries chosen randomly from the image database. The result of each query Q is the set of images that are associated with the BMN, which is the most similar SOM node to Q. By examining the SOM clusters, we found that the number of images associated with any of the SOM nodes is less than 15, which is small enough for manual browsing.

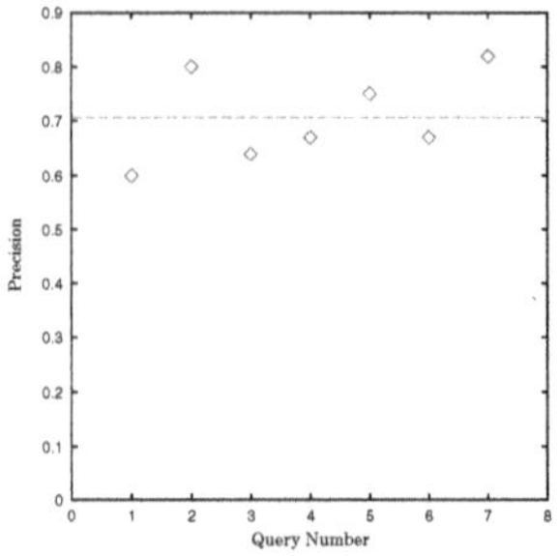

Fig. 9. Precision of the 7 sample queries and their average precision (dotted line)

Table 1. Average time to determine the BMN and average precision of query results

	Average value
Time to determine BMN	4.96 seconds
Precision of query results	70.7%

From the result of each sample query, we calculated the precision p of the query results. Since all the images clustered under, or associated with, the BMN are returned to a user as a result of Q, the computed p is also a measure of precision of the SOM-based clustering method.

To compute the precision of query results, let N_T be the total number of returned images (images associated with BMN) and N_R be the number of relevant images in N_T. Then, the precision p_i of the result of query q_i is computed as follows:

$$p_i = \frac{N_R}{N_T} \tag{8}$$

Figure 9 shows the precision of the 7 sample queries and their average precision (the dotted line). The average precision $\bar{p}$ is computed as follows:

$$\bar{p} = \frac{1}{N_Q} \sum_{i=1}^{N_Q} p_i \tag{9}$$

Where, N_Q is the total number of sample queries, which is equal to 7 in our test. As shown in Table 1, the average precision of query results is about 70.7% (it is also a measure of precision of the SOM-based clustering method). We, also, measured the average query response time, which is equal to the time it take to determine the BMN of a query. Table 1 shows that the average query response time is equal to 4.96 seconds. Even though it is difficult to compare with other systems due to differences in computing environments, our average query response time is comparable to many systems such as [1][5][8] based on the recorded search time in the corresponding papers.

6 Conclusion

In this work, we have implemented a wavelet-based indexing and retrieval system that clusters images into a database and provides query-by-example access to the stored images. We showed that the edge feature is important in indexing and retrieving images. Our edge feature vector is compact, fast to compute and supports similarity retrieval. By normalizing the edge features (high order autocorrelation features), they become invariant to shift, image size and gray level, which are essential properties for similarity-based retrieval in large image database environments. Even though the system currently only supports query-by-example querying, it can be easily extended to support querying by sketch of dominant edges, which is a rough representation of the image. Based on the experimental results, the system shows a high search precision, on the average, which is due to the SOM-based clustering of similar images. Although most of the search time is spent in finding the BMN, the overall search time is comparable to many existing systems.

References

1. M. Flickner, H. Sawhney, W. Niblack, J. Ashley, Q. Huang, B. Dom, M. Gorkani, J. Hafner, D, Lee, D. Perkovic, D. Steele, P. Yanker. *Query by Image and Video Content: The QBIC System.* IEEE Computer Magazine, Sept. 1995. 164, 165, 175
2. J. R.Smith, S. F.Chang. *VisualSEEK: A Fully Automated Content-Based Image Query System.* ACM Multimedia Conference, Boston, pp.87-98, Nov. 1996. 164, 165
3. E.Albuz. E.Kocalar, A. A.Khokhar. *Scalable Image Indexing and Retrieval Using Wavelets.* ICASSAP 1999. 164, 165
4. M.Kobayakawa, M.Hoshi, T.Ohmori, T.Terui. *Interactive Image Retrieval Based on Wavelet Transform and Its Application to Japanese Historical Image Data.* IPSJ Trans. on , Vol.40, No.3, pp.899-911, March 1999. (In Japanese) 164, 165
5. J. Z.Wang, G.Wiederhold, O.Firschein, S. X.Wei. *Content-based Image Indexing and Searching Using Daubechies' Wavelets.* Springer-Verlag Int'l Journal on Digital Libraries. Vol.1, pp.311-328, 1997. 164, 165, 168, 175
6. A.Natsev, R.Rastogi, K.Shim. *WALRUS: A Similarity Retrieval Algorithm for Image Databases.* SIGMOD record, vol.28, no.2, pp.395-406, Philadelphia, PA, 1999. 164, 168
7. T.Kohonen. *Self-Organizing Maps.* Springer-Verlag, 1997. 2nd extended edition. 164, 166, 171
8. C. E.Jacobs, A.Finkelstein, D. H.Salesin. *Fast Multiresolution Image Querying.* Proc. of ACM SIGGRAPH, New York, 1995. 165, 168, 175
9. E.Oja, J.Laaksonen, M.Koskela, S.Brandt. *Self-Organizing Maps for Content-Based Image Database Retrieval.* Published by Elsevier Science B. V., in Kohonen Maps, pp.349-362. 1997. 165
10. T.Kurita, N.Otsu, T.Sato. *A Face Recognition Method Using Higher Order Local Autocorrelation And Multivariate Analysis.* Prod. of 11th Int'l Conf. on Pattern Reconition, pp.213-216, The Hague, 1992. 169

11. M.Kreutz, B.Volpel, H.Janssen. *Scale-Invariant Image Recognition Based on Higher Order Autocorrelation Features.* Pattern Recognition, Vol.29, No.1, pp.19-26, 1996. 169

{tancl,caorn,shenpy}@comp.nus.edu.sg

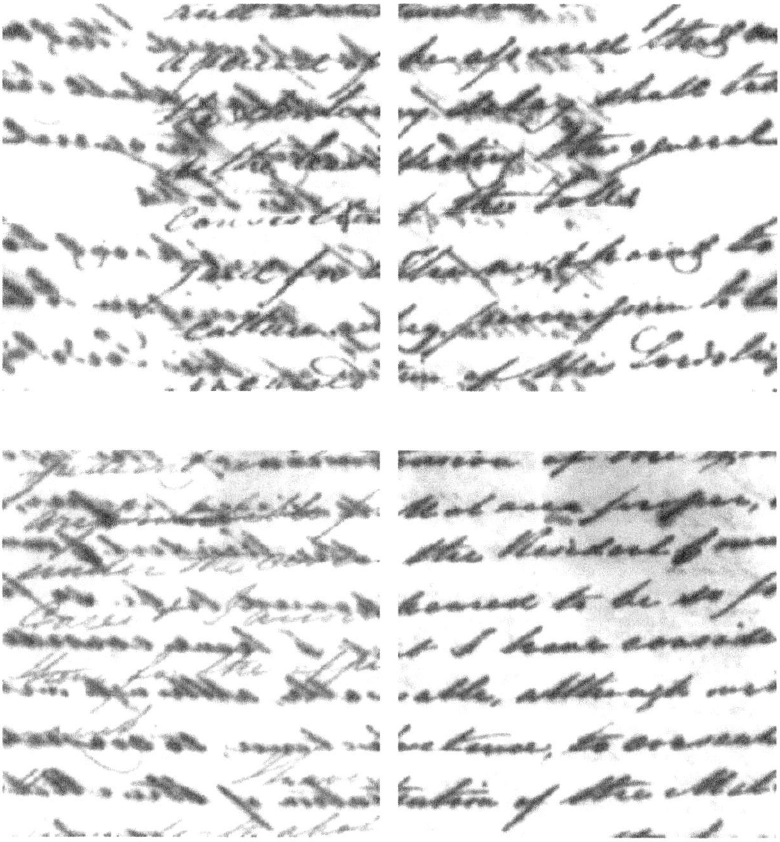

$$= \quad - \quad -$$

$$= \qquad + $$

$$= \qquad - $$

$$= \frac{-}{-} \qquad - $$

. $\times$

$$= \qquad = $$

$= \; < \quad \Phi \quad > \quad \in$

$= \; < \quad \Psi \quad > \quad \in$

$= \; < \quad \Psi \quad > \quad \in$

$= \; < \quad \Psi \quad > \quad \in$

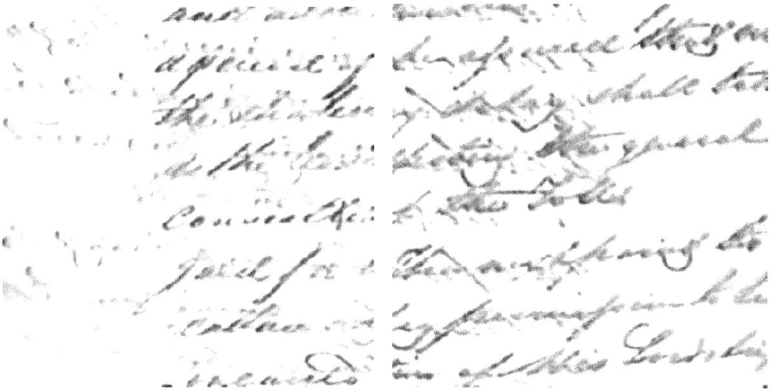

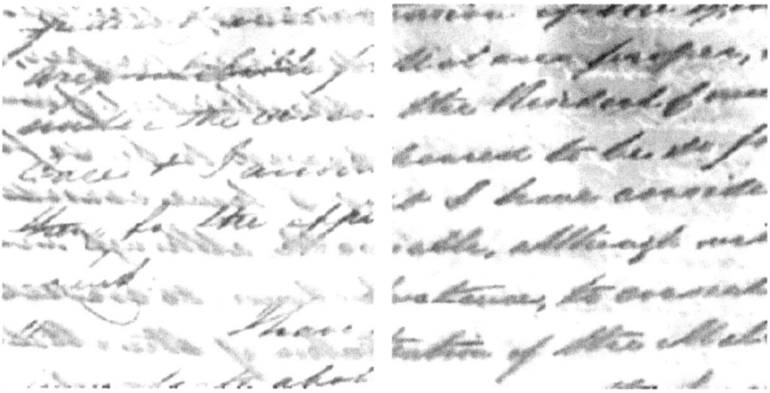

$=$

$\times$

= = =

> > > = =

=

=

= =

’ = ’ ’ = =

’ = =

<=

= >=

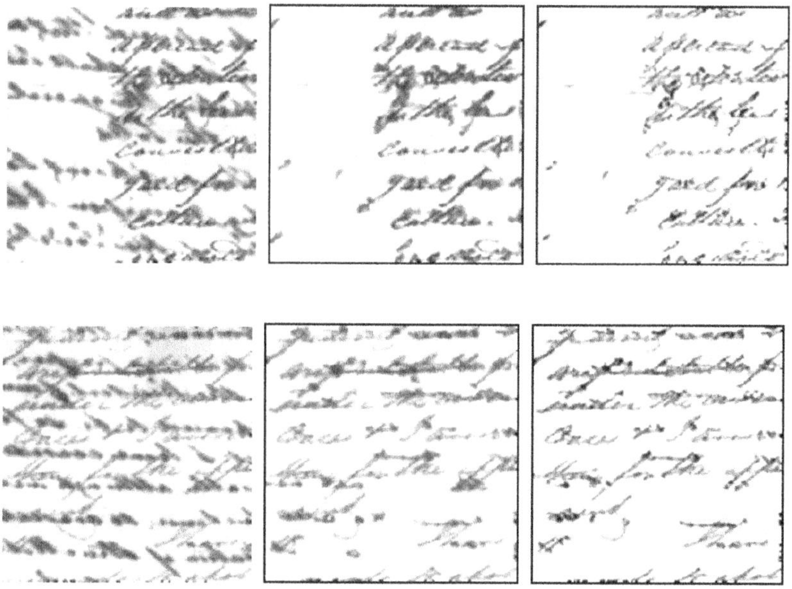

$$= \frac{\qquad}{\qquad}$$

$$= \frac{\qquad}{\qquad}$$

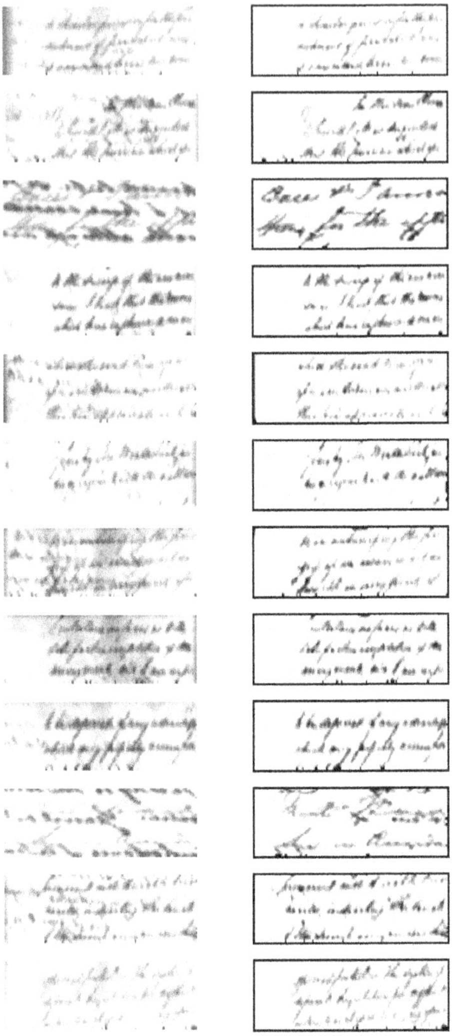

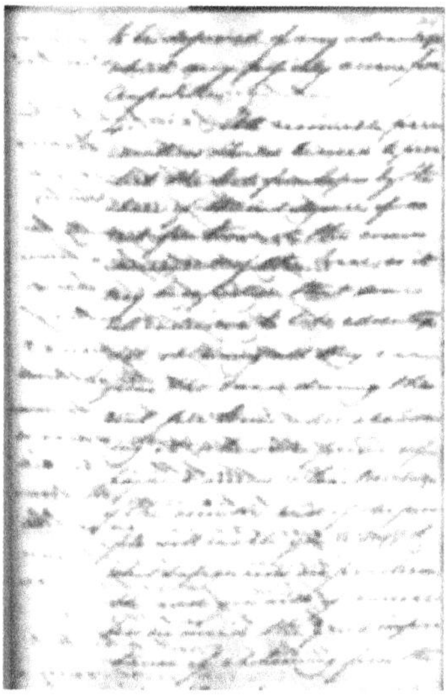

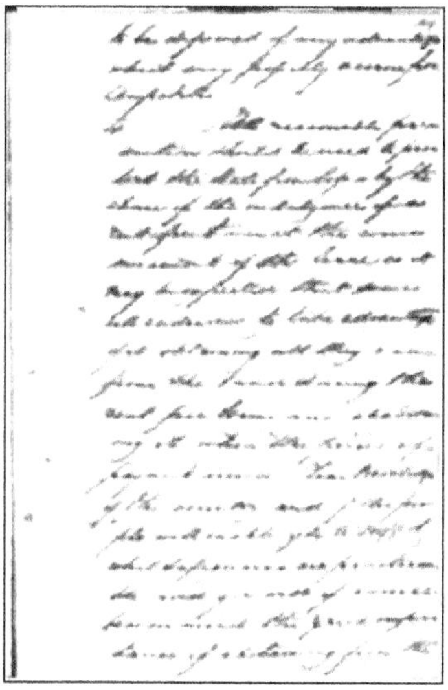

Wavelet Packets for Lighting-Effects Determination

Abbas Z. Kouzani and S. H. Ong

School of Engineering and Technology, Deakin University
Geelong, Victoria 3217, Australia

Abstract. This paper presents a system to determine lighting effects within face images. The theories of multivariate discriminant analysis and wavelet packets transform are utilised to develop the proposed system. An extensive set of face images of different poses, illuminated from different angles, are used to train the system. The performance of the proposed system is evaluated by conducting experiments on different test sets, and by comparing its results against those of some existing counterparts.

1 Introduction

The appearance of a person is highly dependent on the lighting conditions. Often slight changes in lighting produce large changes in the person's appearance. Since the face images in the known face database are taken under front-lit lighting, recognition of a face image taken under a different lighting condition becomes difficult. Determining the lighting effects within a face image is therefore the first crucial step of building a lighting invariant face recognition system.

While there has been a great deal of literature in computer vision detailing methods for face recognition, few efforts have been devoted to image variations produced by changes in lighting. In general, recognition algorithms have either ignored lighting variation, or dealt with it by measuring some properties or features of the image which are at least insensitive to the variability. Yet, features do not contain sufficient information necessary for recognition. Furthermore, faces often produces inconsistent features under different lighting conditions.

In this paper, a hybrid method is proposed based on theories of multivariate discriminant analysis and wavelet packets transform to classify face images based on the lighting effects present in the image. An extensive set of face images of different poses, illuminated from different angles, are used in the training of the system.

The paper is organised as follows. In Section 2, the existing work is reviewed. Section 3 presents the lighting-effects determination system. In Section 4, the experimental results are presented and discussed. Finally, the concluding remarks are given in Section 5.

Y. Y. Tang et al. (Eds.): WAA 2001, LNCS 2251, pp. 188–199, 2001.

2 Review of Existing Methods

To handle image variations that are due to lighting, three main methods have been used in the literature. These methods, used by object recognition systems as well as by systems that are specific to faces, are explained below.

2.1 Shape from Shading

The shape-from-shading method [1] utilises the gray-level information to determine the 3D shape of the object. Most algorithms which attempt to determine shape-from-shading, are designed for images of arbitrary objects with smooth brightness variations [1]. These algorithms estimate shapes from the limited information contained within an image. However, since the knowledge about the surface of human heads is not used by these algorithms, the estimation of head shapes from the limited information of a 2D image restricts the performance of these algorithms in practical applications, and therefore their use is unsuitable for face recognition.

2.2 Image Representation Models

Ideally, an image representation should be invariant to lighting changes. It has been theoretically shown that a representation which is invariant to lighting does not exist for unconstrained 3D objects [2]. However, for certain classes of objects this limitation does not necessary apply. Four image representations are explained below.

1. **Edge Maps:** Intensity edges coincide with gray-level transitions. Gray-level transitions can be due to discontinuities in the surface colour or orientation. Such edges are expected to be insensitive to lighting changes. The advantage of using an edge representation is that it is a relatively compact representation. Such an edge representation is used by several face recognition systems [3].
2. **Gabor-Like Filters:** Physiological evidence indicates that at the early stages of the human visual system the images are processed by local, multiple, and parallel channels that are sensitive to both spatial frequency and orientation. Several face recognition systems filter the gray-level image by a set of 2D Gabor-like functions before attempting to recognise the faces in the image [3,4]. Convolving the image with 2D Gabor-like filters is often similar to enhancing edge contours, as well as valleys and ridge contours from the image.
3. **Derivatives of Gray-Level:** Derivatives of the gray-level distribution were used by several face recognition systems [3] to reduce the effects of changes in lighting conditions on face images. The derivatives used include directional and non-directional first- and second-order derivatives. It can been shown analytically that, under certain conditions, changes in ambient light will affect the gray-level image but not its derivatives. However, this is not the

case in the natural lighting conditions where the direction of the light source is also changed.
4. **Logarithmic Transformation:** Logarithmic transformation is a non-linear transformation of the image intensities used in computer vision [5]. There is physiological evidence that logarithmic transformation approximates the response of cells in the retina of the human eye.

Adini et al. [6] reported that for most image representations considered, the percentage of miss-recognition was above 50 percent. Therefore, the above listed image-representation methods can be used in a lightin-effects determination system.

2.3 Example-Based Models

An example-based method handles image variations that are due to lighting differences by using, as a model, an explicit 3D model or, alternatively, a number of corresponding 2D face images taken under different lighting conditions. A number of 2D images can either be used as independent models or combined into a model-based recognition system such as those described in [7]. In the following, three examples of this method are given.

1. **Independent Image Comparison:** The face model here consists of a large set of images of the same face containing all possible variations. The recognition process involves the comparison of the distances between an input image and all the images comprising the model. A problem with this approach is that the number of images that the model must contain may be very large. Furthermore, this approach has limited generalisation capacity beyond the parameter values that are sampled and stored.
2. **Learning the Lighting Direction:** Learning the input/output mapping from examples is a powerful problem-solving mechanism, once a large number of examples is available. Brunelli [8] used one crude 3D head model to generate computer-generated masks for modulating the intensity of 2D front-view face images in order to produce images illuminated from different angles. The produced images are used for training an HyperBF network in which the lighting direction of the light source is associated with a vector of measurements derived from a front-view face image. The images for which the lighting direction must be computed are very constrained - they are front-view faces with a fixed inter-ocular distance [9]. In addition, the calculation and compensation of the lighting direction are done based on a simple lighting model of the light source that does not represent a variety of complicated lighting conditions which exist in practical situations.
3. **Fisherfaces:** This approach which is reported to perform better than the others, was proposed by Belhumeur et al. [10]. The idea is to produce classes in a low dimensional face image subspace obtained from linearly projecting a high-dimensional image space to the subspace. The multivariate discriminant analysis [11] is used to select most discriminating features. In the most

discriminating feature space, the factors that are not related to classification are discarded or weighted down, and factors that are crucial to classification are emphasised. Belhumeur et al. have conducted experiments on fisherfaces and three standard face recognition methods including the eigenfaces, and have reported lower error rates for the fisherfaces method. A drawback of this approach is that the transformation coefficients of different classes are very close to each other, compared to the other methods. That will cause false recognition [12].

2.4 Discussions

Among the methods described above, the image representation models improve the accuracy of the recognition, but fail to offer a robust invariance to lighting changes [6]. The example-based methods such as Brunelli's method [8] are promising and can produce better results than those of the image representation models. However, the performances of the existing example-based methods are still not satisfactory and there is plenty of room for improvement.

3 Proposed System

In the proposed system, the theories of the multivariate discriminant analysis and the wavelet packets transform are combined to form a learning system for determining the lighting-effects in the input face image. This combination is explained in the following.

3.1 Multivariate Discriminant Analysis

Multivariate discriminant analysis performs dimensionality reduction using linear projection [11]. Each image is considered as a sample point in this high-dimensional space. A problem with this method is that if the within-class scatter matrix [11] is singular in the computation of lighting direction for face images, This stems from the facts that the number of images in the training set is much smaller than the number of pixels in each image.

In order to overcome the complication of the singular within-class scatter matrix, the Principal Component Analysis (PCA) [13] is employed. The PCA builds a low-dimensional face space from a high-dimensional image space using example face images. The face space built by the PCA is an approximation of the real face space. But in order to have a reasonable approximation of the real face space, a large number of face images should be presented to the PCA method. If a large number of face images is not available, the PCA builds a face space that poorly approximate the real face space.

We propose the utilisation of the wavelet packets projection for reducing the dimensionality of face images. In order to overcome the complication of the singular within-class scatter matrix, the training image set is first projected to a lower-dimensional space using the wavelet packets transform. This projection reduces the size of the matrix; therefore, the matrix becomes square which will make it non-singular and invertible. Then, the discriminant analysis projection is performed in the space of the wavelet packets projection.

3.2 Wavelet Packets

The main difference between the wavelet packets transform and the wavelet transform is that, in the wavelet packets, the basic two-channel filter bank can be iterated either over the low-pass branch or the high-pass branch. This provides an arbitrary tree structure with each tree corresponding to a wavelet packets basis. The decision to split or merge is aimed at achieving minimum distortion.

Best Basis Method: The wavelet packets transform offers a choice of optimal bases for the representation of a specific signal [14]. Therefore, it is possible to seek the best basis by a criterion. The chosen basis should carry substantial information about the signal. Since compression is the goal, the basis which minimises the number of significantly non-zero coefficients in the resulting transform is chosen. *Entropy* is a suitable cost function for compression.

3.3 Selection of Best Basis for Face-Image Class

The wavelet packets transform and the best basis selection algorithm find the optimal basis for the representation of a specific signal such as face images. To select the basis for face images, 200 gray-scale front-view 64×64 face images are used as the training set. The training set is divided into four groups; each group consists of 50 face images. The following experiment is separately performed on each group of the face images. For each face image the stat-quadtree of entropy values is first created. For each group, 50 stat-quadtrees are obtained. Next, the entropy values of 50 stat-quadtrees are averaged. This generates four stat-quadtrees, one for each of the groups of the training set. Then, on each stat-quadtree, the best basis selection algorithm is performed to pick out the best basis from all the possible bases. The algorithm minimises the entropy values in the stat-quadtree. After obtaining the four best bases for the three groups of the training set, it is found that the four bases are the same. The maximum depth of splitting is chosen as 6 (explained in the following).

3.4 Selection of Best Filter and Best Decomposition Level

In the wavelet transform, the choice of filters is crucial not only for obtaining satisfactory reconstruction of the original signal, but also for determining the shape of the wavelet used for performing the analysis.

To achieve the best compression of human face images, the best filter and the best decomposition level in the wavelet packets transform must be chosen. The best filter can be chosen by examining different filters and selecting the one with the highest information packing capability. An experiment is carried out to select the best filter and the best decomposition level for the face-image class. Four groups of the training set of the face images are used in each experiment. Each group contains 50 gray-scale front-view 64 × 64 face images.

Six types of orthonormal quadrature mirror filters (Haar, Beylkin, Coiflet, Daubechies, Symmlet, and Vaidyanathan) are examined. Each one of the six types of filters is used with a specific filter parameter. For instance, the Symmlet filter is used with various number of vanishing moments varying from 4 to 10. In addition, together with each particular type of filter and parameter, different levels of decomposition are used. The applicable range of the decomposition level in this experiment is 2-6. A total of 96 filter variants are constructed using all combinations of filter type, parameter, and decomposition level.

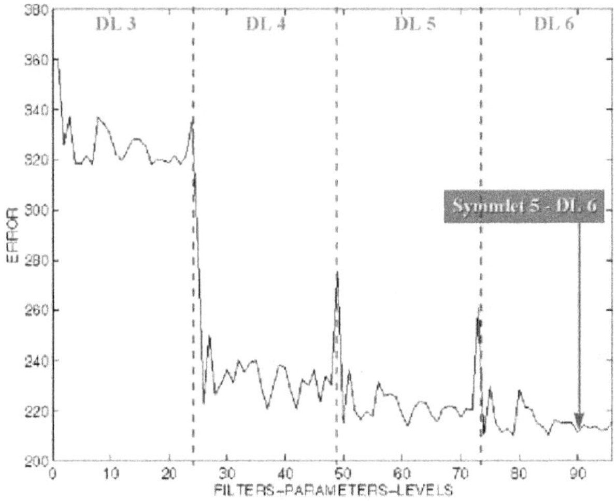

Fig. 1. Best filter and best decomposition level selection results for orthonormal quadrature mirror filters

Each of the above 96 filter variants is applied to each of the four sub-training sets, and the best basis is searched and selected. Compression is then carried out on all the training face images. After compression, the reconstruction is performed on the compressed images. In the reconstruction stage, each face image is reconstructed from 1%, 7.5%, and 15% of the most important information of the transformed coefficients (the coefficients with the highest absolute values). The rest of the coefficients are set to zero before reconstruction. Therefore three images are reconstructed from each compressed image. The errors between the original image and the three reconstructed images are calculated and summed. This is done for all the 50 training face images. The average error is obtained and stored. Figure 1 displays the best filter and the best decomposition level selection results for the orthonormal quadrature mirror filters. Each entry on the horizontal axis represents the measured error for a particular filter and a certain decomposition level. For instance, entry 79 denotes the measured error for the Coiflet filter with parameter 5 and the decomposition level 6.

The results show that the Symmlet filter with 5 vanishing moments and the decomposition level 6 is the best choice for the face image database. The best basis, the best filter, and the best decomposition level selected for the face-images class are employed for reducing the dimensionality of face images. Although the best basis was obtained by using a training set with a limited number of face images, it is experimentally found that adding more face images to the training set does not significantly affect either the structure of the basis nor the compression ratio. A better reconstruction of a face image that is not in the training set is possible using the wavelet packets transform than using the PCA.

3.5 Lighting-Effects Determination System (LEDS)

The LEDS takes an input face image and classifies it into one of possible lighting-effects classes under examination. The LEDS learns to compute the lighting-effects using the multivariate discriminant analysis and the wavelet packets transform. Although the multivariate discriminant analysis has been used by Swets et al. [15] as the most discriminating feature, and later by Belhumeur et al. [10] as the fisherfaces, both the most discriminating feature and the fisherfaces were developed for the purpose of one-step recognition. In the LEDS, however, the utilisation of a **combination** of the multivariate discriminant analysis and the wavelet packets transform is proposed as an example-based scheme for determining the lighting effects, not for recognising faces.

Algorithm 1 ***(Lighting-Effects Determination)*** *The lighting-effects determination process is performed in two stages as described in the following.*

Training: This stage involves the following operations which are performed only once.

1. *A training set of face images of different subjects is acquired. For each possible lighting effect, one image is taken from each subject.*
2. *The face images of the training set are grouped into different lighting-effects classes based on the lighting-effects that they contain.*
3. *An image is manually selected from each class and is named the reference image of the class. Although this selection is an arbitrary choice, the employed principle is that the face should be located in the centre of the image.*
4. *All images of each class are aligned based on the associated reference image using the pixel-based correspondence representation [12].*
5. *The multivariate discriminant analysis and wavelet packets transform is applied to the training set to obtain a dimensionally reduced lighting space.*
6. *The set of weights obtained from projecting each face image of the training set onto the lighting space is stored.*

Determination: This stage involves the following operations to classify the input face image into one of the lighting-effects classes.

1. *The input face image is projected onto the lighting space and a set of weights is calculated.*
2. *The weight pattern is classified into one of the lighting effects classes using the stored weight patterns of the face images in the training set.*

3.6 Training Set

A collection of 63 3D head models are used to generate the training database using computer graphics techniques. The head models have been generated from stereo images obtained using the *C3D* system of the *Turing Institute*. The training database contains 63 sets of 1331 2D full-face images of various poses within $\pm 45°$ rotations about X, Y, and Z directions and with the resolution of $9°$ (see Figure 2).

To quantify the effects of varying lighting, 66 different lighting conditions are considered. For each of the 1331 poses, 66 full-face images are rendered under different lighting conditions. In each image, specific direction and distance of a single light source are implemented. The longitudinal and latitudinal of the light source direction are within $15° - 75°$ of the camera axis. First, the face images are grouped based on the pose of each face. Each group representing a specific pose contains 4158 face images of 63 people. Then, the face images within each

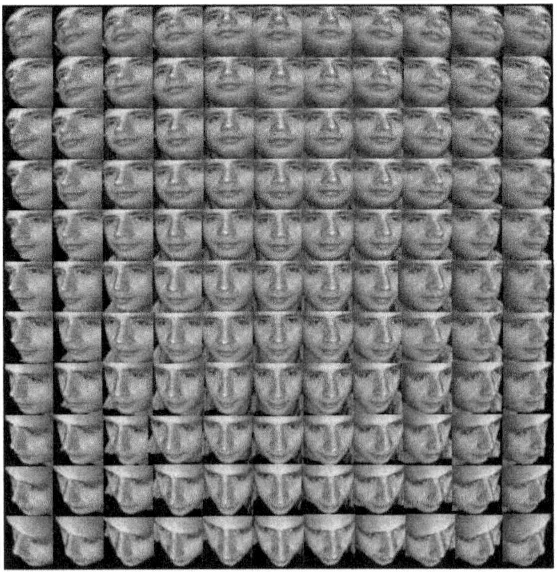

Fig. 2. Face images rendered from a 3D head model of the Turing database

group are divided into 66 different classes. This classification is done based on the lighting direction of each face image. Therefore, each group that represents a specific pose will contain 66 classes of different lighting effects. It should be stated that both the most discriminating feature and the fisherfaces put the face images of one person taken under different lighting conditions into the same class for the purpose of recognition. However, since this aim of this work is to determine the lighting effects, only the face images with similar lighting effects are put into the same class in the face space proposed here.

4 Experimental Results

To evaluate the performance of the LEDS, the results of experiments performed on three different test sets, are presented and discussed below. The test sets used in these experiments are as follows.

- **Test Set 1** contains 411 face images of the Harvard face database. In each image in this database, a subject holds his head steady while being illuminated by a dominant light source. The space of the light source directions is then sampled in $15°$ increments. Figure 3 illustrates sample face images from Test Set 1.
- **Test Set 2** consists of 495 images constructed from the Yale face database. The 165 face images of the Yale face database are first copied into the test set. Then, 330 extra images are produced by rotating each image randomly within the range of $10° - 90°$ twice in the 2D plane. These images are added to the test set. Figure 4 illustrates sample face images from Test Set 2.
- **Test Set 3** is constructed by the author and contains 2710 face images. Face images of ten people were used to build this test set. A set of 271 lighting masks are superimposed on each face image to generate 271 images under different lighting conditions. In each mask, specific direction and distance of a single light source are implemented. Figure 5 illustrates sample face images from Test Set 3.

Face images of Test Sets 1-3 are aligned using the pixel-based correspondence method [12], and are presented to two different systems. The first system uses the multivariate discriminant analysis and the PCA for classification of the lighting effects in the test face images. The PCA method has been trained on 200 front-view face images. The second system, that is the LEDS, uses the multivariate discriminant analysis and the wavelet packets transform for classification of the lighting effects in the test face images. The wavelet packets transform has also been trained on the 200 front-view face images. Table 1 summarises the results obtained from this experiment.

As can be seen from the table, the LEDS achieves a higher correct classification of the lighting effects for all three test sets than that of the method which uses the multivariate discriminant analysis and the PCA. It can be seen that the LEDS achieves a classification rate of 86.7% for Test Set 3 which is not as high as the rate obtained for Test Sets 1-2. The reason for this performance

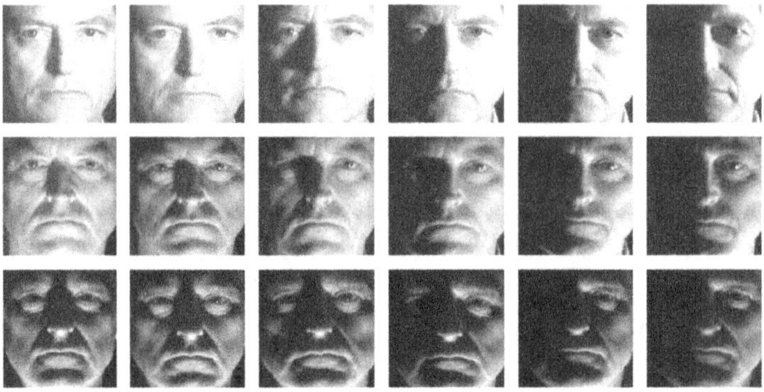

Fig. 3. Test Set 1 sample face images from the Harvard face database

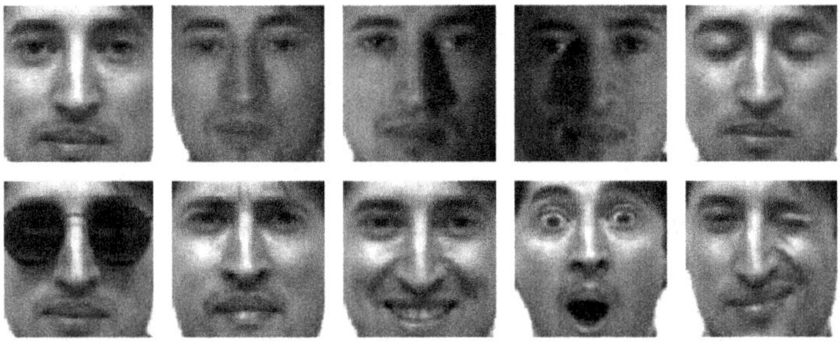

Fig. 4. Test Set 2 sample face images from Yale face database

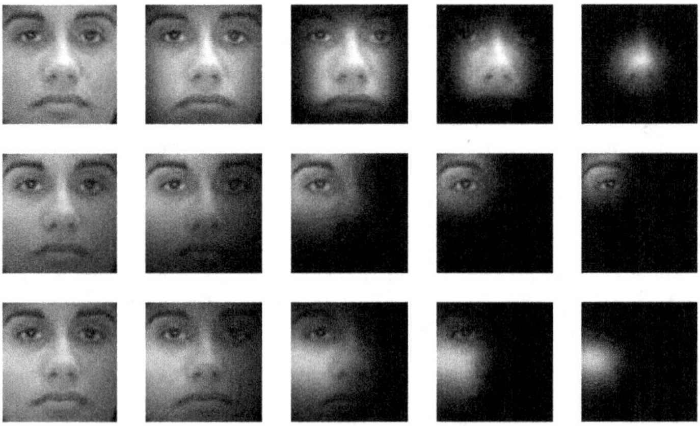

Fig. 5. Test Set 3 sample face images

is that the LEDS is trained on the face images containing real lighting effects, whereas the test images of Test Set 3 contains synthesised lighting effects in which the lighting masks are simply superimposed on face images taken under front-lit lighting. These lighting masks are simple approximations of the real lighting effects. Therefore, the images produced would only be an imitation of a corresponding real illuminated face image. However, training the LEDS on the example images containing synthesised lighting effects can improve the correct classification rate when the system is tested on this kind of face images.

Table 1. Classification of lighting effects for Test Sets 1-3

Method	Test Set	Correct Classification	Classification Rate
Ideal System	1	411	100%
	2	495	100%
	3	2710	100%
Multivariate Discriminant Analysis + PCA	1	377	91.7%
	2	411	83.0%
	3	2043	75.4%
Proposed LEDS	1	396	96.3%
	2	458	92.5%
	3	2349	86.7%

5 Concluding Remarks

A method has been proposed based on theories of multivariate discriminant analysis and wavelet packets transform to classify face images based on the lighting effects present in the image. An extensive set of face images of different poses, illuminated from different angles, are used in the training of the system. The performance of the system has been evaluated by conducting experiments on different test sets and by comparing its results against those of the existing counterparts. The system improves the performances of the existing counterparts because of the utilisation of the combination of the multivariate discriminant analysis and the wavelet packets transform for determination of the lighting effects, and the utilisation of training face images containing realistic lighting effects. The system may fail to determine an lighting effect in the input face image if the image contains a lighting effect that is not covered by the system or the image contains an extreme lighting effect. The performance of the system can be improved by increasing the number of face images in the lighting-effects classes, and by including more lighting-effects classes in the training sets.

References

1. B. K. P. Horn and M. J. Brooks, Eds., *Shape from Shading*, MIT Press, Cambridge, Mass., 1989. 189
2. Y. Moses and S. Ullman, "Limitation of non-model-based recognition schemes," in *Proc. European Conference on Computer Vision*, G. Sandini, Ed., 1992, pp. 820–828. 189
3. R. Brunelli and T. Poggio, "Hyperbf networks for real object recognition," in *Proc. IJCAI*, Sydney, Australia, 1991, pp. 1278–1284. 189
4. J. Buhmann, M. Lades, and F. Eeckman, "Asilicon retina for face recognition," Tech. Rep. 8996-CS, Institute of informatik, University of Bonn, 1993. 189
5. D. Reisfeld and Y. Yeshurun, "Robust detection of facial features by generalised symmetry," in *Proc. International Conference on Pattern Recognition A*, 1992, pp. 117–120. 190
6. Y. Adini, Y. Moses, and S. Ullman, "Face recognition: The problem of compensating for changes in illumination direction," *IEEE Trans. on Pattern Analysis and Machine Intelligence*, vol. 19, no. 7, pp. 721–732, July 1997. 190, 191
7. P. Hallinan, "A low-dimensional representation of human faces for arbitrary lighting conditions," in *Proc. IEEE Conf. Computer Vision and Pattern Recognition*, 1994, pp. 995–999. 190
8. R. Brunelli, "Estimation of pose and illumination direction for face processing," Tech. Rep. TR-AI 1499, Massachusetts Institute of Technology, November 1994. 190, 191
9. R. Brunelli and T. Poggio, "Face recognition: Features versus templates," *IEEE Transaction on Pattern Analysis and Machine Intelligence*, vol. 15, no. 10, pp. 1042–1052, 1993. 190
10. P. N. Belhumeur, J. P. Hespanha, and D. J. Kriegman, "Eigenfaces vs. fisherfaces: Recognition using class specific linear projection," *IEEE Trans. on Pattern Analysis and Machine Intelligence*, vol. 19, no. 7, pp. 711–720, July 1997. 190, 194
11. G. J. McLachlan, *Discriminant Analysis and Statistical Pattern Recognition*, Wiley, New York, 1992. 190, 191
12. A. Z. Kouzani, F. He, and K. Sammut, "Towards invariant face recognition," *International Journal of Information Science*, vol. 123, no. 1-2, pp. 75–101, 2000. 191, 194, 196
13. A. Hyvarinen, J. Karhunen, and E. Oja, *Independent Component Analysis*, John Wiley and Sons, 2001. 191
14. R. R. Coifman and M. V. Wickerhauser, "Entropy-based algorithms for best basis selection," *IEEE Trans. Infor. Theory*, vol. 38, no. 2, pp. 713–718, March 1992. 192
15. D. L. Swets and J. J. Weng, "Shoslif-o: Shoslif for object recognition and image retrieval (phase ii)," Tech. Rep. CPS-95-39, Michigan State University, October 1995. 194

xtang@ie.cuhk.edu.hk

$- \rightarrow -$

$\tau \quad = \quad -\tau$

$\in \quad \tau \quad \tau$

$$\tau = \int_{-\infty}^{+\infty} -\tau \frac{}{\sqrt{}}\psi \frac{-}{}$$

$$= \int_{-\infty}^{+\infty} \frac{}{\sqrt{}}\psi \frac{- \quad -\tau}{} \qquad = -\tau$$

$$= \quad -\tau$$

τ

$$\tau = \int_{-\infty}^{+\infty} -\tau \frac{}{\sqrt{}}\psi \frac{-}{}$$

$$= \int_{-\infty}^{+\infty} \frac{}{\sqrt{}}\psi \frac{- \quad -\tau}{} \qquad = -\tau$$

$$= \quad -\tau$$

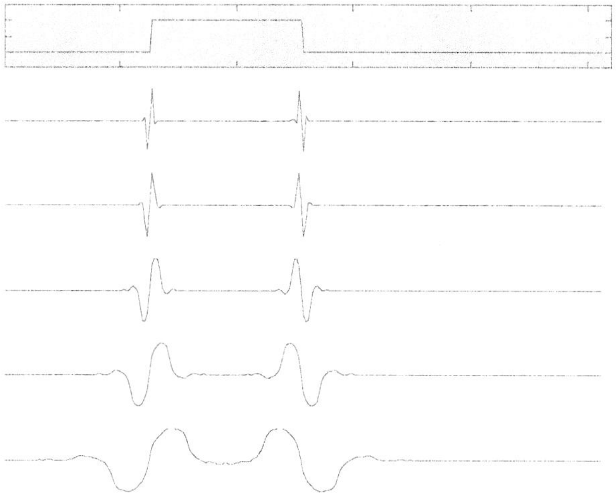

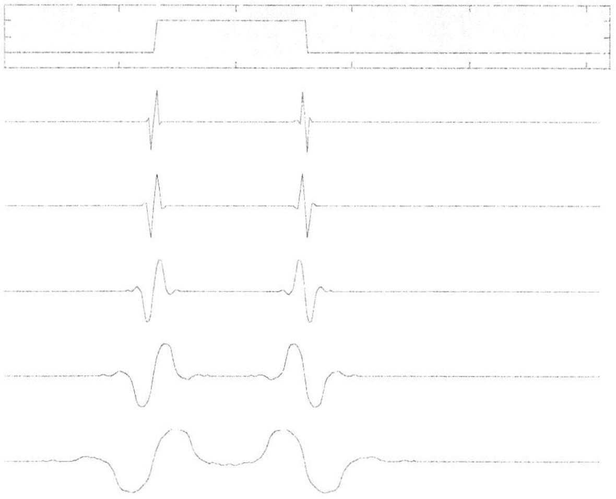

$_{\tau} \quad = \quad -\tau \qquad \in \qquad \tau$

$$\psi \quad = \frac{}{\sqrt{}} \psi \frac{-}{} \qquad \in \mathbb{Z}$$

$$= \qquad = \langle \quad \psi \quad \rangle = \int_{-\infty}^{+\infty} \frac{}{\sqrt{}} \psi \frac{-}{}$$

τ

$$\begin{aligned} = \quad _{\tau} \qquad &= \langle \, _{\tau} \, \psi \quad \rangle \\ &= \int_{-\infty}^{+\infty} -\tau \frac{}{\sqrt{}} \psi \frac{-}{} \\ &= \int_{-\infty}^{+\infty} \frac{}{\sqrt{}} \psi \frac{- \qquad -\tau}{} \qquad = -\tau \\ &= \qquad -\tau \end{aligned}$$

$\tau = \cdot \qquad \in \mathbb{Z} \qquad _{\tau} \qquad = \qquad -$

$= \quad -$

τ

$=$

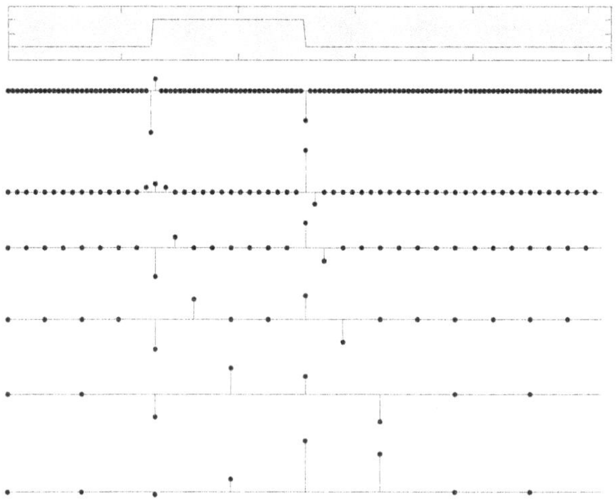

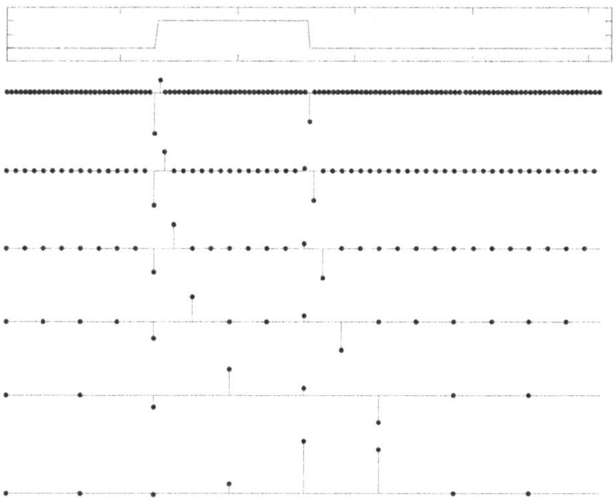

ω

$$\omega \; = \frac{}{} \; \left| < \quad \psi \quad > \right|$$

$=$

$$= \quad \left\| \quad - \quad \right\| < \delta$$

δ

$$= \left\{ \quad \left\| \quad - \quad \right\| < \frac{\delta}{} \right\}$$

$$\Omega = \left| \quad \right|$$

$$\equiv \quad = \Omega \quad \Omega \quad \Omega$$

$$= \frac{\Omega \cdot \Omega}{\sqrt{\Omega \cdot \Omega}}$$

Ω Ω

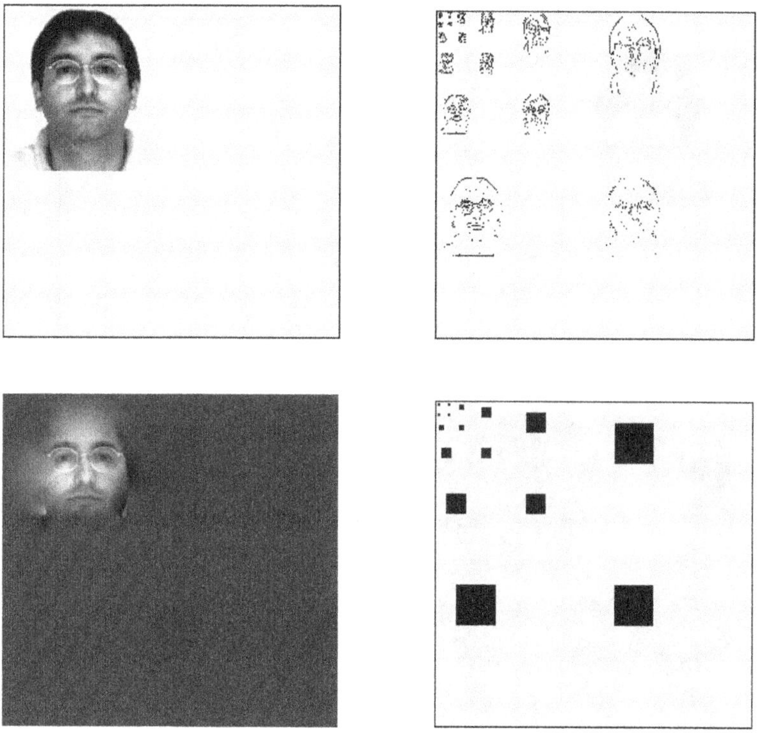

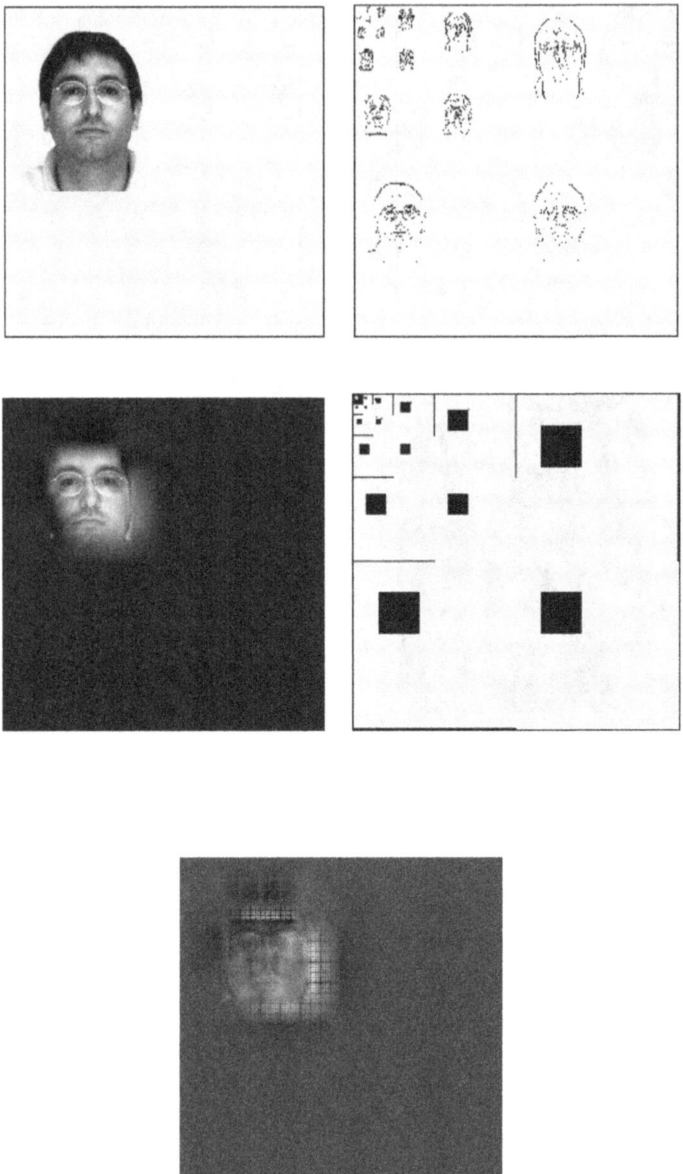

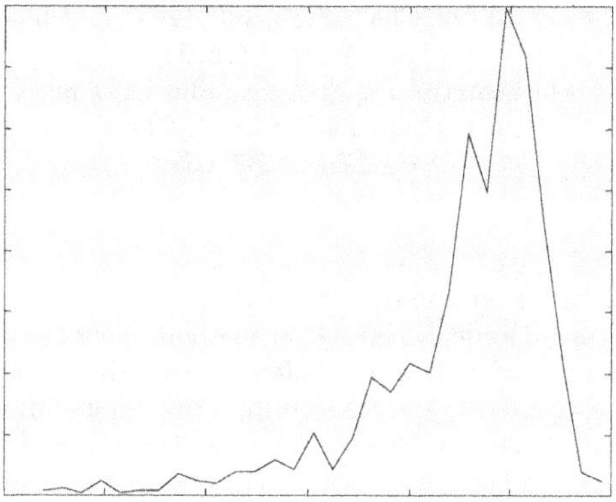

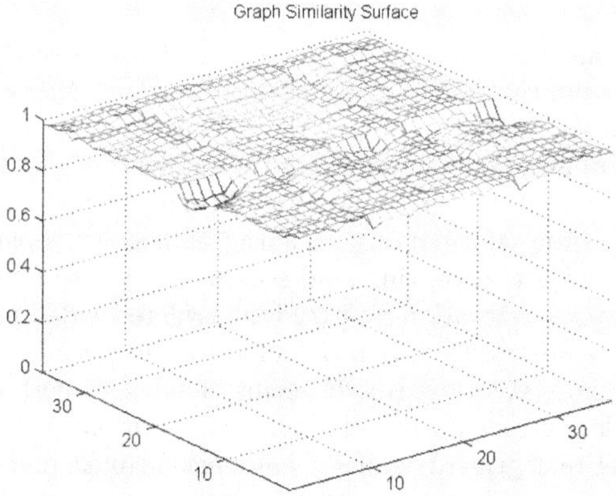
Graph Similarity Surface
1
0.8
0.6
0.4
0.2
0
30
20
10
10
20
30

Text Extraction Based on Nonlinear Frame

Yujing Guan[1] and Lixin Zhang[2]

[1] Jilin University Information Technologies Co. Ltd
Qianjin Rd. 95, Changchun, 130012, P. R. China
yjguan@hotmail.com
[2] Mathematics Department, Jilin University
Changchun, 130012, P. R. China
zhang_lixin@163.com

Abstract. Locating and extracting text in image or video has been studied in recent decade. There is no method robust for all kinds of text, it may be necessary to apply different methods to extract different kinds of text and fuse these results temporarily. So finding new method is important. In this paper, we combine order statistic and frame theory and give a new method, it can extract text of various colors and size once, the experimental result is satisfying.

1 Introduction

In this new era of information explosion, especially because of the development of Multimedia and Internet, a lot of information present themselves as image or video. Problems about how to obtain the information one wants from them become more and more important. Among them, locating and extracting text in image is a very useful and challenging work. The text embedded in image or video usually provide information about the names of people, organization, or about location, subject, date, time and scores, etc. Those texts are powerful resources for indexing, annotation and content-oriented video processing. So a lot of people get to work with this problem in recent decade, many methods are proposed [1,2,3,4,5,10,11,12]. But it seems that each method has its limitation. For example, current optical character recognition(OCR) technology is restricted to finding text printed against clean backgrounds, and can not handle text printed against shaded or textured backgrounds or embedded in images. Even as S. Antani said in [1], none of the proposed text detection and localization methods was robust for detecting all kinds of text, it might be necessary to apply different methods to extract different kinds of text and fuse these results temporarily. This may be induced by the essential complexity of the problem but make it important to provide more methods for people to select according to the problem they face.

In [6] and [7] Dr. Ma and Dr. Tang apply order statistic to detecting step-structure and page segmentation, they get a good result. But their method is only used for binary image and can't be used for gray image. In this paper we combine order statistic and frame theory and apply them to extracting text in

Y. Y. Tang et al. (Eds.): WAA 2001, LNCS 2251, pp. 211–216, 2001.

complex background and the result is satisfying. As we know, they have not been used in this field up to now.

The proposed method first partition the gray image into a number of small adaptive blocks, for example of 16×16 size, and proceed to find text in each block. The value of each pixel in the block is supposed to be a sample observation of a random variable. Sort the samples in ascending order and get an order statistic. According to the text characters represented in the order statistic, if there is text in the block, there will be steps in the values of the order statistic. Frame is used to detect the steps. Of course, it is possible that there is step in the order statistic while there is no text, so at last it is necessary to test whether the step is formed by text by applying text characters.

2 Order Statistic and Text Characters

Definition 1 . *Let* $(X_1, X_2, \cdots, X_n)$ *be a sample. Order statistic is the statistic obtained by replacing* $(X_1, X_2, \cdots, X_n)$ *in ascending order. They are denoted by* $(X_{(1)}, X_{(2)}, \cdots, X_{(n)})$*, which satisfy*

$$X_{(1)} \leq X_{(2)} \leq \cdots \leq X_{(n)}.$$

In fact, we first partition the gray image into several blocks with suitable area, suppose there are n pixels in a block and sort them in ascending order, then we get order statistics. If there is text in this block, the order statistics would have the following characters:

1. The values in text comprise a subsequence of $(x_{(1)}, x_{(2)}, \cdots, x_{(n)})$, which we denote $(x_{(b)}, x_{(b+1)}, \cdots, x_{(e)})$, and its mean is distinct from those of its left subsequence and right subsequence.
2. There are k and K such that $k \leq e - b \leq K$;
3. There exists a positive constant $\varepsilon > 0$ such that $x_{(e)} - x_{(b)} \leq \varepsilon$;
4. There exists a positive constant σ_T such that the variance of the text subsequence is smaller than σ_T^2.
5. All pixels in text form one or more curves.

3 Frame Transform

As we have declared in section 2, the gray values of the text form a subsequence, $(x_{(b)}, x_{(b+1)}, \cdots, x_{(e)})$, in the order statistic $(x_{(1)}, x_{(2)}, \cdots, x_{(n)})$ of the whole gray values in a block. In ideal case, The values in this subsequence are almost equal, but the neighboring values on the left or the right of it are significantly smaller or larger. In other words, there exist singularity points at the two endpoints of the subsequence. If we can find the correct singularity points, we are able to continue to separate the text from the background graphics. Wavelet has been successfully and frequently applied to singularity detection, but it is not adaptive here.

Definition 2 *The sequence $\{\phi_n\}_{n\in\Lambda}$ is called a frame of a Hilbert space $\boldsymbol{H}$, if there exist two constants $A > 0$ and $B > 0$ such that for any $f \in \boldsymbol{H}$,*

$$A\,\|f\|^2 \leq \sum_{n\in\Lambda} |< f, \phi_n >|^2 \leq B\,\|f\|^2 .$$

When $A = B$ the frame is said tight.

Example 1: Let $N = 2n + 1$, be a positive odd number,

$$\psi^N(t) = \begin{cases} \frac{-1}{\sqrt{N}}, 0 \leq t < \frac{N}{2}, \\ \frac{1}{\sqrt{N}}, \frac{N}{2} \leq t < N, \\ 0, \quad otherwise. \end{cases}$$

and $\psi_{j,n}^N(t) = \frac{1}{\sqrt{2^j}}\psi^N(\frac{t}{2^j} - n)$, then $\{\psi_{j,n}^N(t)\}_{j,n\in\mathcal{Z}}$ is a frame.

Let $f_{j,n}^N =< f, \psi_{j,n}^N >$, then the frame coefficient $f_{j,n}^{i,N}$ indicates the difference of two means of f at the left side and the right side of somepoint. In the following of this paper, call it Haar-N frame. It is obvious that this kind of frame vanish for constant. Similar to wavelet, a larger absolute value of frame coefficient indicates a larger step. But wavelet coefficient indicates a sharp change of the value of the function at a point and Haar-N frame coefficient indicates a step of means of the function at the left side and right side of a point. So, the Haar-N frame is not sensitive to noise while high frequency wavelet coefficient is very sensitive to noise. This character of Haar-N frame is very adaptive to detect change point in order statistic. To simplify discussion, denote the gray value of text, background, text noise and background noise by X,Y,$W1$,$W2$ respectively, suppose $W1$ and $W2$ are zero mean, the sample of $X + W1$ is less than the sample of $Y + W2$. Because we do not know any other statistic property of the above statistic than $E(X+W1)$ is less than $E(Y+W2)$, so we want to detect $E(Y+W2)-E(X+W1)$ to make sure where the samples of text order statistic are and where the samples of background order statistic are.

The existence of noise usually makes the difference of adjoining points in the order statistic decrease or vanish, even makes it smooth, thus it is difficult to detect step with wavelet because wavelet coefficient is generally a linear combination of difference of adjoining points in the order statistic. For example, if we use haar wavelet, the wavelet coefficient at the step point is $\min(Y + W2) - \max(X + W1)$, obviously it is very sensitive to noise and different from $E(Y) - E(X)$. But frame is not sensitive to noise,

From the propostion [8, pp.139], we know the variance of noise becomes $\frac{1}{N}$ times. But for our order statistic, we can not get such a good result, because we do not know where the samples of text are and where the samples of background are. On the other hand, though $E(X+W1)$ and $E(Y+W2)$ are unobtainable, we can calculate the mean of some samples with greater values for the text and that of some samples with smaller values for the background graphics respectively. Lastly we calculate the difference of these two means and base our detection on it instead of $E(Y+W2)-E(X+W1)$. Sometimes, there is error, but it is better than wavelet.

In fact, theoretically we have

$$E(Y+W2)-E(X+W1)$$
$$=\lim_{N\longrightarrow\infty}\left\{\frac{\sum_{i=0,N-1}Y_i+W2_i}{N}-\frac{\sum_{i=0,N-1}X_i+W1_i}{N}\right\}.$$

So the previous difference of two means is just an approximate estimate of $E(Y+W2)-E(X+W1)$ using finite samples, and at the same time it is also the Haar-N frame transform. Since the number of text sample is unknown, we must choose an adaptive N.

A large absolute value of the frame coefficient indicates a step, but a step will make one or more coefficients' absolute values large. Naturally, we should choose the coefficient with greatest absolute value to make sure where the step happens. In wavelet theory, these points are called Maxima points, more details see [9] and [8, ch.6], in this paper we also use this concept. Moreover, we should also notice that there may be large step in the order statistic induced by the complexity of background graphics. Thus we need a threshold τ, to indicate whether a step is large enough, since our supposition that the gray values of text are distinct from those of background graphics has guaranteed that the step at the correct change point should not be trival. By the way, we should notice that our frame coefficients for the order statistic are all nonnegtive in this paper. More precisely, after we get the frame coefficients, we proceed to let those smaller than τ be neglected, and we only test whether those points corresponding to the left frame coefficients are from text using the text characters presented in section 2.

4 Algorithms

After partitioning the gray image into a number of continuity regions with suitable area, such as into squares with $m \times n$ pixels, we replace the values in one block in ascending order and get the order statistic $(x_{(1)}, x_{(2)}, \cdots, x_{(mn)})$. We use the stationarity of the gray values of the text and their distinction from the other values from the background graphics to reduce the separation of text to step points detection. Wavelet is not adaptive here, and we use Haar-N frame. Those step points separate the order statistic into several subsequences, and we will use the text characters to decide which one or several are from text. Here $N > 0$ is an integer, and let τ be a threshold, if a frame coefficient f_i satisfy $|f_i| < \tau$ then f_i must not be a Maxima point formed by any step. Suppose the image has M blocks whose size is m $\times$ n, ε is the difference of the maximum and minimum of text sample, σ_T^2 is the maximum variance of text, k and K are minimum and maximum number of text point in one block if there is text in this block. We give our algorithm as follows:

Algorithm : For every block do

1. Get order statistic: Get the samples from current block, and sort them in ascending order to get the order statistic, $X_{(0)}, X_{(1)}, \cdots, X_{(mn-1)}$.

2. Calculate frame coefficients: for i=0,1,$\cdots$,mn-1, calculate

$$f_i = \frac{1}{\sqrt{N}} \left(\sum_{j=i}^{i+\frac{N}{2}-1} X_{(j)} - \sum_{j=i-\frac{N}{2}}^{i-1} X_{(j)} \right).$$

   Here when $j < 0$, $X_{(j)} = X_{(0)}$, when $j > mn - 1$, $X_{(j)} = X_{(mn-1)}$.
3. Find Maxima points:
   (a) For i=0,1,$\cdots$,mn-1, if $f_i < \tau$, set $f_i = 0$.
   (b) Find Maxima points, suppose the number of the Maxima points is a, sort the Maxima points in ascending order and denote them as $\alpha_1, \alpha_2, \cdots, \alpha_a$; So there are $a + 1$ subsequences of the order statistic as follows:

$$\{X_{(\alpha_0)}, \cdots, X_{(\alpha_1 - 1)}\}, \{X_{(\alpha_1)}, \cdots, X_{(\alpha_2 - 1)}\}, \cdots, \{X_{(\alpha_a)}, \cdots, X_{(\alpha_{a+1})}\},$$

   where we let $\alpha_0 = 0$, $\alpha_{a+1} = mn - 1$.
4. For every subsequence of the order statistic, $i = 0, 1, \cdots, a$, decide whether it is text according to the text charecters in setion 2.

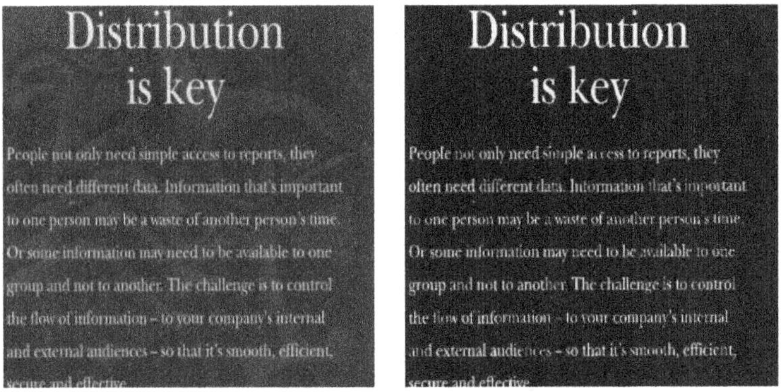

Fig. 1. A scanned image and the extracted text image

5 Examples

We applied our new method to some pictures and got a satisfying result. Fig.1 left was a scanned image, the background is a purple flower, Fig.1 right is the result of the extracted text image. The text of Fig.2 left was added by computer, 3 color text, black, blue and red were added, and the image was conversed to gray image, Fig.2 right is the extracted text image from the background.

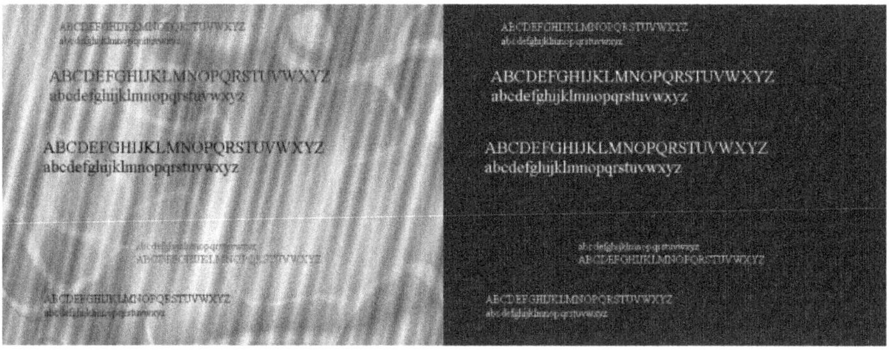

Fig. 2. ext image added by computer and the extracted text image

References

1. S. Antani, D. Crandall, A. Narasimhamurthy, V. Y. Mariano, R.Kasturi, Evaluation of Methods for Detection and Location of Text in Video, In Proc. 4th IAPR International workshop on document analysis systems - DAS '2000, Rio Othon Palace Hotel - Rio de Janevio, 10-13 December 2000. 211
2. A. Antonacopoulos and D. Karatzas, An Anthropocentric Approach to Text Extraction from WWW Images, In Proc. 4th IAPR International workshop on document analysis systems - DAS '2000, Rio Othon Palace Hotel - Rio de Janevio, pp 515-525, 10-13 December 2000. 211
3. U. Gargi, S. Antani, R. Kastui, Indexing Text Events in Digital Video Databeses, In Proc. International conference on pattern Recognition, Vol. 1, pages 916-918, Aug. 1998. 211
4. Yassin M. Y. Hasan and Lian J.Karam, Morphological Text Extraction from Images, IEEE Transaction on Image Processing, Vol. 9, No. 11, pp 1978-1983, Nov. 2000. 211
5. Huiping Li, David Doermann and Omid Kia, Automatic Text and Tracking in Digital Video, IEEE Transaction on Images processing, Vol. 9, No. 1, pp 147-156, Jan. 2000. 211
6. Hong Ma, Yong Yu, Li Ma, M. Umeda, Detection of Step-Structure Edge Base on Order Statistic Filter, preprint. 211
7. Hong Ma, Zhou Jie, Yuanyang Tang, Nonlinear Stochastic Filtering Methods of Adaptive Page Segmentation, preprint. 211
8. Stephne Mallat, A Wavelet Tour of Signal Processing, Academic Press, San Diego, 1998. 213, 214
9. Stephne Mallat and W. L. Hwang, Singularity detection and processing with wavelets. IEEE trans. on info. theory, (38):617-643, March, 1992. 214
10. Anil K. Jain and Bin Yu, Automatic Text Location in Images and Video Frames, Pattern Recognition, Vol. 31, No. 12, pp 2055-2076, 1998. 211
11. Victor Wu, Raghvan Manmatha, and Edward M. Riseman, TextFinder: An Automatic System to Detect and Recognize Text in Images, IEEE Transaction on Patter Analysis and Machine Intelligence, Vol. 21, No. 11, pp 1224-1229, Nov. 1999. 211
12. Yu Zhong, Hongjiang Zhang, and Anil K. Jain, Automatic Caption Localization in Compressed Video, IEEE Transaction on Patter Analysis and Machine Intelligence, Vol. 22, No. 4, pp 385-392, Apr. 2000. 211

qwang@public.glptt.gx.cn

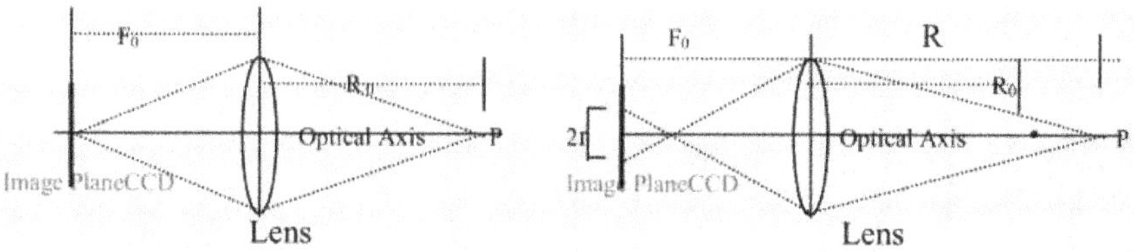
F0
R0
Optical Axis
P
Image PlaneCCD
Lens
F0
R
R0
2r
Optical Axis
P
Image PlaneCCD
Lens

•

$$+ \quad = \quad *$$

$$+ \quad = \quad *$$

$$+ \quad = \quad *$$

$$= +$$

$$= \sqrt{\quad + \quad}$$

$$= \quad \frac{\quad}{\quad}$$

•

$$\leq \quad ^{\alpha -} \quad = \sqrt{\quad + \sigma}$$

α σ

$$- \quad - \quad -\frac{\alpha -}{} \quad \sigma \quad +$$

=

α

σ

α σ

					α		σ

α

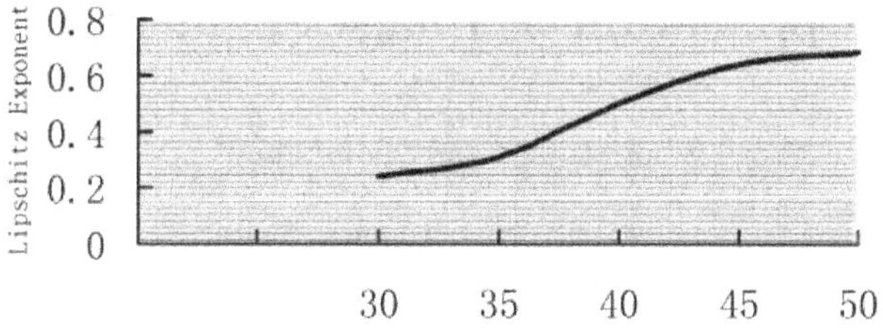

α

α

Construction of Finite Non-separable Orthogonal Filter Banks with Linear Phase and Its Application in Image Segmentation

Hanlin Chen[1] and Silong Peng[2]

[1] Inst. of Math., Academia Sinica
100080, Beijing, PRC
chen@math03.math.ac.cn
[2] NADEC, Inst. of Automation, Academia Sinica
100080, Beijing, PRC
pengsl@nadec.ia.ac.cn

Abstract. In [7], a large class of bi-variate finite orthogonal wavelet filters was constructed. In this paper, we propose a more general expression of the filter bank with linear phase which is called standard method. Beside this, a non-standard method is also presented. A interesting example is also given. By using this non-separable wavelet filter bank, we present a novel method of segmenting a image into two parts: one part is texture with special property and another part is image of piecewise smooth in some sense.

1 Introduction

Recently, many researchers are working on non-separable wavelets (see [1,2,3,4,7,8] and the references therein). In [7], a large class of bivariate compactly supported orthogonal symmetric wavelet filters (low-pass and high-pass) with arbitrary length are presented in explicit expression.

In this paper, we give another two methods of constructing bivariate compactly supported orthogonal symmetric wavelet filters. The standard method in this paper is similar to that of [7], but it's result is more general. We prove that non-standard method is included in the standard method. The standard method is introduced in next section. The non-standard method will be given in section 3. A simple image segmentation method by using the filters is given in section 4.

2 Standard Method

Let $\{V_j\}$ be a two dimensional MRA, then there exists a function $m_0(\xi,\eta)(\xi,\eta \in R)$ such that $\hat{\varphi}(2\xi,2\eta) = m_0(\xi,\eta)\hat{\varphi}(\xi,\eta)$, where $\hat{\varphi}$ is the Fourier transform of φ, and m_0 is called **Symbol Function** of the scaling function φ. The orthogonality of $\{\varphi(x-j,y-k)\}_{j,k\in\mathbb{Z}}$ implies that m_0 satisfies

$$|m_0(\xi,\eta)|^2 + |m_0(\xi+\pi,\eta)|^2 + |m_0(\xi,\eta+\pi)|^2 + |m_0(\xi+\pi,\eta+\pi)|^2 = 1. \quad (1)$$

Y. Y. Tang et al. (Eds.): WAA 2001, LNCS 2251, pp. 223–229, 2001.

If a trigonometric polynomial $m_0(\xi,\eta)$ satisfies (2.1) and $m_0(0,0) = 1$, we call m_0 a **Orthogonal Lowpass Wavelet Filter**. Assume that $m(x,y)$ is a polynomial of x and y with real coefficients. Rewrite $m(x,y)$ into its polyphase form as

$$m(x,y) = f_1(x^2,y^2) + xf_2(x^2,y^2) + yf_3(x^2,y^2) + xyf_4(x^2,y^2). \tag{2}$$

It is easy to see that $m(e^{i\xi}, e^{i\eta})$ satisfies (1) is equivalent to

$$\sum_{\nu=1}^{4} |f_\nu(e^{i\xi}, e^{i\eta})|^2 = \frac{1}{4}, \quad \xi,\eta \in R. \tag{3}$$

In this paper, all polynomials are with real coefficients in default.

In some applications, it is better to use a filter with linear phase than a filter with nonlinear phase. It is well known that in one dimension case, there does not exist a orthogonal filter with linear phase beside Haar filter. But in high dimension case, we can find many filters with linear phase.

Definition 1. *Given a polynomial $m(x,y)$, if*

$$\overline{m(e^{i\xi}, e^{i\eta})} = \pm e^{-iM_1\xi} e^{-iM_2\eta} m(e^{i\xi}, e^{i\eta}), \tag{4}$$

where M_1 and M_2 are positive integers, then we say that $m(e^{i\xi}, e^{i\eta})$ has **Linear Phase**.

To our purpose, we introduce a kind of matrix transform. For a matrix A of size $m \times m$, define $A^S := H_m A H_m$, where $H_m = (h_{kl})_{k,l=1}^m$ is matrix of size $m \times m$, with $h_{kl} = 1$ when $k + l = m + 1$ and 0 otherwise. Moreover, denote $\mathcal{U}_2$ to be the set of all real unitary matrices with size 4×4. The following theorem give a large class of symmetric wavelet filters.

Theorem 1. *Let*

$$m(x,y) = \frac{1}{4} X \cdot \prod_{\mu=1}^{N} (U_\mu D_\mu(x^2,y^2) U_\mu^T) \cdot V_0, \tag{5}$$

where $U_\mu \in J_2 := \{U | U \in \mathcal{U}_2, \;\; U = \pm U^S\}$, and $D_\mu(x,y) = diag\{1,\ x,\ 1,\ x\}$, or $diag\{1,\ y,\ 1,\ y\}$, for $\mu = 1, \cdots, N$, and $V_0 = (1\ 1\ 1\ 1)^T$, $X = (1\ x\ y\ xy)$, then $m(e^{i\xi}, e^{i\eta})$ is a symmetric filter and satisfies (1).

Proof. The proof is direct.

Remark 1. Although the form of (5) is similar to that of [7], but we can see that this form is more general. In fact, the filters given by non-standard method later are included in this form, but not included in the expression of [7].

3 Non-standard Method

In this section, we will introduce a new method to construct finite orthogonal symmetric wavelet filters. Given a polynomial with real coefficients $m(x, y)$, and let

$$m(x,y) = f_1(x^2,y^2) + xf_2(x^2,y^2) + yf_3(x^2,y^2) + xyf_4(x^2,y^2). \tag{6}$$

If

$$\overline{m(e^{i\xi},e^{i\eta})} = \pm e^{-i(2M_1+1)\xi}e^{-i(2M_2+1)\eta}m(e^{i\xi},e^{i\eta}), \tag{7}$$

that is $m(x,y) = \pm x^{2M_1+1}y^{2M_2+1}m(\frac{1}{x},\frac{1}{y})$, then we obtain

$$f_1(x,y) = \pm x^{M_1}y^{M_2}f_4(\frac{1}{x},\frac{1}{y}), \qquad f_2(x,y) = \pm x^{M_1}y^{M_2}f_3(\frac{1}{x},\frac{1}{y}). \tag{8}$$

If $m(e^{i\xi}, e^{i\eta})$ satisfies (1), then f_1, f_2, f_3 and f_4 satisfy (3). Substitute (8) into (3) to obtain

$$|f_1(e^{i\xi},e^{i\eta})|^2 + |f_2(e^{i\xi},e^{i\eta})|^2 = \frac{1}{8}. \tag{9}$$

Conversely, if we have two polynomials f_1 and f_2 satisfy (9), then (8) will give f_3 and f_4, such that we can get a finite orthogonal symmetric wavelet filter. The following theorem give a large class of the solutions of (9).

Theorem 2. *Let*

$$(f_1(x,y) \;\; f_2(x,y))^T = \frac{1}{4}\prod_{\mu=1}^{N}(A_\mu E_\mu A_\mu^T)\cdot(1 \;\; 1)^T, \tag{10}$$

where A_μ is any real unitary matrix of size 2×2, $E_\mu = diag(1,\ x)$ or $diag(1,\ y)$, for $\mu = 1,\cdots,N$, then f_1 and f_2 satisfy (17).

Proof. Since A_μ's are unitary matrices, and E_μ's are paraunitary matrices, the conclusion is followed immediately.

The non-standard method looks like different from the standard method, but in fact, all filters result from non-standard method can be constructed by standard method, which is the following theorem.

Theorem 3. *Let*

$$m(x,y) = f_1(x^2,y^2) + xf_2(x^2,y^2) + yf_3(x^2,y^2) + xyf_4(x^2,y^2),$$

where $f_1(x,y)$ and $f_2(x,y)$ satisfy (10), $m(x,y)$ satisfies (7), then we have

$$m(x,y) = \frac{1}{4}X\cdot\prod_{\mu=1}^{N}(U_\mu D_\mu(x^2,y^2)U_\mu^T)\cdot V_0, \tag{11}$$

where

$$U_\mu = \begin{pmatrix} A_\mu & 0 \\ 0 & A_\mu^S \end{pmatrix}, \quad D_\mu(x,y) = \begin{pmatrix} E_\mu(x,y) & 0 \\ 0 & E_\mu(x,y) \end{pmatrix}.$$

By using this construction method, we can construct the following non-separable filter banks. The lowpass filter is:

$$\frac{1}{8}\begin{pmatrix} 1 & -1 & 1 & 1 \\ 1 & 1 & 1 & -1 \\ -1 & 1 & -1 & 1 \\ 1 & 1 & -1 & 1 \end{pmatrix}.$$

The three highpass filters are

$$\frac{1}{8}\begin{pmatrix} 1 & -1 & 1 & 1 \\ 1 & 1 & 1 & -1 \\ 1 & -1 & -1 & -1 \\ -1 & -1 & 1 & -1 \end{pmatrix}, \frac{1}{8}\begin{pmatrix} -1 & 1 & 1 & 1 \\ -1 & -1 & 1 & -1 \\ 1 & -1 & 1 & 1 \\ -1 & -1 & -1 & 1 \end{pmatrix}, \frac{1}{8}\begin{pmatrix} 1 & -1 & -1 & -1 \\ 1 & 1 & -1 & 1 \\ 1 & -1 & 1 & 1 \\ -1 & -1 & -1 & 1 \end{pmatrix}.$$

4 Image Segmentation with Non-separable Symmetric Filter

The filters given in previous section is very good in some sense: their element are 1 or -1 (if we omit the factor $\frac{1}{8}$) which will be useful in computation; they are all with linear phase, two of them are symmetric, the other two are anti-symmetric. These filters have bad regularity in contrast with the well known biorthogonal 9/7 wavelets. These filters act as derivative operators such as Sobel operator. The following examples illustrate this fact.

Fig. 1. Original image

Image segmentation is important in many applications such as image compression and computer vision. In some applications, it will be useful to segment an image into two parts: one part is include regions with dense edges, and the other regions are with few edges. In general case, a sub-area of a image with dense edges maybe texture, which will be difficult to processed in applications such as compression. By using the derivative property of the filters, we present a novel method to do segmentation.

In the area which may be texture, the distance between edges are short. In addition, there are always all direction of edges, this means that for each channel

Fig. 2. Filtering result of the first high-pass filter. Left: bigger than 16. Right: smaller than -16

Fig. 3. Filtering result of the second high-pass filter. Left: bigger than 16. Right: smaller than -16

of the above filter, various edges will appear in the texture region. On the other hand, we only have one kind of edge near the edge of piecewise smooth area. Therefore we utilize this fact to segment a image.

Here we have three high-pass filters, we call them three channels. In this algorithm, we do not do the down-sampling, just convolute the image with the filters.

Suppose B is the currently processed channel. Let $BP1 = B > th1$ and $BP2 = B > th2$, where $th1 > th2$ are two positive numbers. In $BP2$, we can find all the areas which contain at least one point of $BP1$, all these areas put together to obtain BP. Similarly we can get the BM in which every point is negative number in B.

Let DBP is a matrix with same size of B, its elements is

$$DBP(p) = \begin{cases} 0, & BP(p) = 0 \\ d(p, BM), & otherwise \end{cases}$$

where $p = (i, j)$ is a point, $d(p, BM)$ is the distance of p and the nonzero point in BM. Similarly, we can define DBM.

If a point in B is located in the texture area, then at least one of the corresponding value in DBP and DBM are small. Let SB is a matrix which indicates that at each point, both values in DBP and DBM are smaller than a given threshold th. Of course, the selection of th depend on the scale of the texture one prefer, and the segment result depends on the selection of the threshold.

Let $S1$, $S2$ and $S3$ be the corresponding segment results of three channels respectively. Then we can segment roughly as $S = S1 + S2 + S3 > 1$, which

Fig. 4. Filtering result of the third high-pass filter. Left: bigger than 16. Right: smaller than -16

means that, if a point is located in the texture area, then it will appear in at least two of the channels.

At last to obtain a true area, we need to do some small dilation and erosion. A segmentation example is given as follows by using the above filters and the well known image.

Fig. 5. Original Image

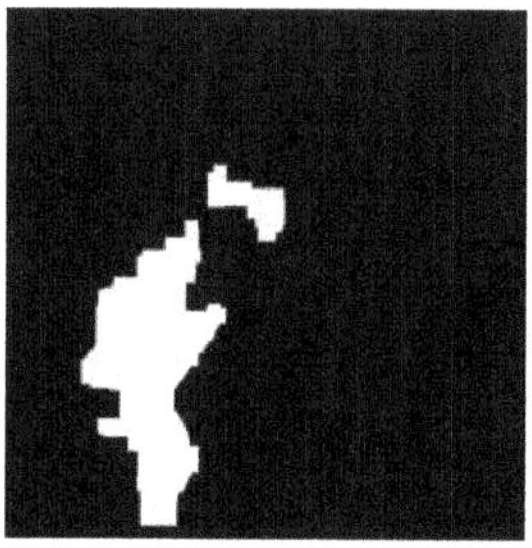

Fig. 6. Segmentation result ($th = 20$)

References

1. I. Daubechies, Ten Lectures on Wavelets, CBMS, 61,SIAM, Philadelphia, 1992. 223
2. Wenjie He and Mingjun Lai, Construction of Bivariate Compactly supported Biorthogonal Box Spline Wavelets with Arbitrarily High Regularities, Applied Comput. Harmonic Analysis, **6**(1999) 53-74. 223
3. Wenjie He and Mingjun Lai, *Examples of Bivariate Nonseparable Compactly Supported Orthonormal Continuous Wavelets*, Wavelet Applications in Signal and Image Processing IV, Proceedings of SPIE, **3169**(1997) 303-314. 223
4. J. Kovacevic and M. Vetterli, Nonseparable multidimensional perfect reconstruction filter banks and wavelet bases for R^n, IEEE Tran. on Information Theory, 38, **2**(1992) 533-555. 223
5. S. Mallat, Review of Multifrequency Channel Decomposition of Images and Wavelet Models, Technical report 412, Robotics Report 178, New York Univ., (1988).

6. Y. Meyer, Principe d'incertitude, Bases hilbertiennes et algebres d'oper-ateurs, Seminaire Bourbaki 662,1985-86, Asterisque (Societe Mathematique de France).
7. Silong Peng, *Construction of Two Dimensional Compactly Supported Orthogonal Wavelet Filters with Linear Phase*, (to appear in ACTA Mathematica Sinica), (1999). 223, 224
8. Silong Peng, *Characterization of Separable Bivariate Orthonormal Compactly Supported Wavelet Basis*, (to appear in ACTA Mathematica Sinica), (1999). 223
9. Silong Peng, *N dimensional Compactly Supported Orthogonal Wavelet Filters*, (to appear in J. of Computational Mathematics), (1999).

yytang@comp.hkbu.edu.hk

houyuhua@mail.henu.edu.cn

ρ

$= \quad =$

$= \qquad =$

$=$

$= \quad \mu \quad \sigma \quad + \quad =$

$\Theta \quad \Theta \quad \Theta$

$= \{\Theta \quad \Theta \quad \Theta\} \quad \beta \quad = \quad = \quad \Theta$

$\Theta = \quad \beta \quad = \quad \Theta$

$=$

$= \quad \Theta \quad \Theta \quad \Theta$

$=$

$\in \{ \quad \cdots \quad \}$

$= \qquad =$

$\mu \qquad \sigma$

$$= \qquad\qquad =$$

$$= \quad \mu \quad \sigma \qquad + \qquad + \qquad =$$

$$\left\{ \; \rho \; \right\}$$

$$\begin{matrix} \varepsilon^{\rho} & \varepsilon^{\rho} & \varepsilon^{\rho} \\ \varepsilon^{\rho} & \varepsilon^{\rho} & \varepsilon^{\rho} \\ \varepsilon^{\rho} & \varepsilon^{\rho} & \varepsilon^{\rho} \end{matrix}$$

$$= \begin{matrix} -\varepsilon^{\rho} \; -\varepsilon^{\rho} & \varepsilon^{\rho} & \varepsilon^{\rho} \\ \varepsilon^{\rho} & -\varepsilon^{\rho} \; -\varepsilon^{\rho} & \varepsilon^{\rho} \\ \varepsilon^{\rho} & \varepsilon^{\rho} & -\varepsilon^{\rho} \; -\varepsilon^{\rho} \end{matrix}$$

$$\Theta = \quad \beta \qquad = \Theta$$

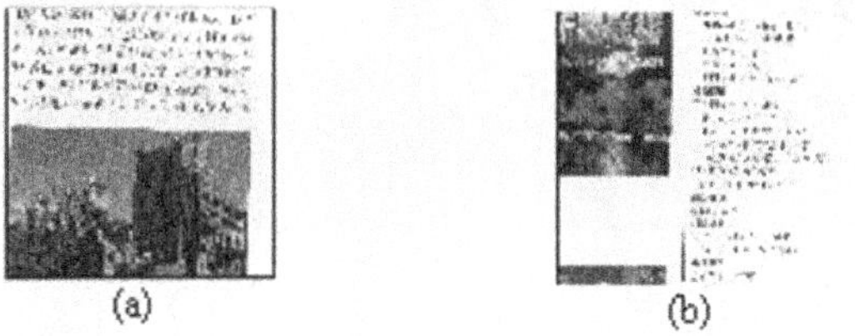

(a) (b)

$\times$

$\times$

(c)

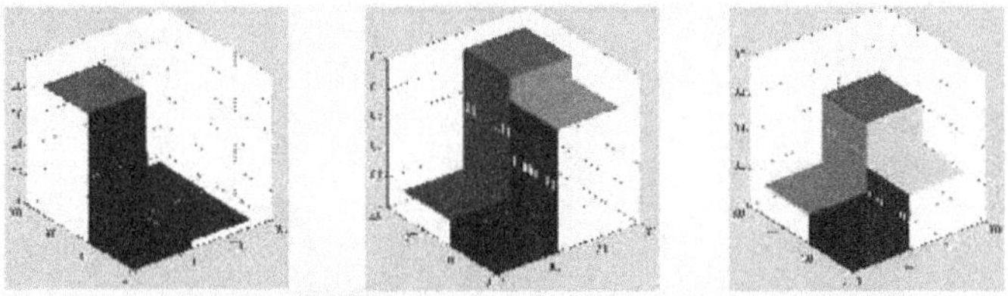

background text image

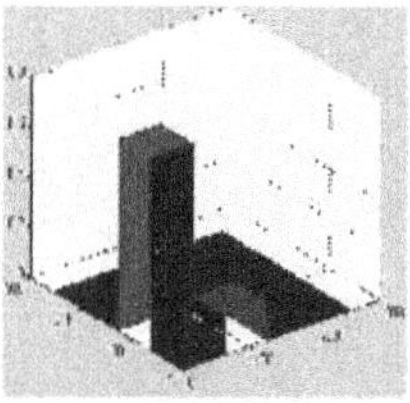

background

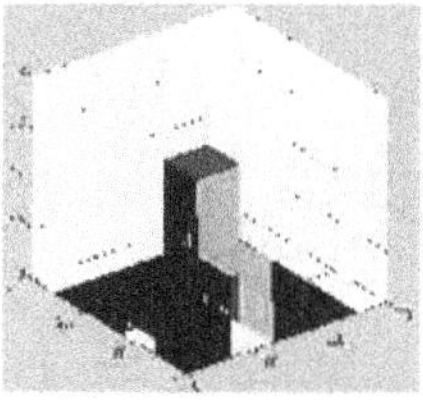

text

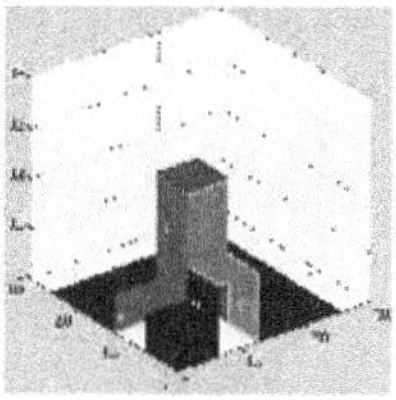

image

—

=

=

$\Theta = \quad \beta \quad = \quad \Theta$

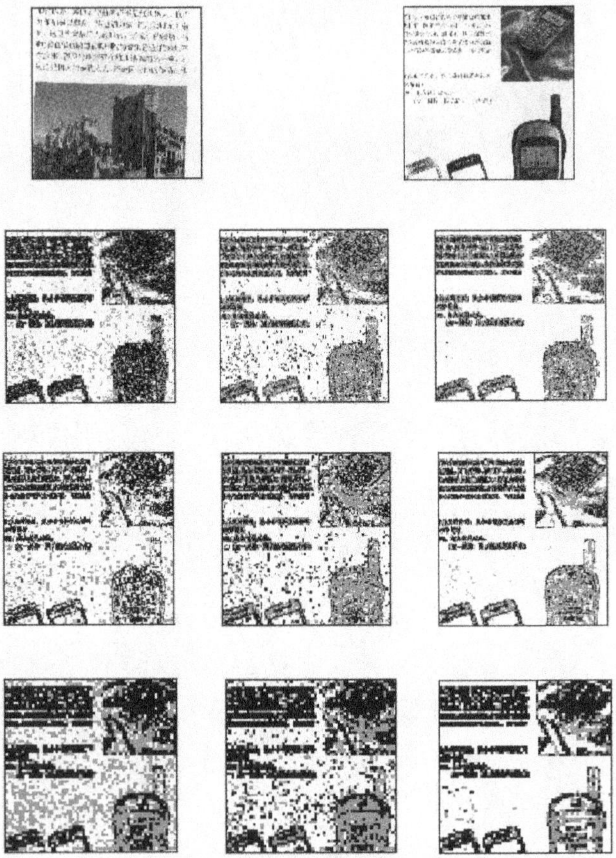

tmlaw@vtc.edu.hk

	橫	豎	撇	捺	點
	一	丨	丿	㇏	丶

絲

∠

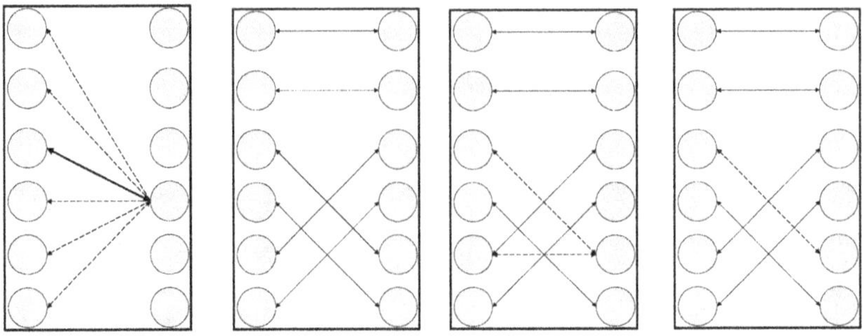

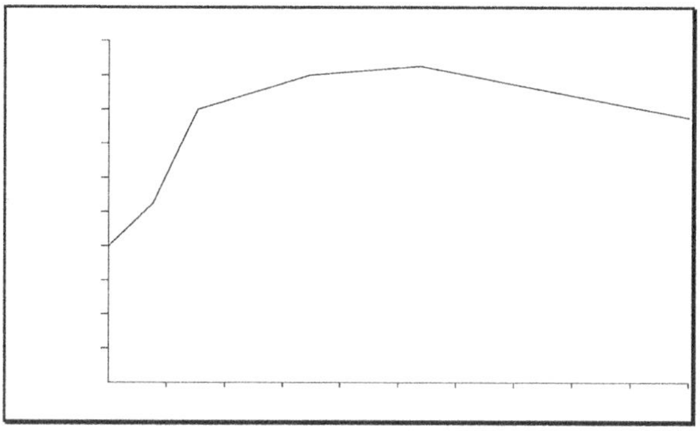

Automatic Detection Algorithm of Connected Segments for On-line Chinese Character Recognition

Tak Ming Law

Hong Kong Institute of Vocational Education (Morrison Hill)
Department of Computing, 6 Oi Kwan Road, Wan Chai, Hong Kong.
Email: tmlaw@vtc.edu.hk

Abstract. This paper presents a very easy way to detect the improper connected strokes by simply breaking all the strokes into pieces of segments. Once the strokes of the character decomposed into segments, as the basis of recognition, the connected stroke problem is no longer exists anymore.

1 Introduction

One of the most popular multimedia devices for people to enter Chinese characters into the system is on-line Chinese character recognition system.

There are so many ways to perform on-line Chinese character recognition. For examples, some researchers utilize individual classifiers [1] to derive the best final decision from the statistical point of view [2] and others classify characters by feature extraction [3] or structural [4]. The simplest method used for the recognition is template matching [5]. Some works emphasized on characters searching look up [6]. Relaxation is a well-known matching method, which has been employed for the recognition of Chinese character [7]. Some other methods like attributed string matching by split-and-merge and segment-order free techniques [8] are also applied in on-line characters and numeric digit recognition. Now, some researchers are developing on-line Chinese signature verification by using some advanced character recognition techniques [9]. However, some products in the market do generate some incorrect results. The problems may be due to the inefficiency of database structure and retrieval methods.

On-line Chinese character recognition algorithms are usually based on comparing the similarities of the individual stroke segments between the input and reference characters. However, the accuracy of the measurement always hindered by the improper connected strokes caused by running handwriting on the electronic tablet. We have found a very easy way to detect the improper connected strokes by simply breaking all the strokes into pieces of segments. Once the strokes of the character decomposed into segments, as the basis of recognition, the connected stroke problem is no longer exists anymore. Lets start by looking at the foundation that our recognition system based on in section 2.

Y. Y. Tang et al. (Eds.): WAA 2001, LNCS 2251, pp. 242-247, 2001.

2 Basic Stroke Types (Segment Type)

In our database, for simplicity and performance, we only consider five types of stroke, which is the basic stroke. Basic stroke can be divided as five strokes presented in the following table [10].

Table 1. Five basic types of segment

Stroke Name	Horizontals 橫	Verticals 豎	South-West Slanting 撇	South-East Slanting 捺	Dot 點
Symbol	h	v	p	n	d
Stroke Shape					

The above five kinds of basic strokes are in their own unique directions without turning points. In our system, we call the above basic stroke types as segment types and which as the basic elements of the whole Chinese character database. Our system breaks all the compound-segment strokes into segments that used as the elements to represent the entire character. The features of each segment have been analyzed and placed into a single vector.

Segments compose a compound-segment stroke. The amount of stroke count is different from segment count for a particular character. For example, The character (絲) is counted as 12 strokes in the regular Chinese database but 16 segments in our system since the character contains 4 compound-segment strokes (∠) each consisting of 8 segments. Our system breaks all the compound-segment strokes into segments that are used as the elements to represent the entire character. The features of each segment have been analyzed and placed into a single vector.

3 Connected Segments Handling

A stroke segment is measured from pen down to pen up. A connected segment is a segment with freeman code 1, 2, 3, 4, and is located between standard segments. In this system, we can easily detect all segments with the freeman code 1, 2, 3 and 4 from the input characters. The system counts them as the end of the segments; otherwise, they will be counted as connected segments. Fig. 1 shows an example of eliminating hooks.

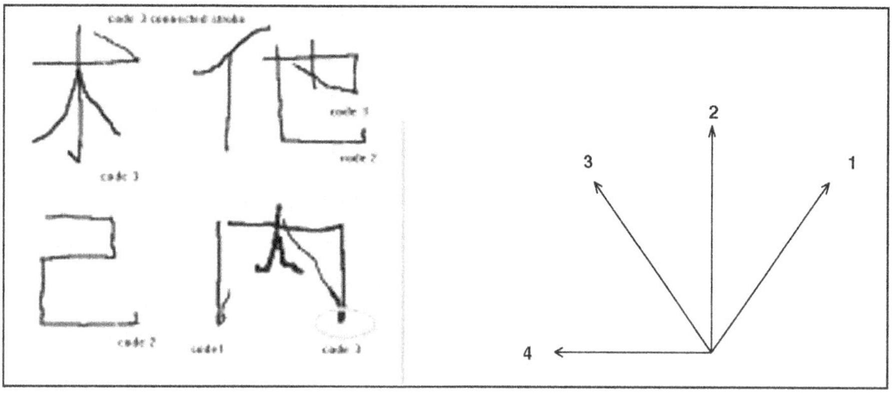

Fig. 1 Connected segment detection Algorithm utilizes the Freeman Code 1, 2, 3 and 4 to eliminate hooks. Segments will be ignored if detected as the code within 1, 2, 3 or 4. This algorithm can also be applied in dehooking (eliminating the necessary segments of characters).

If two standard segments without intermediate line segments are connected by tail to head, it is the head-tail type connection. For instance ( ) and ( ), connected as (辶), is a head-tail connection. On the other hand, if two standard segments are connected with an intermediate line segment, it is the backward type connection. For instance, (—) and (|) with an intermediate (丶), connected as (十), is a backward connection. If a character has several connected segment radicals, but there is no connected segments within, then the writing is running hand writing. Obviously, the key problem for recognizing running hand writing is to solve the problem of connections between segments. The knowledge-based approach is applied for decomposing a connected segment into separate segments. Some rules based upon the knowledge of segment connection are summarized as follows:

1. If a connected segment is classified as a standard segments, e.g., (矢) ---> (矢), then the segment is decomposed into line segments for recognizing radicals because the number of segments as well as corresponding directions are the same.
2. When one writes characters, one usually starts writing character by character from left to right, line by line from top to bottom. If a compound-segment segment consists of several line segments, then the direction of the segment is between "h", "v", "p", "n" segment type. Other than that, We can say it is a connected segment and the linkage can be cut off. That is, if the freeman code of that line segment is within 1, 2, 3 and 4, then our system will not consider it as a standard segment and will automatically ignore it. Some cases of connected segments are shown in the following table. (The above explanations are originated from [11])

Table 2. Examples of connected segments (Cited from [11])

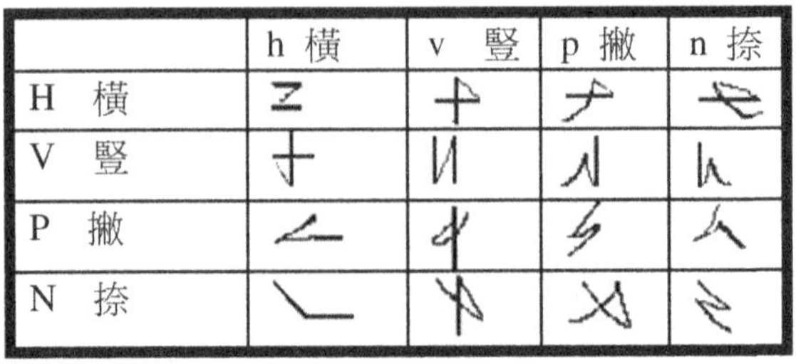

	h 橫	v 豎	p 撇	n 捺
H 橫				
V 豎				
P 撇				
N 捺				

Since our system will ignore those stroke segments with freeman code 1, 2, 3 and 4, around 95% of the connected segments shown on the above table are automatically detected.

When people write characters in free hand, the frequency of segment number variation will occur as shown in Fig. 2 This shows a 37.4% frequency of missing segments and the maximum missing segment count is seven at a time. Moreover, there is only 1.4% frequency of additional segments and the maximum number is two segments at a time.

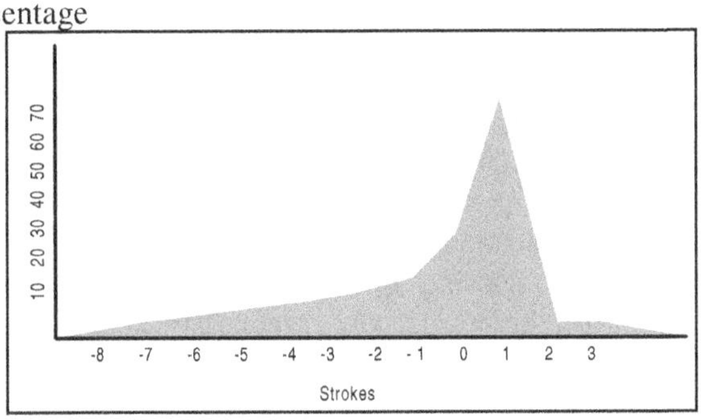

Fig. 2 Segment variation for Chinese script writing. (Cited from [10])

The range of segment number variation is from -2 to +2 as usual. As in section 2, we assume the input character X () and the reference character G (乞) have the same segment number (Fig. 3). The system matches them by calculating the segment distance between the two characters. The result is the same as our assumption. As mentioned before in section 2, each vector in the data dictionary is dedicated for one segment of the character. The segmentation process will break the character into segments during the preprocessing stage. Although the input character X has one head-tail connected segment; it does not affect the amount of the total stroke segment of the character. Thus, the connected segment does not have any influence on the recognition stage. (The above example was inspired from [10])

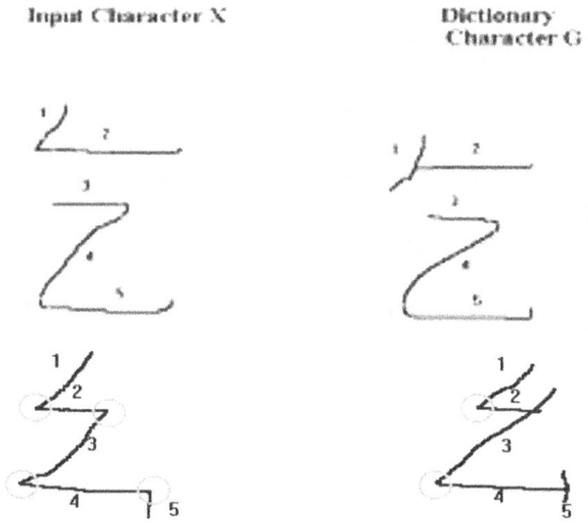

Fig 3 The connected segments do not affect the amount of segment number after segmentation because the reference characters in dictionary are in segment based. (Inspired from [10]).

4 Concluding Remarks

The above connected stroke detection algorithm has been tested by implementations and shows satisfactory results on improving the accuracy of overall Chinese character recognition. In order to recognize the contribution of the automatic detection algorithm, we perform the experiment by combining the techniques mentioned in [12], [13], [14] and [15] as a whole system.

4.1 Experiment Results

To perform the practical experiment, a database composed of 1100 Chinese characters was constructed as the database, which had been trained for five times during the signal learning stage. The segment numbers of the characters ranged from 1 to 31. In order to include the variations of segment features in the database, each time of the signal learning was trained with a different writing style. It is a closed result tested by the author himself, with limited cursive writing, in his own laboratory. Although the result is writer dependent, it still shows that an integrated recognition system using the proposed algorithm is very promising.

Reference

1. K.Yamamoto And A. Rosenfeld, Recognition Of Hand-Printed Kanji Characters By Relaxation Method, Proc. 6th ICPR, (1982) 395-398.

2. Eveline J. Bellegarda, Jerome R. Bellegarda, David Nahamoo, And Krishna S. Nathan, A Fast Statistical Mixture Algorithm For On-Line Handwriting Recognition, IEEE Transactions On Pattern Analysis And Machine Intelligence, Vol. 16, No. 12, (1994).
3. J.-W. Tai, T.-J. Liu And L.-Q. Zhang, A New Approach For Feature Extraction And Feature Selection Of Handwritten Chinese Character Recognition, From Pixels To Features III: Frontiers In Handwriting Recognition S. Impedovo And J.C. Simon (Eds.) (1992) 479-491.
4. Yih-Tay Tsay And Wen-Hsiang Tsai, Attributed String Matching By Split-And-Merge For On-Line Chinese Character Recognition, IEEE Transactions On Pattern Analysis And Machine Intelligence, Vol. 15, No.2, (1993) 180-185.
5. M.Nakagawa, K. Aoki, T. Manable, S. Kimura, And N. Takahashi, On-Line Recognition Of Hand-Written Japanese Characters In JOLIS-1, Proc. 6th ICPR,(1982) 776-779.
6. K.S. Leung, Y. Fan And F.Y. Young, A Chinese Dictionary System Based On Fuzzy Logic And Object-Oriented Approach, Computer Processing Of Chinese And Oriental Languages, Vol. 6, No. 2, (1992) 205-219.
7. C. H. Leung, Y. S. Cheung And Y. L. Wang, A Knowledge-Based Stroke Matching Method For Chinese Character Recognition, IEEE Trans. Man Cybern. 17, (1987) 993-999.
8. S.-L.Shiau, S.-J. Kung, A.-J. Hsieh, J.-W.Chen And M.-C.Kao, Stroke-Order Free On-Line Chinese Character Recognition By Structural Decomposition Method, From Pixels To Features III:Frontiers In Handwriting Recognition, (1992) 117-127.
9. Ke Jing, Qiao Yi Zheng, A Local Elastic Matching Method For On-Line Chinese Signature Verification, Journal Of Chinese Information Processing, Vol.12, No.1, (1998) 57-63.
10. Chi Chung Zhang, Chinese Recognition Techniques, Chinese Signal Processing, Tsing Hwa University Press, (1992).
11. Y.J.Liu And J.W. Tai, An On-Line Chinese Character Recognition System For Handwritten In Chinese Calligraphy, From Pixels To Features III: Frontiers In Handwriting Recognition, S. Impedovo And J.C. Simon (Eds), (1992) 87-99.
12. Tak-Ming Law, The Decision Path Classification For A Segment-Based On-Line Chinese Character Recognition, Proceedings Of The Conference On Applications Of Automation Science And Technology, Hong Kong (1998) 227-231.
13. Tak-Ming Law, Signal Smoothing, Sampling, Interpolation And Stroke Segmentation Algorithm For On-Line Chinese Character Recognition, Proceedings Of The Second International Conference On Information, Communications & Signal Processing, Singapore (1999),.
14. Tak-Ming Law, Signal Learning Algorithms And Database Architecture For On-Line Chinese Characters Recognition, Proceedings Of The 2000 International Workshop On Multimedia Data Storage, Retrieval, Integration And Applications, Hong Kong (2000) 68-74.
15. Tak-Ming Law, Segmentation Analysis And Similarity Measure For Online Chinese Character Recognition, Proceedings Of The International Conference On Chinese Language Computing, Chicago, Illinois, USA (2000).

{huwp,bobling}@ee.uwa.edu.au

huwp@mailbox.gxnu.edu.cn

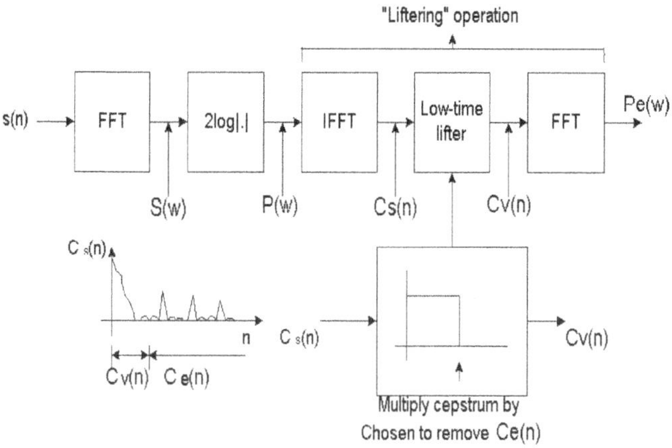

"Liftering" operation
s(n)
FFT
2log|.|
IFFT
Low-time lifter
FFT
Pe(w)
S(w)
P(w)
Cs(n)
Cv(n)
C s(n)
n
C v(n)
C e(n)
C s(n)
Cv(n)
Multiply cepstrum by
Chosen to remove Ce(n)

$\in$

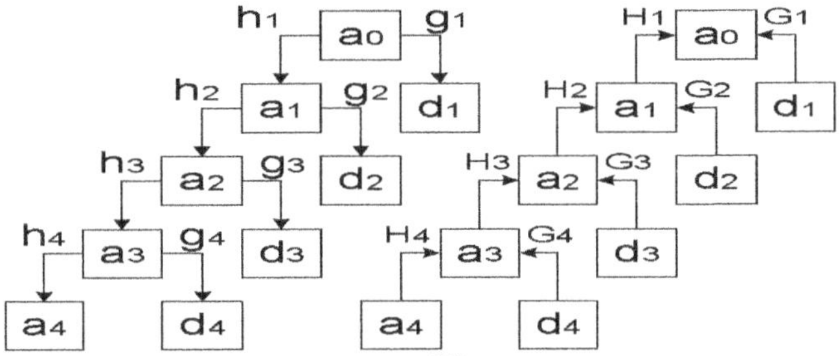

h1
a0
g1
h2
a1
g2
d1
h3
a2
g3
d2
h4
a3
g4
d3
a4
d4
H1
a0
G1
H2
a1
G2
d1
H3
a2
G3
d2
H4
a3
G4
d3
a4
d4

2a 2b

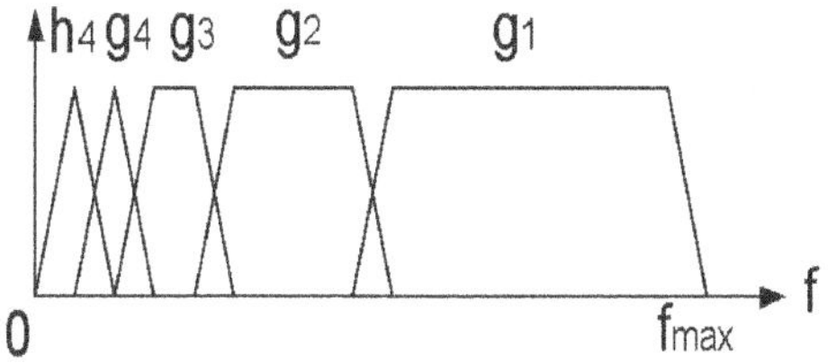
h4 g4 g3 g2 g1
0
fmax
f

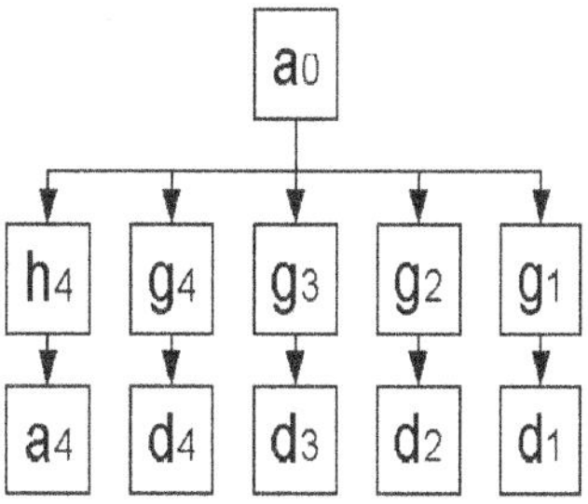
a0
h4 g4 g3 g2 g1
a4 d4 d3 d2 d1

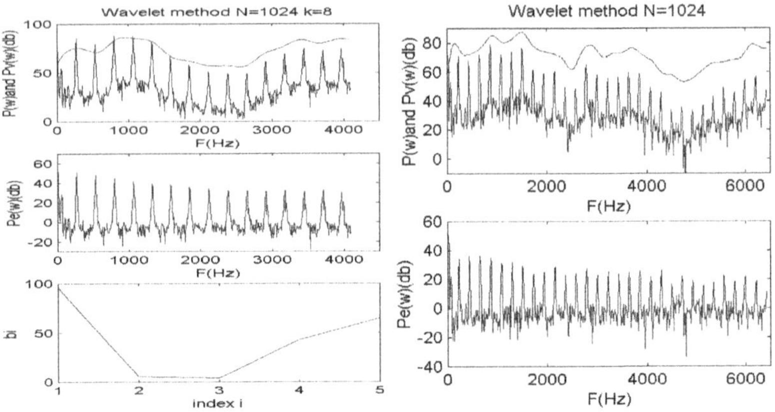

Wavelet method N=1024 k=8
P(w)and Pv(w)(db)
F(Hz)
Pe(w)(db)
F(Hz)
bi
index i
Wavelet method N=1024
P(w)and Pv(w)(db)
F(Hz)
Pe(w)(db)
F(Hz)

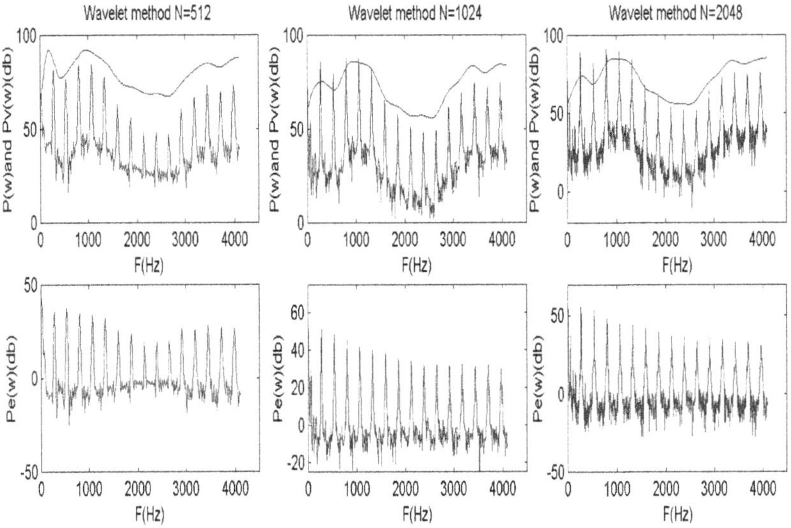

Wavelet method N=512
Wavelet method N=1024
Wavelet method N=2048
P(w)and Pv(w)(db)
F(Hz)
Pe(w)(db)
F(Hz)

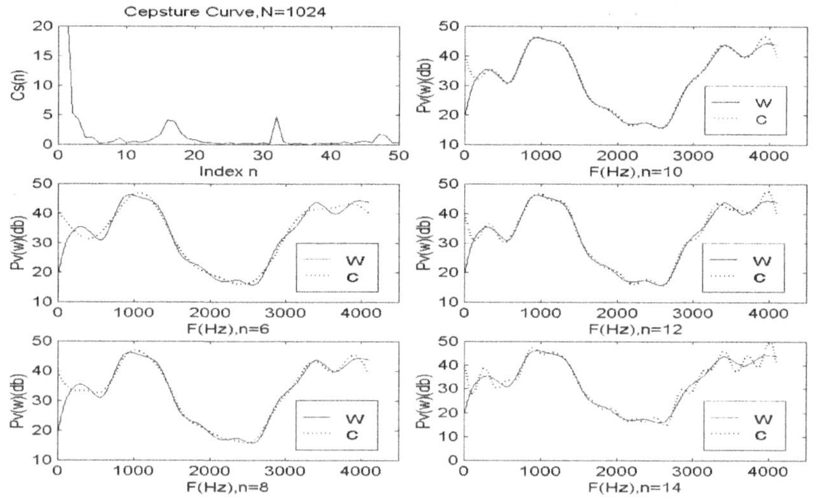
Cepsture Curve,N=1024
Cs(n)
Index n
Pv(w)(db)
F(Hz),n=10
W
C
F(Hz),n=6
F(Hz),n=12
F(Hz),n=8
F(Hz),n=14

A Proposal of Jitter Analysis Based on a Wavelet Transform

Jan Borgosz and Boguslaw Cyganek

Electronic Engineering and Computer Science Department, Academy of Mining and Metallurgy, Mickiewicza 30, 30-059 Kraków, Poland
{borgosz,cyganek}@uci.agh.edu.pl

Abstract. The paper puts forth a proposal for a new jitter measurement method based on a wavelet transform usage. There are many problems associated with the generation jitter free reference clock for measurements. The proposed method does not need a reference clock which is its main advantage over known methods that rely on reference clock usage. Additionally, presented wavelet transform applied to the jitter signals allows for more detailed analysis than offered by other methods. Comparison of classic and wavelet approach is presented. Problems like wavelet function type, order and post processing methods are also indicated.

1 Introduction

Estimating the jitter of a transmission clock is an important problem in telecommunication measurements. The classic approach to jitter measurement analysis usually consists of processing steps that use a reference clock [2][6][7]. The most troublesome part of the measurement process is to correlate slopes of the reference and received clocks. The purpose of this paper is to present a totally different wavelet based approach to jitter measurement analysis as compared to the aforementioned methods. Possibility of the usage different post processing methods is also shown (e.g. neural networks, fuzzy logic).

2 Jitter Theory

A jitter is an unwanted, spurious transmission clock phase modulation that orginates from the physics of semiconductor device [2][6]. Modeling this phenomenon using a modulation scheme allows us to describe it with a multitone technique [5]. A single tone modulation case can be described:

$$y(t) = A \cdot cos(2 \cdot \pi \cdot f_n \cdot t + \varphi_m(t)) \tag{1}$$

Y. Y. Tang et al. (Eds.): WAA 2001, LNCS 2251, pp. 257-268, 2001.

where y is the jittered clock signal with amplitude A [V] and base frequency f_n [Hz]. Phase modulating function $\varphi_m(t)$ can be described as follows:

$$\varphi_m(t) = 2 \cdot \pi \cdot k \sin \cdot (2 \cdot \pi \cdot f_{jitt} \cdot t) \quad \textbf{(2)}$$

where f_{jitt} is jitter frequency [Hz], $k \geq 0 \wedge k \in R$ the jitter amplitude in telecommunication UI units (UI means *Unit Interval* which is equal to one cycle of transmission clock). Results of more extended simulations are shown in Fig. 1. For view clarity some assumptions were made: simulation time 0.001s, base clock frequency f_n =1kHz, jitter amplitude A_{jitt} =1UI, jitter frequency f_{jitt} =100 Hz.

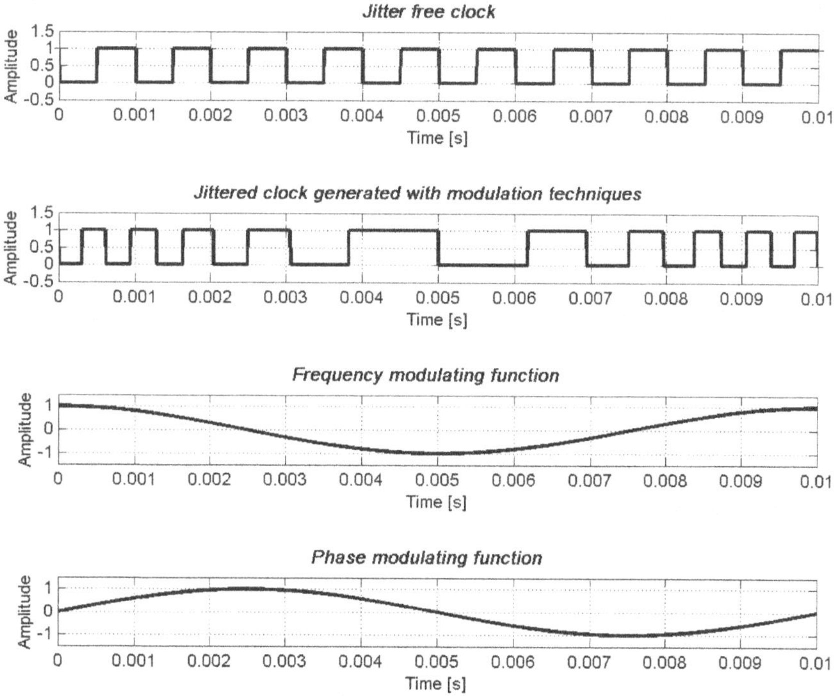

Fig. 1. Example of jitter simulation results

All calculations will be presented for the sinus function, which easily can be changed to other wave shapes. Note that the sinus waves after the comparator will be square wave with 50% duty cycle – ideal clock signal [3][6][7].

3 Classic Measurement Environment

Jitter test equipment is used with Equipment Under Test (EUT). Generated test signals are transmitted over telecom line into Equipment Under Test. EUT retransmits received data to the meter over telecom line again. Test equipment processes all received information and calculates results.

In classic approach, the implementation of the jitter meter is part of the structure shown in Fig. 2. Data received from the telecommunication line is being transformed by line interfaces. Simultaneously to data processing and Bit Error Ratio calculations, transmission clock is recovered and passed to the FPGA meter input. Output from this module is connected to the Digital Signal Processor. DSP may bypass data straight to the host processor or improve processing by additional calculations (e.g. FFT or other filtering methods). Host processor (Fig. 2) helps to visualize measurement results to the end user.

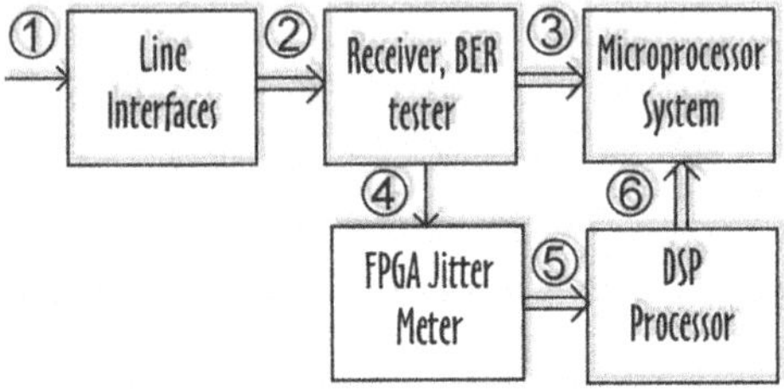

Fig. 2. Jitter meter structure 1) Signal from the telecommunication line 2) Signal from the line interfaces. 3) BER information 4) Jittered clock 5) Jitter measure results 6) Jitter measure results after DSP

Here is example of digital jitter meter. There are three main components of FPGA implementation presented in Fig. 3: jitter-free clock generator, phase comparator and FIR filter [5][6].

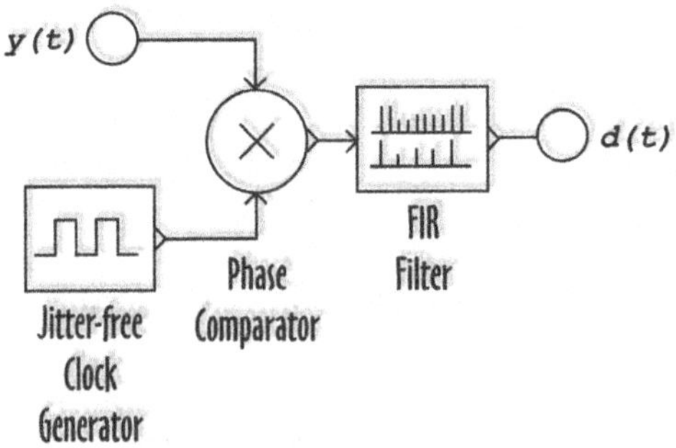

Fig. 3. FPGA jitter meter implementation

Digital phase detector forms series of pulses in accordance with the phase differences between jittered clock signal *y(t)* described by equations (1), (2) and reference clock with frequency f_n. This way formed signal (extended by sign bit, that describes phase shift direction) is provided into the FIR filter input. It can be seen, that time resolution (phase measure quantization) depends on sampling clock. An appropriate

FIR structure is selected due to measurement type and range. Filtered signal with jitter information *d(t)* is available at the FIR output. It may be shown, that *d(t)* is equal to:

$$d(t) = k \cdot sin(2 \cdot \pi \cdot f_{jitt} \cdot t) \quad (3)$$

4 Jitter Measurement with Wavelet Transform

4.1 Bessel representation of a single tone modulated signal

A single tone modulated signal can be written as follows:

$$y_{PM}(t) = A \cdot cos(\Omega \cdot t + \Delta\Theta_{PM} \cdot sin\varpi \cdot t) \quad (4)$$

As shown in [5] this representation can be replaced by a more appropriate form that makes use of a Bessel function:

$$y_{PM}(t) = A \cdot \sum_{n=-\infty}^{\infty} J_n(\Delta\Theta_{PM}) \cdot cos(\Omega + n \cdot \varpi) \cdot t \quad (5)$$

where J_n is an n-th order Bessel function of the first kind. In this case equations (1) and (2) can be rewritten as follows ($\Omega_n = 2 \cdot \pi \cdot f_n$ and $\varpi_j = 2 \cdot \pi \cdot f_{jitt}$):

$$y(t) = A \cdot \sum_{n=-\infty}^{\infty} J_n(2 \cdot \pi \cdot k) \cdot cos(\Omega_n + n \cdot \varpi_j) \cdot t \quad (6)$$

4.2 Jittered signal integration and RMS calculations

A jittered sine signal can be integrated by a circuit with a much higher cut-off frequency than the maximum. In this case, the integrator like that of an accumulator. A jittered signal given by (6) after the integration will be equal to:

$$y_{INT}(t) = \frac{A}{\Delta T} \cdot \int_{t}^{t+\Delta T} \sum_{n=-\infty}^{\infty} J_n(2 \cdot \pi \cdot k) \cdot cos(\Omega_n + n \cdot \varpi_j) \cdot t \quad (7)$$

where ΔT is the integration period. Another operation that can be performed on the jittered signal is the RMS calculation:

$$y_{RMS}(t) = A \cdot \frac{\sqrt{\int_{t}^{t+\Delta T} \left(\sum_{n=-\infty}^{\infty} J_n(2 \cdot \pi \cdot k) \cdot cos(\Omega_n + n \cdot \varpi_j) \cdot t \right)^2}}{\Delta T} \quad \textbf{(8)}$$

Because a direct analysis of (7) and (8) were somewhat cumbersome, therefore some numerical computations were performed and are presented in Fig. 4. The relationship between frequency changes in the jittered signal as well as the amplitude changes in the integrated jittered signal or RMS can be observed in Fig.4b and Fig.4c.

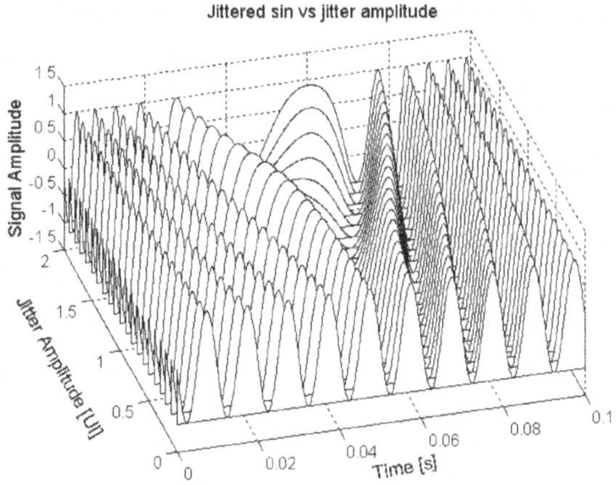

(a)

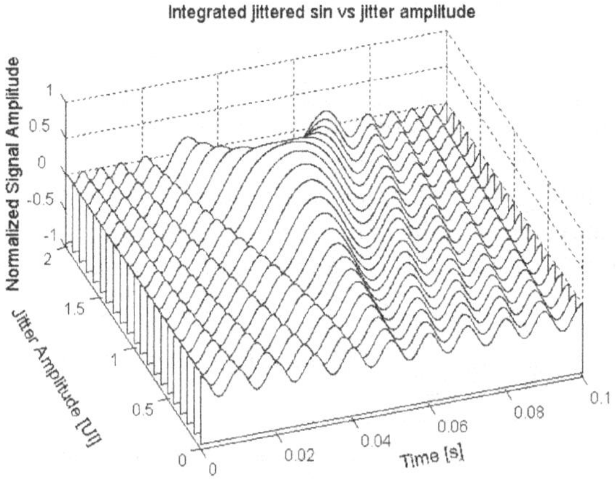

(b)

Fig. 4. a) Set of jittered sinusoids with jitter frequency f_{jitt}=10Hz, carrier frequency f_n = 100 Hz for different jitter amplitudes, b) the same sinusoids after integration

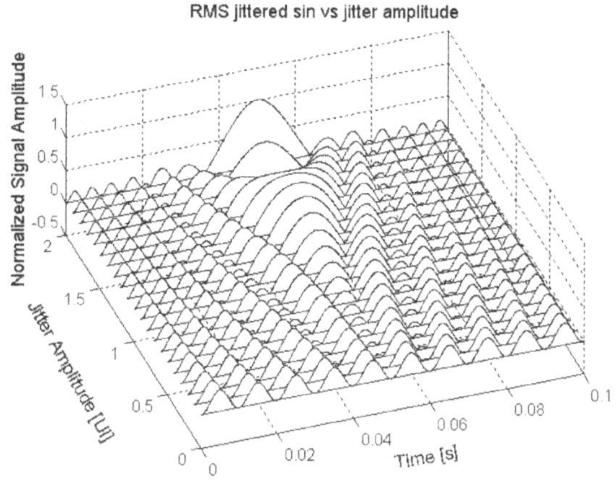

(c)

Fig. 4 - continuation. c) the same sinusoids after RMS calculation

Also numerical computations were performed for square waves – sinus waves after the comparator and are presented in Fig. 5. Note that, there is no need for reference clock usage for jitter detection. Results of practical experiments are shown in Fig. 6. They are confirmation of presented here calculations.

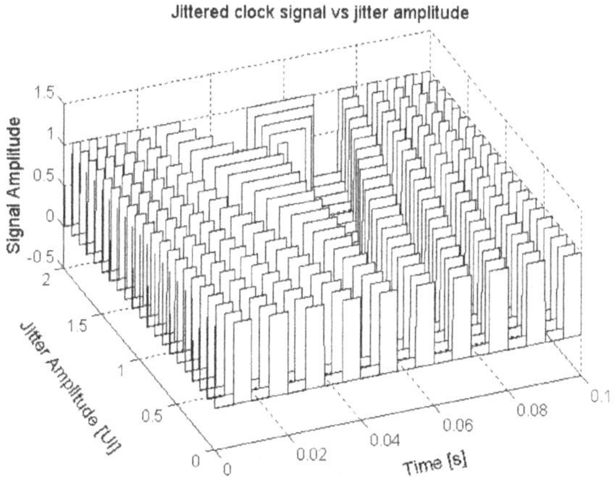

(a)

Fig. 5. a) Set of jittered clock signals with jitter frequency f_{jitt}=10Hz, carrier frequency f_n = 100 Hz for different jitter amplitudes

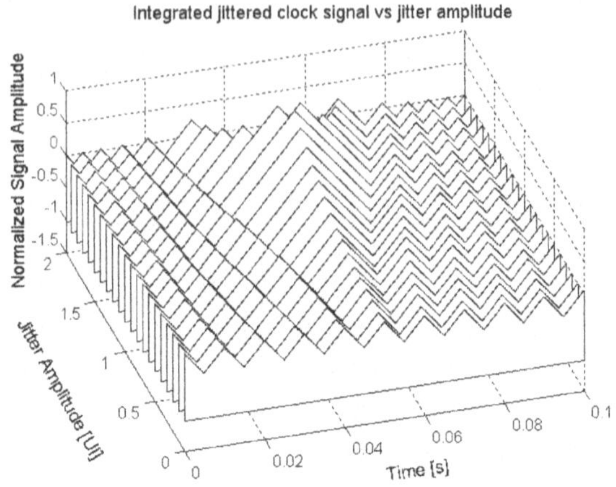

(b)

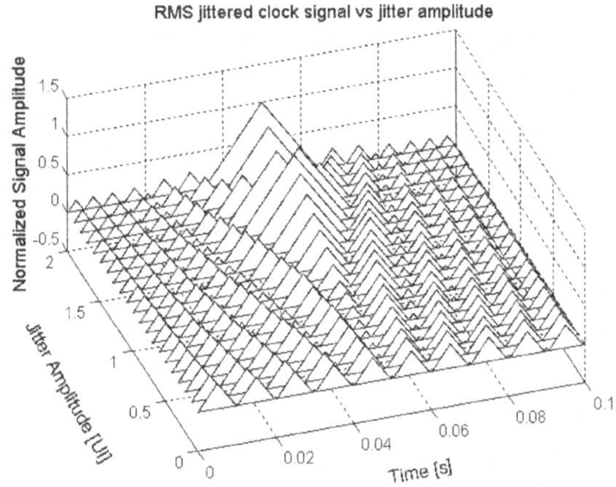

(c)

Fig. 5 - continuation. b) the same clock signals after integration, c) the same clock signals after RMS calculation

4.3 Wavelet transform applied to jitter analysis

High quality jitter measurements involve: amplitude, frequency and time - changes in time of both parameters. According to the authors opinion, the best tool for such an analysis is the Continuous Wavelet Transform (CWT) [1] [4], which is represented by the following equation:

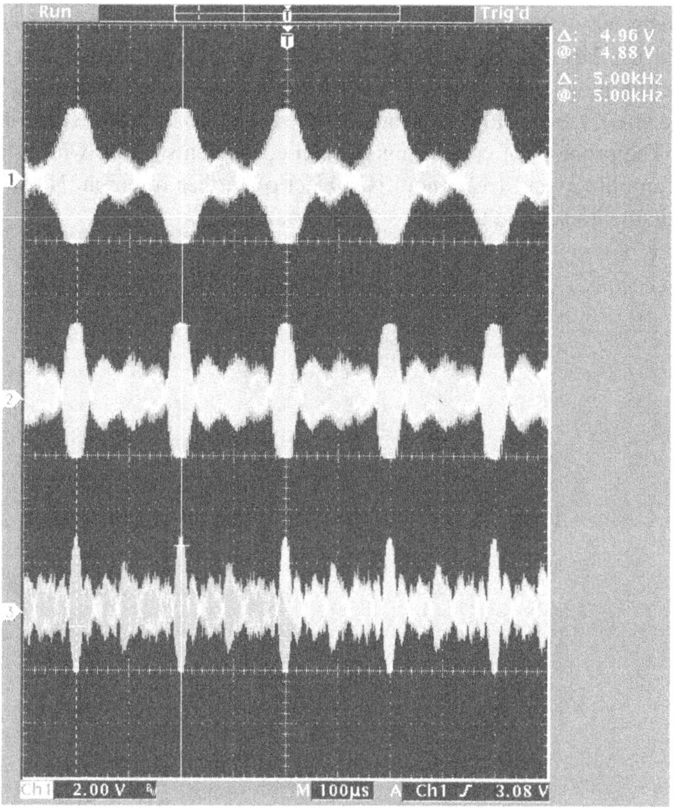

Fig. 6. Practical tests. Integrated jittered signals of E1 standard – 2,048 MHz base clock. Jitter frequency 5kHz, cursor positions: 1) jitter amplitude 0.5UI, 2) jitter amplitude 1UI, 3) jitter amplitude 1.5UI

$$C(s,p) = \int_{-\infty}^{\infty} f(t)\cdot\Psi(s,p,t)dt \tag{9}$$

where $\Psi(s,p,t)$ is mother wavelet, s the scale and p the position. Inserting (7) into (9) provides us the following formula:

$$C(s,p) = \int_{-\infty}^{\infty}\left(\left(\frac{A}{\Delta T}\cdot\int_{t}^{t+\Delta T}\left(\sum_{n=-\infty}^{\infty} J_n(2\cdot\pi\cdot k)\cdot cos(\Omega_n + n\cdot\varpi_j)\cdot t\right)dt\right)\cdot\Psi(s,p,t)\right)dt \tag{10}$$

As can be observed in (10), there is no easy way to find a relationship between jitter parameters and *CWT* coefficients for a signal after integration. This problem is the subject of research, as is also a selection of a proper mother wavelet function.

During tests the authors decided to use the following wavelets (Fig.7): *Mexican Hat*, *Morlet*, *Coiflets 2-5*, *Biorthogonal 2.6 2.8 4.4 5.5 6.8*, because their shapes appeared to be the most appropriate [1] [4] to analyze the signals shown in Fig.4b and Fig.4c. As can be seen in Fig. 8. changes of wavelet coefficients carry information about jitter. The problem of correlating wavelet coefficients *C(s,p)* with jitter parameters (i.e. its amplitude and frequency) is subject of further research. Neural networks or fuzzy logic methods seem to be most appropriate.

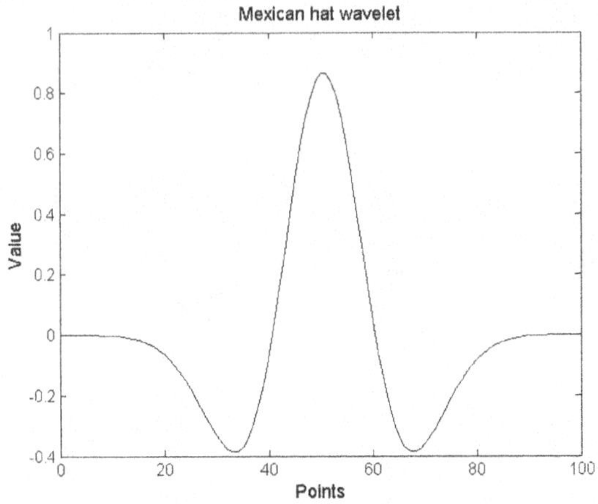

(a)

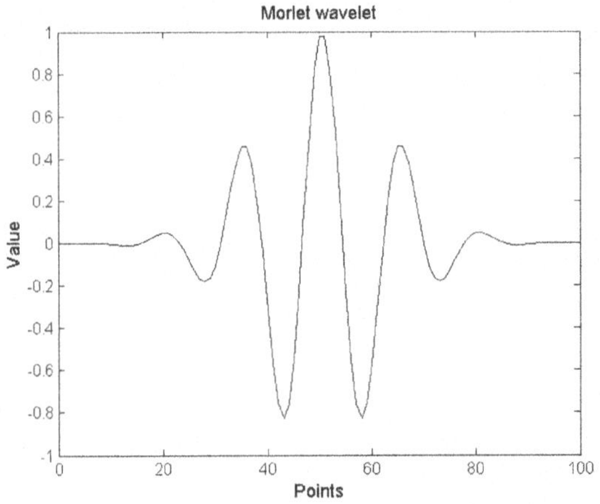

(b)

Fig. 7. Example of used wavelets a) Mexican Hat, b) Morlet

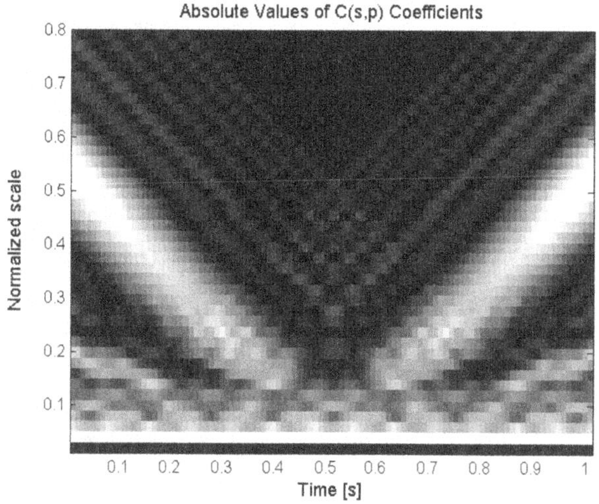

(a)

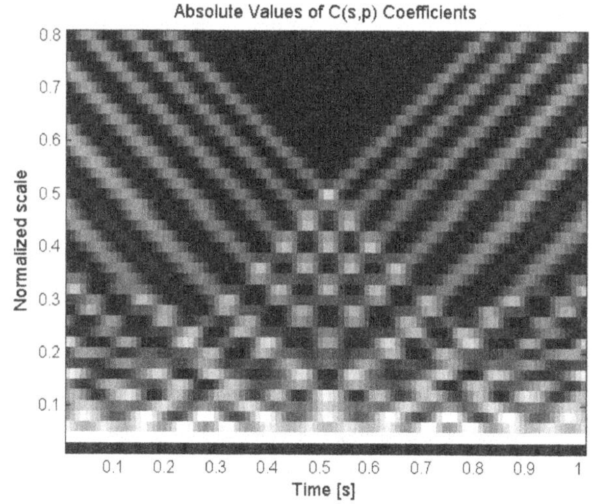

(b)

Fig. 8. Set of *Mexican Hat* wavelet coefficients for integrated jittered signals shown in Fig.4b: a) jitter amplitude equals 0UI, b) jitter amplitude equals 0.5 UI

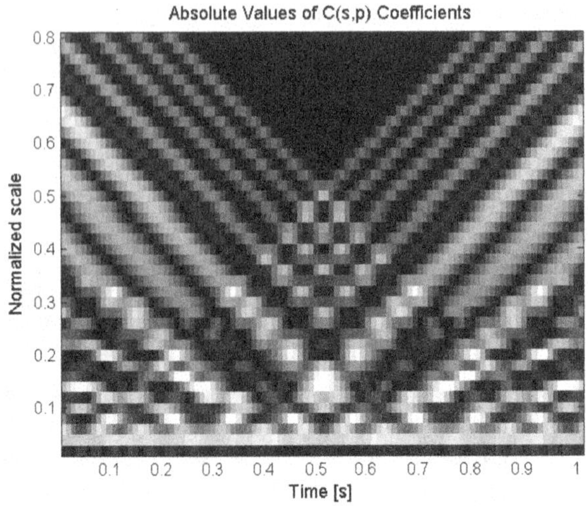

(c)

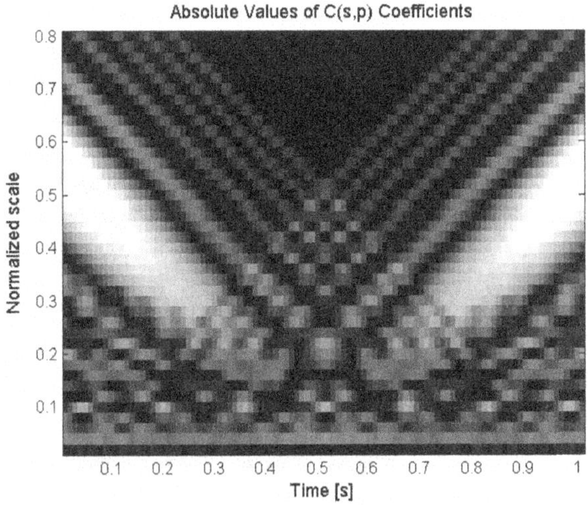

(d)

Fig. 8 - continuation. Set of *Mexican Hat* wavelet coefficients for integrated jittered signals shown in Fig.4b: c) jitter amplitude equals 1 UI, d) jitter amplitude equals 1.5 UI

5 Conclusions

A new jitter measurement method using the wavelet transform was developed. In this paper we found correlation between jitter in signals and *CWT* coefficients of integrated jittered signals. This method provides a novel way of jitter measurement without reference clock usage. Wavelet transform with all benefits and lack of reference clock make this method very attractive.

An analytic description of the method was presented. Furthermore wavelet types selection, operations performed on jittered signal (other than integration and RMS), as well as relations between wavelet coefficients *C(s,p)* and jitter amplitude, given by analytic equations (10), are under continuous research.

This paper is a result of research work registered in KBN The State Committee for Scientific Research of Poland at number *7 T11B 072 20* and its publication is sponsored from KBN founds.

References

1. Bia•asiewicz J.: Falki i aproksymacje (in Polish). WNT (2000)
2. Feher and Engineers of Hewlett-Packard: Telecommunication Measurements Analysis and Instrumentation, Hewlett-Packard(1991)
3. Glover I. A., Grant P.M.: Digital Communications. Prentience Hall (1991)
4. Prasad L., Iyengar S.S.: Wavelet Analysis. CRC Press (1997)
5. Szabatin J.: Podstawy teorii sygnalów (in Polish). WKL (2000)
6. Trischitta P.R., Varma E.L.: Jitter in Digital Transmission System. Artech House Publishers (1989)
7. Takasaki Y., Personick S.D.: Digital Transmission Design and Jitter Analysis. Artech House Publishers (1991)

Skewness of Gabor Wavelets and Source Signal Separation

Weichuan Yu[1], Gerald Sommer[2], and Kostas Daniilidis[3]

[1] Dept. of Diagnostic Radiology, Yale University
weichuan@noodle.med.yale.edu
[2] Institut für Informatik, Universität Kiel
gs@ks.informatik.uni-kiel.de
[3] GRASP Laboratory, University of Pennsylvania
kostas@grasp.cis.upenn.edu

Abstract. Responses of Gabor wavelets in the mid-frequency space build a local spectral representation scheme with optimal properties regarding the time-frequency uncertainty principle. However, when using Gabor wavelets we observe a skewness in the mid-frequency space caused by the spreading effect of Gabor wavelets. Though in most current applications the skewness does not obstruct the sampling of the spectral domain, it affects the identification and separation of source signals from the filter response in the mid-frequency space. In this paper, we present a modification of the original Gabor filter, the skew Gabor filter, which corrects skewness so that the filter response can be described with a sum-of-Gaussians model in the mid-frequency space. The correction further enables us to use higher-order moment information to separate different source signal components. This provides us with an elegant framework to deblur the filter response which is not characterized by the limited spectral resolution of other local spectral representations.

1 Introduction

According to the well known uncertainty principle, the product of the spatial and the spectral support of a filter has a lower bound. Because Gabor filters [1] can achieve such a lower bound they are very useful in many spectral analysis tasks such as image representation (e.g. [2]) and the spatio-temporal analysis of motions in image sequences (e.g. [3,4]). Besides, Gabor filters were shown to approximate biological models of vision (e.g. [5,6,7]). In the spatio-temporal models for motion estimation [3,8], the energy spectrum of a constant translational motion can be characterized as an oriented plane passing through the origin in the spectral domain. Sampling the spectrum with a set of Gabor filters at different frequencies and orientations [4] may help us to estimate the orientation of the spectral plane. Grzywacz and Yuille [9] further argued that the spectral support of a Gabor filter is a measure of uncertainty and the angle between two tangential lines of the support, which pass through the spectral origin, represents the uncertainty of orientation estimation (see figure 1). This angle is desired to be

Y. Y. Tang et al. (Eds.): WAA 2001, LNCS 2251, pp. 269–283, 2001.

the same for filters at different frequencies. Thus, the spectral support should be proportional to the distance between the origin and the support center.

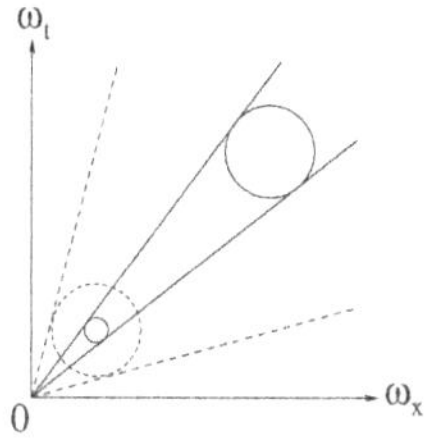

Fig. 1. The motivation of applying 2D Gabor wavelets (redrawn from [9]). We represent the spectral support of a 2D Gabor filter with a circle. Applying a set of filters with constant scale may cause larger angular uncertainty at lower frequencies (as shown by the angle between two dashed lines). Thus, the spectral support of filters should be directly proportional to the mid-frequency

In Gabor filters, impulse responses have the same support in low and high frequencies. However, we would prefer the support to be inversely proportional to the mid-frequency. The coupling of the bandwidth with the mid-frequency yields Gabor wavelets which are extensively used in signal analysis and image representation (e.g. [10,2]).

In applying Gabor wavelets we observe a positive skewness in the mid-frequency space [9]. This skewness did not draw considerable attention in the computer vision community because most applications of Gabor wavelets are classification tasks. Being aware of the non-symmetrically spreading effect of Gabor wavelets in the mid-frequency space, we argue that an isotropic dissemination of the mid-frequency representation of the filter response (we call this local spectral representation the *mid-spectrum*) may facilitate the deblurring of filter responses so that we no more suffer from the limited resolution of frequency-based approaches. This is especially useful in source signal separation and multiple spectral orientation analysis. Based on this motivation we design a new filter to correct the skewness effect (section 2). In section 3 we further describe the 1D corrected mid-spectrum with a sum-of-Gaussians model and use higher-order moments to identify different source components. The deblurring of the mid-spectrum is also demonstrated. In section 4 we extend the analysis to 2D spectral orientation analysis. This paper is concluded in section 5.

2 The Skewness of Gabor Wavelets

We first explain the positive skewness of Gabor wavelets. For simplicity we begin with a 1D Gabor filter whose impulse response reads

$$g_1(x;\omega_0,\sigma_x) := \frac{1}{\sqrt{2\pi}\sigma_x} e^{-\left(\frac{x^2}{2\sigma_x^2}\right)} e^{j\omega_0 x}. \tag{1}$$

Here ω_0 denotes the mid-frequency and σ_x is the scale parameter. The spectrum of $g_1(x;\omega_0,\sigma_x)$ is a Gaussian centered at ω_0

$$G_1(\omega;\omega_0,\sigma_x) = e^{-\frac{\sigma_x^2(\omega-\omega_0)^2}{2}} \tag{2}$$

with bandwidth inversely proportional to σ_x. In applications, we usually calculate the spatial convolution between $g_1(x;\omega_0,\sigma_x)$ and the input signal $i(x)$

$$h_1(x;\omega_0,\sigma_x) := i(x) * g_1(x) = \int_{\xi=-\infty}^{\infty} i(\xi) g_1(x-\xi) d\xi. \tag{3}$$

At a fixed position x_0, the filter response is simplified as an inner product

$$h_1(x_0;\omega_0,\sigma_x) = \int_{\xi=-\infty}^{\infty} i(\xi) g_1(x_0-\xi) d\xi. \tag{4}$$

Using the facts that $g_1(x_0-x) = g_1^\star(x-x_0)$ and $G_1^\star(\omega) = G_1(\omega)$ (here $\star$ denotes conjugation) the above inner product can also be represented in the spectral domain according to the Parseval theorem ([11], pp.113-115) as

$$h_1(x_0;\omega_0,\sigma_x) = \int_{\omega=-\infty}^{\infty} I(\omega) G_1(\omega) e^{j\omega x_0} d\omega. \tag{5}$$

Here $I(\omega)$ is the spectrum of $i(x)$. Thus, for $x = x_0$ (for simplicity we set $x_0 = 0$) we obtain a local spectral representation which is a function of the mid-frequency ω_0 and the scale σ_x. We call this representation the mid-spectrum of the signal.

The mid-spectrum $h_1(\omega_0,\sigma_x)$ spreads every spectral Dirac component of the source signal into a function of ω_0. Assume that the spectrum of a source signal is a Dirac function: $I(\omega) = \delta(\omega-\omega_i)$ originating from a complex harmonic. Equation (5) then turns out to be

$$h_1(\omega_0,\sigma_x) = G_1(\omega_i;\omega_0,\sigma_x) = e^{-\frac{\sigma_x^2(\omega_0-\omega_i)^2}{2}}. \tag{6}$$

When the parameter σ_x is a constant, $h_1(\omega_0,\sigma_x)$ is a Gaussian spreading of $\delta(\omega-\omega_i)$ and there is no skewness. But if the wavelet property is preferred, i.e. if σ_x is inversely proportional to ω_0

$$\sigma_x = \frac{C}{\omega_0} \tag{7}$$

with C as a constant. Then, we observe the positive skewness of ω_0 [9] (see also figure 2)

$$h_1(\omega_0,C) = e^{-\frac{C^2(\omega_0-\omega_i)^2}{2\omega_0^2}}. \tag{8}$$

We may straightforwardly extend the above analysis for n-dimensional Gabor wavelets with isotropic envelope. For 2D Gabor wavelets in the spatio-temporal domain we have the following relation

$$\sigma_x = \sigma_t = \frac{C}{\sqrt{\omega_{x0}^2+\omega_{t0}^2}}. \tag{9}$$

The mid-spectrum of a 2D spectral impulse $\delta(\omega_{x0}-\omega_{xi},\omega_{t0}-\omega_{ti})$ reads

$$h_2(\omega_{x0},\omega_{t0},C) = \exp\{-\frac{C^2[(\omega_{x0}-\omega_{xi})^2+(\omega_{t0}-\omega_{ti})^2]}{2(\omega_{x0}^2+\omega_{t0}^2)}\}. \tag{10}$$

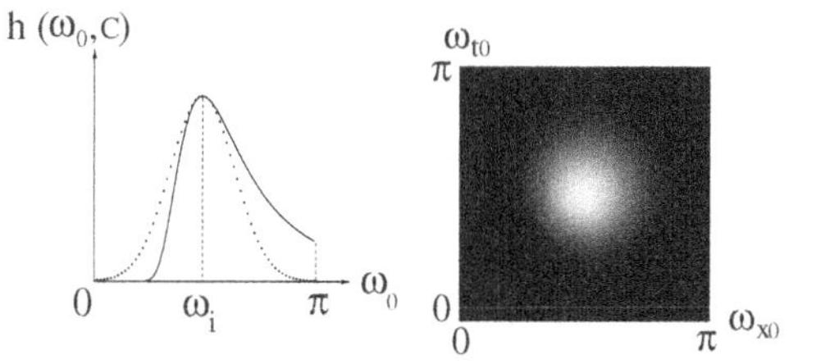

Fig. 2. The skewness of Gabor wavelets. **Left:** The solid curve denotes $h_1(\omega_0, C)$ and the dotted curve is a Gaussian function centered at ω_i with the scale parameter $\frac{\omega_i}{C}$. $C = 3.5, \omega_i = \frac{\pi}{2}$. **Right:** 2D skewness $h_2(\omega_{x0}, \omega_{t0}, C)$. $C = 3.5, \omega_{xi} = \omega_{ti} = \frac{\pi}{2}$

In many Gabor wavelet approaches, this skewness seems harmless because it does not obstruct the description of different signals with a set of samples [12,13]. The main attention was attracted to the efficient covering/sampling of the spectrum as well as the coefficient estimation of the Gabor basis [10,2]. But we should keep in mind that the spreading effect of Gabor wavelet filtering (see equation (8)) really blurs the input signal non-symmetrically in the mid-frequency space. For the sake of source signal identification and separation, we prefer to have a symmetric spreading. In the following we present a new filter to correct this positive skewness.

2.1 Correcting the Skewness

In order to achieve symmetry in the mid-spectrum, we introduce a new skew Gabor filter whose spectral definition reads

$$SG_1(\omega; \omega_0, C) := \exp\{-(\frac{C^2}{2})(\frac{\omega - \omega_0}{\omega})^2\}. \tag{11}$$

There exists no analytical expression of the skew Gabor filter in the spatial domain because there is no closed-form representation of the inverse Fourier transform of $SG_1(\omega)$. But we may obtain an FIR version of both the real and the imaginary part of the skew Gabor filter $sg_1(x)$ using filter-design in the Fourier domain and discrete Fourier transform. In figure 3 we display one example of the skew Gabor filter. It is similar to a Gabor filter with subtle shape differences inside the Gaussian envelope.

Replacing $G_1(\omega; \omega_0, \sigma_x)$ in equation (5) with $SG_1(\omega; \omega_0, \sigma_x)$ yields a mid-spectrum with an ideal Gaussian shape

$$sh_1(\omega_0, C) = \exp\{-\frac{C^2(\omega_0 - \omega_i)^2}{2\omega_i^2}\}. \tag{12}$$

Similarly, we may correct the skewness of 2D Gabor wavelets by using a 2D skew Gabor filter

$$SG_2(\omega_x, \omega_t; \omega_{x0}, \omega_{t0}, C) = \exp\{-(\frac{C^2}{2})[\frac{(\omega_x - \omega_{x0})^2 + (\omega_t - \omega_{t0})^2}{\omega_x^2 + \omega_t^2}]\}. \tag{13}$$

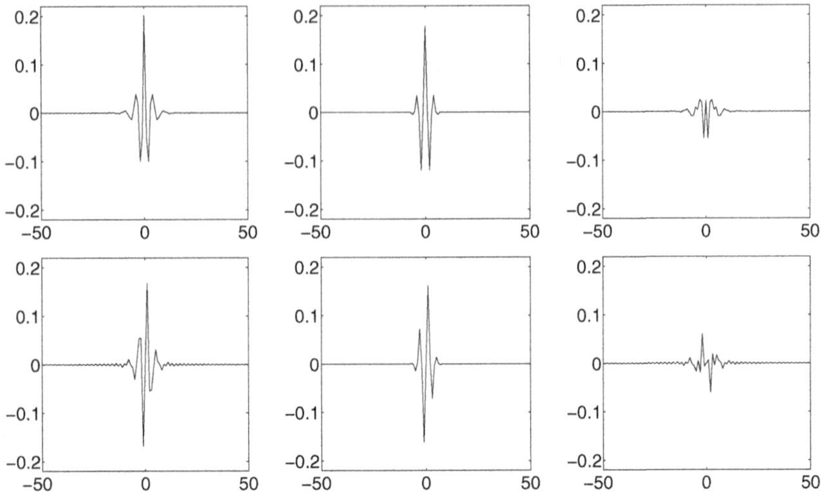

Fig. 3. Top: The real parts of a 1D skew Gabor filter (left) and of a Gabor filter (middle) as well as their even-symmetric difference (right). **Bottom:** The imaginary parts of both filters (left: skew Gabor; middle: Gabor) and their odd-symmetric difference (right). The parameters are $C = 3.5$ and $\omega_0 = \frac{\pi}{2}$

The mid-spectrum corresponding to $\delta(\omega_{x0} - \omega_{xi}, \omega_{t0} - \omega_{ti})$ is then an ideal 2D Gaussian (cf. equation (10))

$$sh_2(\omega_{x0}, \omega_{t0}, C) = \exp\{-(\frac{C^2}{2})[\frac{(\omega_{x0} - \omega_{xi})^2 + (\omega_{t0} - \omega_{ti})^2}{\omega_{xi}^2 + \omega_{ti}^2}]\}. \tag{14}$$

In figure 4 we display a 1D cosine sequence and the correction of the skewness in the mid-spectrum. Here we use only one constant C to keep the Gaussian envelope isotropic. It is also possible to apply two different constants (i.e. $C_x \neq C_t$) in order to form a mid-spectrum with an elongated Gaussian shape. But this is beyond the scope of this paper.

3 1D Source Signal Separation

In the following we demonstrate the merit of correcting the positive skewness. We start with 1D source signal separation. We assume that the spectrum of an input signal is composed of two Dirac components

$$S(\omega) = a_1\delta(\omega - \mu_1) + a_2\delta(\omega - \mu_2), \tag{15}$$

where their amplitudes (a_1 and a_2) and offsets (μ_1 and μ_2) are unknown. Our goal is to estimate these amplitudes and offsets from the mid-spectrum so that

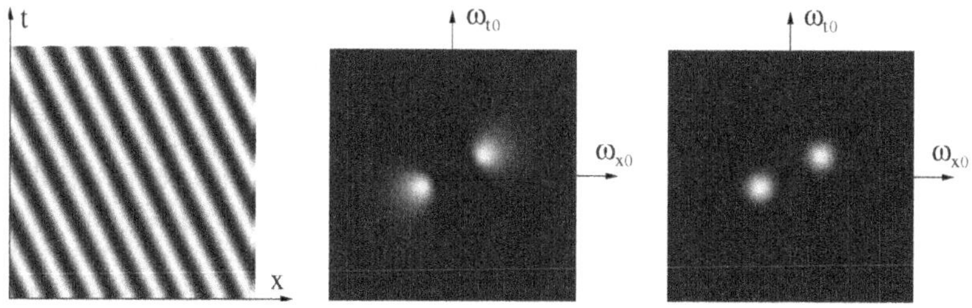

Fig. 4. Left: Cosine sequence $f(x,t) = 5\cos(\frac{\pi}{4}(x+0.5t))$. **Middle:** Mid-spectrum using Gabor wavelets with $C = 3.5$. **Right:** Mid-spectrum using 2D skew Gabor filters with the same C. The skewness in the middle image is corrected

the source components can be identified and separated. Here we do not discuss the traditional Fourier analysis, but focus on the comparison with Gabor wavelets.

If we apply plain Gabor wavelets for filtering, the mid-spectrum is an overlap of two skewness curves (cf. equation (8)). Though iterative algorithms (e.g. [14]) or learning methods (e.g. [15]) may be used to extract the desired parameters, such non-analytic approaches are computationally inefficient and are sensitive to initial values and related parameters in the cost function. Besides, they are susceptible to local minima in the regression procedure. Thus, we prefer to use an analytic framework for parameter regression.

The correction of skewness makes this idea possible. Under the same assumption as that in equation (15), the mid-spectrum of skew Gabor filters is then a sum of two differently weighted and shifted Gaussian functions (for simplicity we omit the coefficient term $\frac{1}{\sqrt{2\pi}\sigma}$ of Gaussian)

$$g(\omega_0) = g_1(\omega_0) + g_2(\omega_0) \tag{16}$$

with

$$\begin{cases} g_1(\omega_0) = a_1 e^{-\frac{(\omega_0-\mu_1)^2}{2(\frac{\mu_1}{C})^2}} \\ g_2(\omega_0) = a_2 e^{-\frac{(\omega_0-\mu_2)^2}{2(\frac{\mu_2}{C})^2}} \end{cases} . \tag{17}$$

The scale parameters in above Gaussians are proportional to the mean values. In figure 5 we demonstrate the mid-spectrum of plain Gabor wavelet filtering as well as the mid-spectrum of skew Gabor filtering.

The sum-of-Gaussians model is well studied from statistic aspect and is widely used in neural network approaches (e.g. [15,14]). One benefit of this model is that we are able to use higher-order moment information to extract parameters. According to Appendix A we obtain the following system of equations

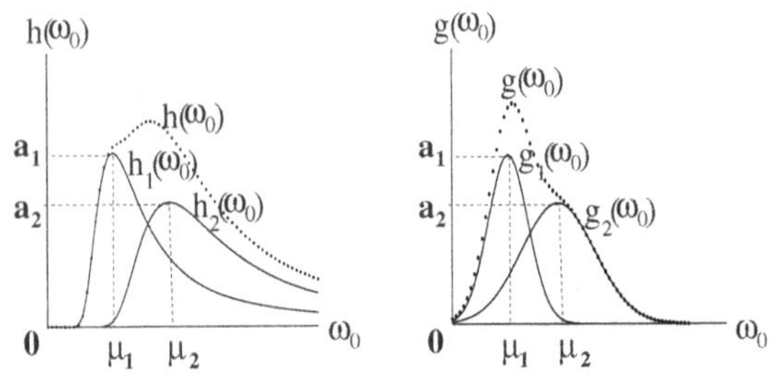

Fig. 5. Left: The mid-spectrum of plain Gabor wavelet filtering. **Right:** The superposition of two Gaussians after 1D skew Gabor filtering. The scale parameters of these two Gaussians are determined by $\frac{\mu_1}{C}$ and $\frac{\mu_2}{C}$, respectively

in a_1, a_2, μ_1, and μ_2

$$\begin{cases} a_1\mu_1 + a_2\mu_2 = m_0\frac{C}{\sqrt{2\pi}} & := b_1 \\ a_1\mu_1^2 + a_2\mu_2^2 = m_1 b_1 & := b_2 \\ a_1\mu_1^3 + a_2\mu_2^3 = \frac{1}{\frac{1}{C^2}+1} m_2 b_1 & := b_3 \\ a_1\mu_1^4 + a_2\mu_2^4 = \frac{1}{\frac{3}{C^2}+1} m_3 b_1 & := b_4 \end{cases} . \tag{18}$$

Here m_0 denotes the integration of $g(\omega_0)$ and m_1, m_2, and m_3 denote the first three order moments of $g(\omega_0)/m_0$. Without loss of generality we assume $0 < \mu_1 \leq \mu_2$. Solving these equations (Appendix B) yields

$$\begin{cases} a_1 = \frac{a(2ab_2+bb_1+b_1\sqrt{b^2-4ac})}{b^2-4ac-b\sqrt{b^2-4ac}} \\ a_2 = \frac{a(2ab_2+bb_1-b_1\sqrt{b^2-4ac})}{b^2-4ac+b\sqrt{b^2-4ac}} \\ \mu_1 = \frac{-b+\sqrt{b^2-4ac}}{2a} \\ \mu_2 = \frac{-b-\sqrt{b^2-4ac}}{2a} \end{cases} , \tag{19}$$

where b_1, b_2, b_3, and b_4 are defined in (18) and the variables a, b, and c are defined in (B.6) (Appendix B). The term $b^2 - 4ac$ is guaranteed to be no less than zero (see Appendix B). If $b^2 - 4ac = 0$, there is only one single Gaussian (i.e. $\mu_1 = \mu_2$) and we can estimate its mean value and amplitude directly using equations (A.1) and (A.2).

In figure 6 we display an example of source signal separation. The input signal is composed of two cosine functions

$$s(x) = 2\cos(\frac{\pi}{4}x) + \cos(\frac{3\pi}{8}x) \tag{20}$$

with the spectrum $S(\omega) = \delta(\omega \pm \frac{\pi}{4}) + \frac{1}{2}\delta(\omega \pm \frac{3\pi}{8})$. Now we sample the positive spectral space with Gabor wavelets and skew Gabor filters. We start the mid-frequency at $\omega_0 = \frac{\pi}{128}$ and increase it with a step of $\frac{\pi}{128}$ to get a dense sampling. Here we set the highest mid-frequency as $\omega_0 = \frac{7\pi}{8}$ so that we do not need to consider the boundaries in the mid-spectrum. Using higher-order moments we

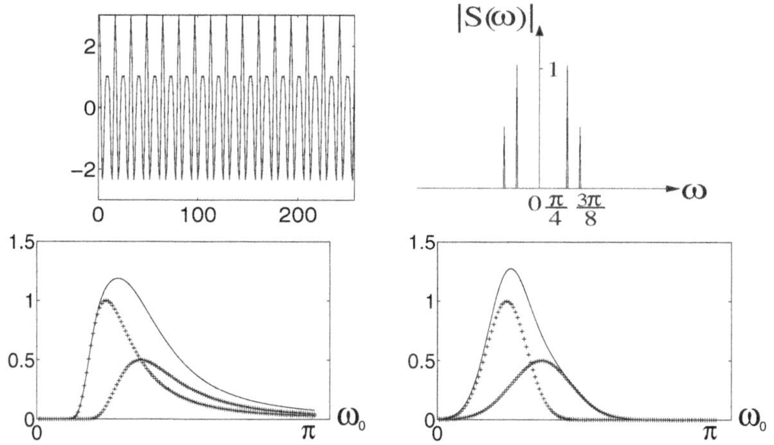

Fig. 6. Top: The source signal and its energy spectrum. **Bottom:** The positive mid-spectra (solid lines) using plain Gabor wavelets (left) and using skew Gabor filters (right). These curves are actually overlapping of the spreading responses of two Dirac functions (shown as crosses)

estimate the amplitudes and the locations of two positive Dirac components

$$\begin{cases} a_1 &= 0.9976 \approx 1 \\ a_2 &= 0.4825 \approx \frac{1}{2} \\ \mu_1 &= 0.8130 \approx \frac{\pi}{4} \\ \mu_2 &= 1.2079 \approx \frac{3\pi}{8} \end{cases} . \tag{21}$$

In the negative frequencies, we may perform a similar procedure to extract the desired parameters. Then, we are able to identify the source signal components in spite of the blurring in the mid-spectrum. In other words, this method can "deblur" the mid-spectrum. Taking into account that a lot of efforts had to be made in filter design so that the blurring after filtering does not significantly affect the identification of signals or orientations (e.g. [16,17]), this framework provides an elegant solution to increases the spectral resolution.

4 Orientation Analysis in 2D Spectral Space

In this section, we analyze the appearance of multiple orientations in 2D spectral space using skew Gabor filters. An important application of this analysis is multiple motion analysis in xt-space. According to [18,19,8], both 1D occlusion and transparency may be modeled as multiple lines in the spectral domain, with some distortion in case of occlusion and without distortion in case of transparency. Thus, the problem of motion estimation turns out to be an issue of orientation analysis in the spatio-temporal space. As the angle between two spectral lines can be arbitrary, eigen-analysis (e.g. [20,21]) cannot properly determine the orientation of multiple lines. Sampling the spectrum with Gabor filters [4] provided a good motivation, but suffered under the limited resolution. Here we prove that this limitation may be overcome using skew Gabor filters.

As the energy spectrum of either occlusion or transparency is mainly a superposition of two spectral lines, the corresponding mid-spectrum after skew Gabor filtering is then the sum of differently weighted 2D Gaussians centered on two spectral lines. Along each spectral line, these Gaussians have the same angular uncertainty due to wavelet property (cf. figure 1). Though the angular distribution $sh_a(\theta)$ of a 2D Gaussian is no more an exact 1D Gaussian, we still can approximate $sh_a(\theta)$ using a Gaussian with appropriate parameters, especially if C is adequately large (e.g. $C \geq 3$). Due to the space limitation we won't delve into the mathematic derivation, but use an example in figure 7 as an intuitive proof. The reader is referred to [22] for more details. After this approximation, all 2D Gaussians centered on the same spectral line have the same angular mean value and the same angular scale parameter σ_a. Consequently, after polar integration [1] we obtain one 1D Gaussian from all 2D Gaussians centered on the same spectral line and the angular distribution of the mid-spectrum is the superposition of two 1D Gaussians. Thus, we are able to extract the exact orientation of the spectral lines from the blurring mid-spectrum using the framework introduced in section 3. The only difference here is that the parameter σ_a is no more proportional to θ_i, but a constant determined by C.

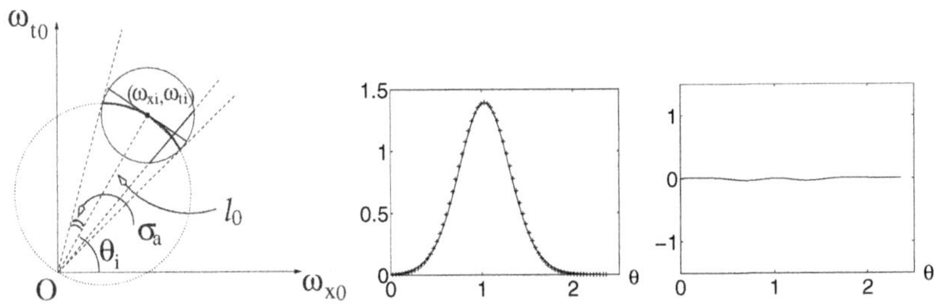

Fig. 7. Left: Polar integration of an isotropic 2D Gaussian centered at $(\omega_{xi}, \omega_{ti})$ can be approximated by an ideal Gaussian function with mean value θ_i and scale parameter σ_a. The solid circle represents the support of the Gaussian. The pencil of lines passing through the origin denotes the integration paths. The middle point of the intersection between a integration path and the solid circle lies on the dotted circle with a diameter l_0. **Middle:** The solid curve is the plot of $sh_a(\theta)$. For comparison we plot an ideal Gaussian with crosses as well. The scale of this Gaussian is $\sigma_a = \sin^{-1}(1/C)$ with $C = 3.5$. **Right:** The maximal difference between the normal Gaussian and $sh_a(\theta)$ is less than 2% of $\max(sh_a(\theta_i))$

[1] This integration is well known as Radon Transform [23].

4.1 Examples

To evaluate the performance of our framework properly, we use synthesized examples. The first example demonstrates the deblurring ability of our framework. We use a 2D signal whose spectrum is composed of two spectral lines passing through the origin (figure 8). The angles between these two lines and the ω_x axis are 15 degrees and 30 degrees, respectively. The mid-spectrum after skew Gabor filtering is strongly blurred due to the spreading effect of filtering and the overlapping of two neighboring Gaussians. The source signal components are hardly to observe in this mid-spectrum. Using higher-order moments, we are still able to determine the orientation of the original spectral lines: $\mu_1 = 13.37°$ and $\mu_2 = 30.07°$. The relative large error in μ_1 is caused by the discrete approximation of the polar integration (e.g. at 0 degree we have more grid points than at 15 degrees). We may reduce this error by increasing the grid density or by interpolation. But we will not enter this topic here.

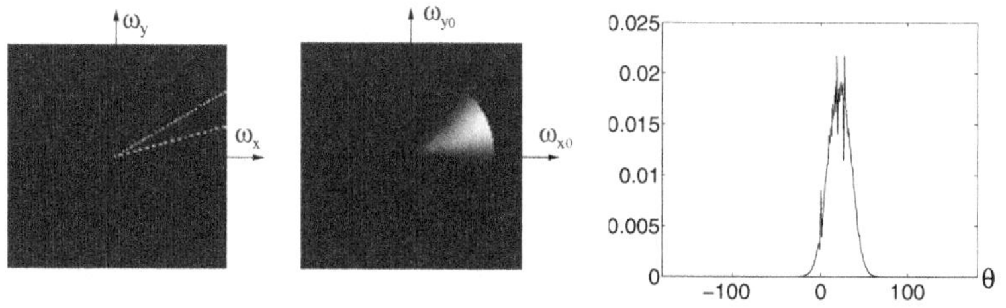

Fig. 8. Left: The spectrum of a 2D signal is composed of two spectral lines passing through the origin with an angle of 15 degrees between them. **Middle:** Mid-spectrum using 2D skew Gabor filters with $C = 6$. The mid-frequency satisfies $\pi/16 \leq \sqrt{\omega_{x0}^2 + \omega_{t0}^2} \leq 3\pi/4$. **Right:** The angular distribution of the mid-spectrum

The second example is to estimate multiple motions in a transparency sequence (figure 9). In this sequence we have one random dot sequence moving at 1.00 [pixel/frame] and one sum-of-cosines sequence moving at 0.40 [pixel/frame], respectively. For clarity of displaying we arrange the maximal amplitude of the cosine sequence to be twice the maximal amplitude of the random dot sequence so that the corresponding spectral lines have the same amplitudes. The spectrum displays the superposition of two motions clearly. The mid-spectrum spreads this distribution. This is clearly to see in the plot after the polar integration. As the spectral lines are symmetric with respect to the origin, we only need one half for estimation. Here we use the higher-order moments in the angular space between 90 degrees and 180 degrees to determine the orientation of spectral lines: $\mu_1 = 134.43°$ and $\mu_2 = 158.62°$. The normal vector of these two lines indicate the velocities: $u_1 = \cot(\mu_1 - 90°) = 1.02$ and $u_2 = \cot(\mu_2 - 90°) = 0.39$.

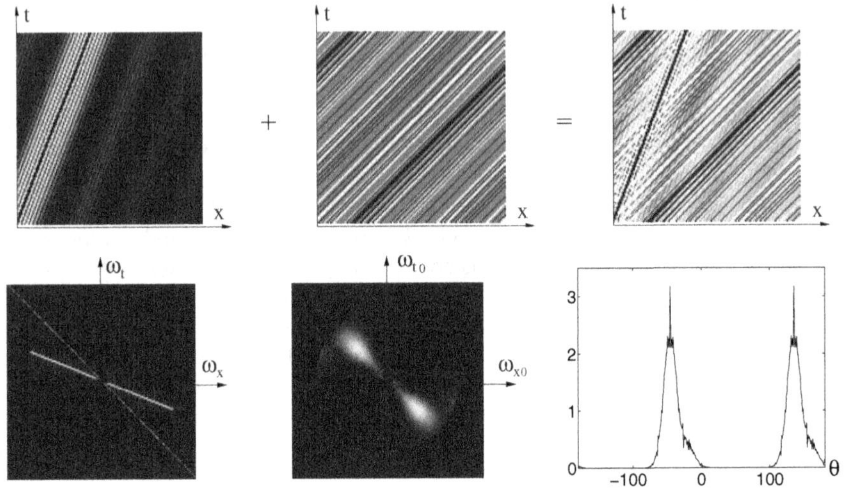

Fig. 9. Top: The transparency sequence (right) $f(x,t) = \sum_{k=1}^{89} A_k \cos(\omega_k(x - 0.4t) + \phi_k) + ran(x - t)$. In the sum-of-cosine sequence (left) ω_k varies from $\pi/16$ to $3\pi/4$ with a step of $\pi/128$. The amplitudes A_k and phase components ϕ_k are randomly chosen. The random dot sequence (middle) $ran(x - t)$ moves with 1 [pixel/frame]. **Bottom Left:** The spectrum of the transparency sequence. **Bottom Middle :** Mid-spectrum using 2D skew Gabor filters with $C = 6$. The middle frequency satisfies $\pi/16 \leq \sqrt{\omega_{x0}^2 + \omega_{t0}^2} \leq 3\pi/4$. **Bottom Right:** The angular distribution of the mid-spectrum

5 Discussions

The skewness correction of Gabor wavelets results in a Gaussian spreading of the input signal in the mid-frequency space. After the correction we are able to model the distribution in the parameter space with a sum-of-Gaussians model. Comparing with the non-symmetric skewness curve, the benefit of using Gaussian functions for distribution description is obvious: Gaussians have good localization ability and are capable of providing simple yet rich descriptions of signals. From the point of view of probabilistic signal processing and pattern recognition, this correction simplifies the tasks of signal analysis significantly. For example, the analytical framework for source signal separation benefits from the statistical simplicity of Gaussians in calculating higher-order moments.

Higher-order moment information is also used in independent component analysis (ICA) approaches [24,25]. In ICA approaches we need a numerical solution (e.g. singular value decomposition (SVD)) because the distribution is unknown. In our framework, however, the sum-of-Gaussians model makes an analytic solution possible. It is also worth mentioning that we need only one superposition of the source signals to separate them (In [25], for example, two linearly independent superposition are needed to separate source signals).

Another point of our source signal separation framework is that most frequency-based methods suffer from low resolution due to spreading and overlapping. By achieving the spreading to have a Gaussian shape, we can separate two overlapping Gaussians in the mid-frequency space. This enables us to reach very fine resolution in the spectral domain and therefore solve the aliasing problem.

In the future work the following points are worth studying:

- Extend the framework to 2D multiple motion analysis, where the source signal itself is a sum of 2D Gaussians.
- Reduce the computational load by using elongated filter masks and by studying how sparsely we can sample the spectrum without affecting the parameter regression.
- Develop efficient estimation algorithms for the spectrum with multiple harmonics (more than two Dirac impulses).
- Study the sensitivity to noise.

Appendix A

For convenience we change the variable in equations (16) and (17) to x and normalize $g_1(x)$, $g_2(x)$, and $g(x)$ to obtain the corresponding distribution density functions $f_1(x)$, $f_2(x)$, and $f(x)$:

$$m_0 = \int g(x)dx = \frac{\sqrt{2\pi}}{C}(a_1\mu_1 + a_2\mu_2), \tag{A.1}$$

$$\begin{cases} f_1(x) = \frac{g_1(x)}{\int g_1(x)dx} = \frac{1}{\frac{\sqrt{2\pi}}{C}\mu_1} e^{-\frac{(x-\mu_1)^2}{2(\frac{\mu_1}{C})^2}} \\ f_2(x) = \frac{g_2(x)}{\int g_2(x)dx} = \frac{1}{\frac{\sqrt{2\pi}}{C}\mu_2} e^{-\frac{(x-\mu_2)^2}{2(\frac{\mu_2}{C})^2}} \\ f(x) = \frac{1}{m_0} g(x) = \frac{1}{a_1\mu_1+a_2\mu_2}[a_1\mu_1 f_1(x) + a_2\mu_2 f_2(x)] \end{cases}.$$

The first three order moments of $f(x)$ read

$$m_1 = \int x f(x)dx = \frac{1}{a_1\mu_1 + a_2\mu_2}(a_1\mu_1^2 + a_2\mu_2^2), \tag{A.2}$$

$$m_2 = \int x^2 f(x)dx = \frac{\frac{1}{C^2}+1}{a_1\mu_1 + a_2\mu_2}(a_1\mu_1^3 + a_2\mu_2^3), \tag{A.3}$$

$$m_3 = \int x^3 f(x)dx = \frac{\frac{3}{C^2}+1}{a_1\mu_1 + a_2\mu_2}(a_1\mu_1^4 + a_2\mu_2^4). \tag{A.4}$$

Reformulate equations (A.1), (A.2), (A.3), and (A.4) yields the equation system (18).

Appendix B

After defining $x_1 = a_1\mu_1$, $x_2 = a_2\mu_2$, we get an equation system of variables x_1, x_2, μ_1, and μ_2 from (18)

$$x_1 + x_2 = b_1, \tag{B.1}$$

$$x_1\mu_1 + x_2\mu_2 = b_2, \tag{B.2}$$

$$x_1\mu_1^2 + x_2\mu_2^2 = b_3, \tag{B.3}$$

$$x_1\mu_1^3 + x_2\mu_2^3 = b_4. \tag{B.4}$$

From (B.1) and (B.2) we obtain

$$x_1(\mu_1 - \mu_2) = b_2 - b_1\mu_2, \tag{B.1-1}$$

$$x_2(\mu_1 - \mu_2) = b_1\mu_1 - b_2. \tag{B.2-1}$$

We multiply both sides of (B.3) and (B.4) with $(\mu_1 - \mu_2)$ and simplify them as

$$(b_2 - b_1\mu_1)\mu_2 = b_3 - b_2\mu_1, \tag{B.3-1}$$

$$(b_2 - b_1\mu_1)\mu_2^2 + (b_2 - b_1\mu_1)\mu_1\mu_2 + b_2\mu_1^2 - b_4 = 0. \tag{B.4-1}$$

Submitting (B.3-1) into (B.4-1) yields

$$a\mu_1^2 + b\mu_1 + c = 0 \tag{B.5}$$

with

$$\begin{cases} a := b_2^2 - b_1 b_3 \\ b := b_1 b_4 - b_2 b_3\,. \\ c := b_3^2 - b_2 b_4 \end{cases} \tag{B.6}$$

This is a standard one variable, two order equation whose discriminator reads

$$\begin{aligned} b^2 - 4ac &= (b_1 b_4 - b_2 b_3)^2 - 4(b_2^2 - b_1 b_3)(b_3^2 - b_2 b_4) \\ &= [a_1 a_2 \mu_1 \mu_2 (\mu_1 - \mu_2)^3]^2 \geq 0 \end{aligned} \tag{B.7}$$

The equality is attainable only when $\mu_1 = \mu_2$, i.e. when we have only one single Gaussian. Then we only need to use (A.1) and (A.2) directly to extract parameters. In case of $b^2 - 4ac > 0$, we have two real roots (without loss of generality we assume $\mu_1 < \mu_2$)

$$\begin{cases} \mu_1 = \frac{-b+\sqrt{b^2-4ac}}{2a} \\ \mu_2 = \frac{-b-\sqrt{b^2-4ac}}{2a} \end{cases}. \tag{B.8}$$

Here $a < 0$ (cf. (B.6)). Submitting μ_1 and μ_2 into (B.1-1) and (B.2-1) and further taking into account that $x_1 = a_1\mu_1$, $x_2 = a_2\mu_2$ we solve a_1 and a_2

$$\begin{cases} a_1 = \frac{a(2ab_2+bb_1+b_1\sqrt{b^2-4ac})}{b^2-4ac-b\sqrt{b^2-4ac}} \\ a_2 = \frac{a(2ab_2+bb_1-b_1\sqrt{b^2-4ac})}{b^2-4ac+b\sqrt{b^2-4ac}} \end{cases}. \tag{B.9}$$

References

1. Gabor, D.: Theory of communication. Journal of the IEE **93** (1946) 429–457 269
2. Lee, T. S.: Image representation using 2D Gabor wavelets. IEEE Trans. Pattern Analysis and Machine Intelligence **18(10)** (1996) 959–971 269, 270, 272
3. Adelson, E. H., Bergen, J. R.: Spatiotemporal energy models for the perception of motion. Journal of the Optical Society of America **1(2)** (1985) 284–299 269
4. Heeger, D. J.: Optical flow using spatiotemporal filters. International Journal of Computer Vision **1(4)** (1987) 279–302 269, 277
5. Daugman, J. G.: Uncertainty relation for resolution in space, spatial frequency and orientation optimized by two-dimensional visual cortical filters. Journal of the Optical Society of America **2(7)** (1985) 1160–1169 269
6. Koenderink, J., Doorn, A. V.: Representation of local geometry in the vision system. Biological Cybernetics **55** (1987) 367–375 269
7. Heitger, F., Rosenthaler, L., der Heydt, R. V., Peterhans, E., Kuebler, O.: Simulation of neural contour mechanisms: from simple to end-stopped cells. Vision Research **32(5)** (1992) 963–981 269
8. Beauchemin, S., Barron, J.: The frequency structure of 1d occluding image signals. IEEE Trans. Pattern Analysis and Machine Intelligence **22** (2000) 200–206 269, 276
9. Grzywacz, N., Yuille, A.: A model for the estimate of local image velocity by cells in the visual cortex. Proc. Royal Society of London. **B 239** (1990) 129–161 269, 270, 271
10. Bovik, A. C., Clark, M., Geisler, W. S.: Multichannel texture analysis using localized spatial filters. IEEE Trans. Pattern Analysis and Machine Intelligence **12(1)** (1990) 55–73 270, 272
11. Bracewell, R. N.: The Fourier Transform and Its Applications. McGraw-Hill Book Company (1986) 271
12. Jain, A., Farrokhnia, F.: Unsupervised texture segmentation using Gabor filters. Pattern Recognition **24(12)** (1991) 1167–1186 272
13. Manjunath, B., Ma, W.: Texture features for browsing and retrieval of image data. IEEE Trans. Pattern Analysis and Machine Intelligence **18(8)** (1996) 837–842 272
14. Poggio, T., Girosi, F.: Networks for approximatation and learning. Proceedings of the IEEE **78** (1990) 1481–1497 274
15. Daugman, J.: Complete discrete 2-d Gabor transforms by neural networks for image analysis and compression. IEEE Trans. Acoustics, Speech, and Signal Processing **36(7)** (1988) 274
16. Simoncelli, E. P., Farid, H.: Steerable wedge filters for local orientation analysis. IEEE Trans. Image Processing **5(9)** (1996) 1377–1382 276
17. Yu, W., Daniilidis, K., Sommer, G.: Approximate orientation steerability based on angular Gaussians. IEEE Trans. Image Processing **10(2)** (2001) 193–205 276

18. Fleet, D. J.: Measurement of Image Velocity. Kluwer Academic Publishers (1992) 276
19. Fleet, D., Langley, K.: Computational analysis of non-Fourier motion. Vision Research **34** (1994) 3057–3079 276
20. Shizawa, M., Mase, K.: A unified computational theory for motion transparency and motion boundaries based on eigenenergy analysis. In: IEEE Conf. Computer Vision and Pattern Recognition, Maui, Hawaii, June 3-6 (1991) 289–295 277
21. Jähne, B.: Spatio-Temporal Image Processing. Springer-Verlag (1993) 277
22. Yu, W., Sommer, G., Daniilidis, K.: Skewness of Gabor wavelets and source signal separation. submitted to IEEE Trans. Signal Processing (2001) 277
23. Radon, J., translated by P. C. Parks: On the determination of functions from their integral values along certain manifolds. IEEE Trans. Medical Imaging **5(4)** (1986) 170–176 277
24. Cardoso, J.: Source separation using higher order moments. In: IEEE International Conf. on Acoustics, Speech and Signal Processing. (1989) 2109–2112 279
25. Farid, H., Adelson, E. H.: Separating reflections from images using independent components analysis. Journal of the Optical Society of America **16(9)** (1999) 2136–2145 279

m.maccallum@napier.ac.uk
a.almaini@napier.ac.uk

μ

- δ
- θ
- α
- β

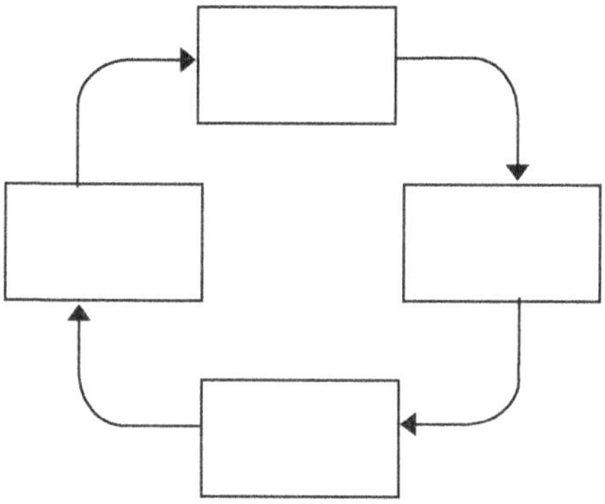

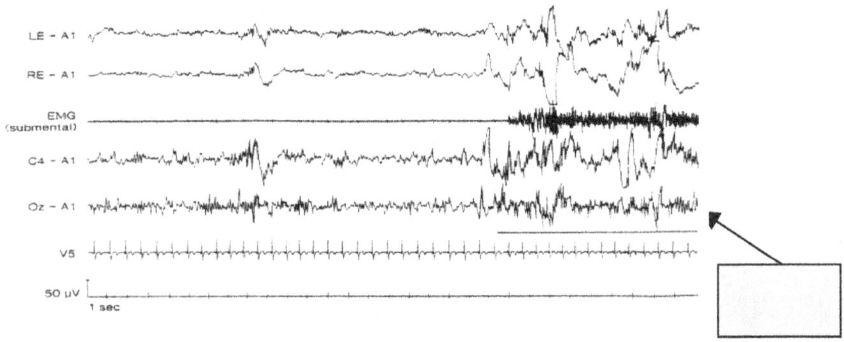

LE – A1
RE – A1
EMG
(submental)
C4 – A1
Oz – A1
V5
50 µV
1 sec

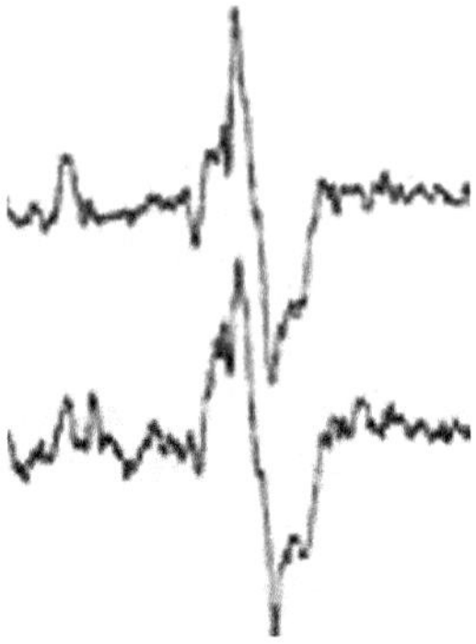

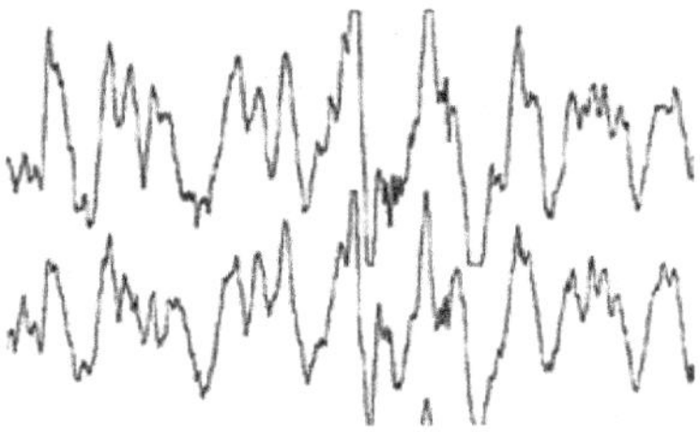

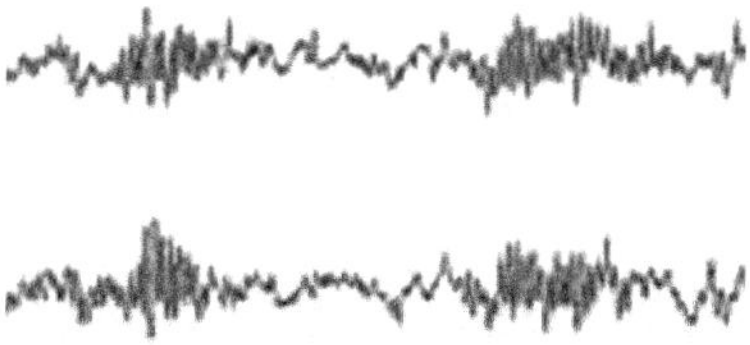

δ

β

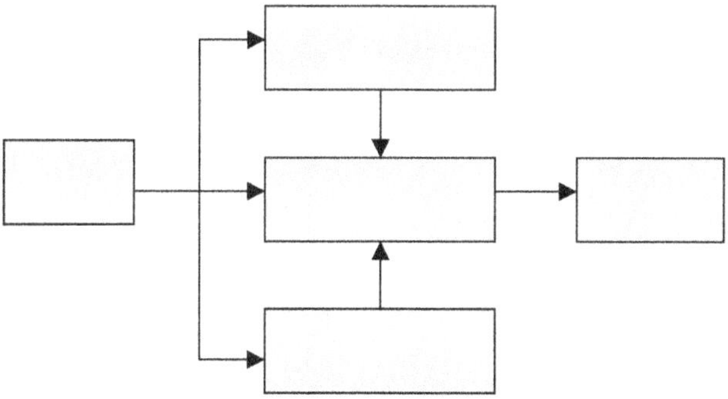

- μ
- δ

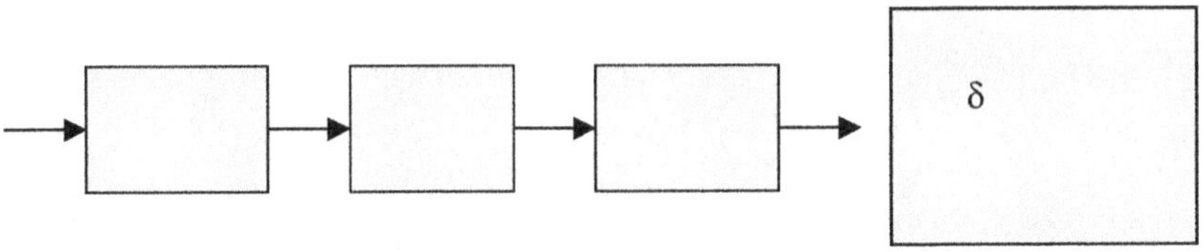

- μ
- μ

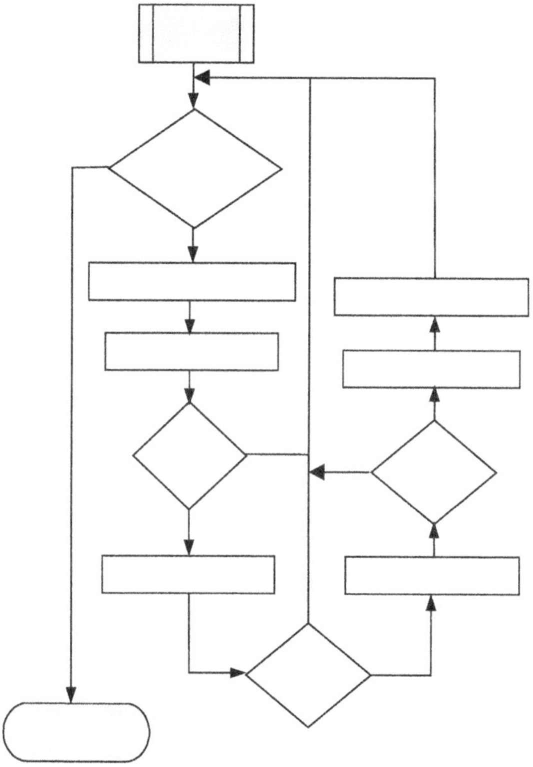

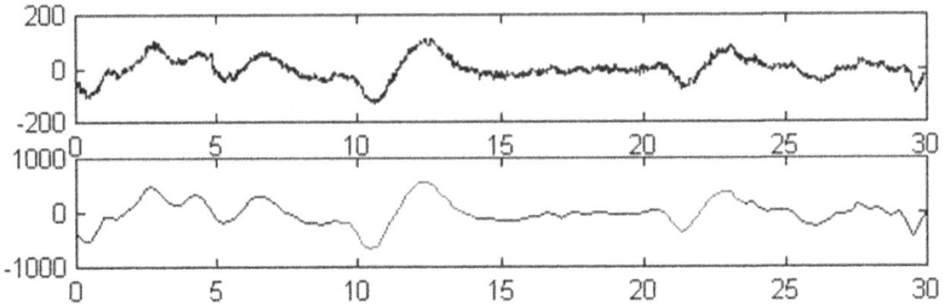
200
0
-200
0 5 10 15 20 25 30
1000
0
-1000
0 5 10 15 20 25 30

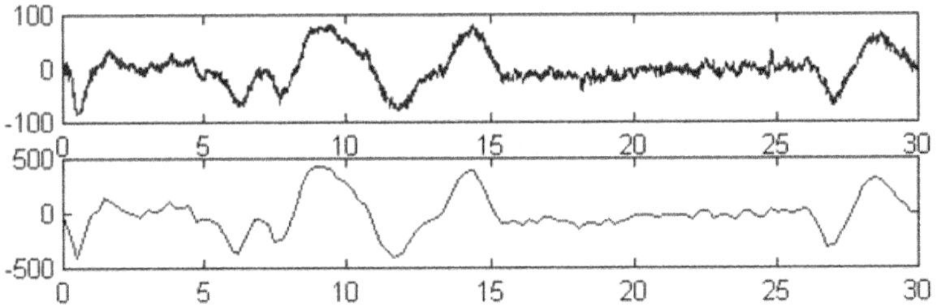
100
0
-100
0 5 10 15 20 25 30
500
0
-500
0 5 10 15 20 25 30

θ α

]

ϕ

ψ

]

ϕ ψ

$< \quad <$

$$= \frac{\sigma}{\quad} \,|\,| \quad + |\,| \quad -| - |$$

$$= \sigma \; |\,|$$

$$\omega \;= \frac{\sigma}{|\omega| \quad +}$$

$$= \quad * \frac{\quad}{\pi} = \int_{-\infty}^{\infty} \frac{\tau}{\pi \quad - \tau} \quad \tau$$

$$(\quad) = {}^{-} \quad (\quad)$$

$$(\quad) = {}^{-} \quad (\quad) \quad >$$

$$(\quad) = \frac{}{\pi \quad -} = {}^{-} \left[\quad \right]$$

$$\left[\quad \right] = \frac{}{\pi \quad -} \quad \frac{}{\pi \quad -}$$

$$= {}^{-} \left[\quad \right]$$

σ

σ

$$\left[\quad \right] = \frac{\sigma}{} \left[|\tau| + |\tau| - |\tau - \tau| \right] \frac{}{\pi - \tau} \frac{}{\pi - \tau} \tau \quad \tau$$

$$= {}^{-}/ \sum_{=-\infty}^{\infty} [\ ] \varphi({}^{-} \ - \ ) + \sum_{=-\infty} / \sum_{=-\infty}^{\infty} [\ ] \phi \ {}^{-} \ -$$

ϕ φ ϕ

$$= {}^{-}/ \int_{-\infty}^{\infty} \phi({}^{-} \ - \ ) \qquad = {}^{-}/ \int_{-\infty}^{\infty} \varphi \ {}^{-} \ -$$

$\in$

$\in$

ϕ ϕ ϕ

$$\phi \quad = \phi \quad * \frac{}{\pi}$$

ϕ

$$\phi \quad = \Phi \; \omega \cdot \quad \omega$$

$$\frac{\left| \quad \phi \quad \right|}{|\omega|} \; \omega = \frac{|\Phi \;\, \omega|}{|\omega|} \; \omega < \infty$$

ϕ

$$\psi \quad \tau = {}^{-}/ \psi \; {}^{-} \; \tau - \quad = \quad {}^{-}/ \phi \; {}^{-} \; - \quad \cdot \frac{}{\pi \quad - \tau}$$

$$= - \; {}^{-}/ \phi \; {}^{-} \; \tau - \quad * \frac{}{\pi\tau}$$

$$= \quad {}^{-}/ \psi \; {}^{-} \; \tau -$$

$$\psi \; \tau - \quad = -\phi \; \tau - \quad * \frac{}{\pi\tau}$$

$$= {}^{-}/ \int_{-\infty}^{\infty} \quad \psi\left({}^{-} \quad - \right) \quad \in$$

$$= \big[\; [\; + \;] \; [\;]\big] = \frac{\sigma}{} - \quad {}_{\psi} \quad \tau - \; |\tau| \quad \tau \quad {}^{+}$$

$$_{\psi}(\;\; \tau) = \;\; \psi \;\;\; \psi \;\;\; - \tau$$

$=$ $\qquad$ $[\;]$

$$[\;] = \;\; (\;\; [\;]) = \frac{\sigma}{} \;\; _{\psi}(\;\;)(\;\;) \;\; ^{+} = \sigma \;\;\; ^{\gamma}$$

$$_{\psi} \;\;\; = - \;\;\; _{\psi} \;\; \tau \, |\tau| \;\; \tau$$

$(\;)$ $\qquad$ $(\;)$

ψ $\qquad$ ϕ

$(\;)$ $\qquad$ $(\;)$

ψ

$(\;)$

$(\;)$

$$= \; ^{-}/ \; ^{/ \; -} \;\; _{=} \;\; \varphi \; ^{-} \;\; - \;\; + \;\; _{=} \; ^{-}/ \; ^{/ \; -} \;\; _{=} \;\; \phi \; ^{-} \;\; - \qquad = \qquad -$$

$$= \qquad +$$

σ

$$= \qquad +$$

σ

$[\ ]=\ (\ )$

$[\ ]=\ [\ ]+\ [\ ]\ =$

$[\ ]\qquad[\ ]$

$\in\qquad=\qquad\sigma$

$(\ )=\ \ /\ \ /\ \ [\ ]\phi(\ -\ -\ )\qquad=\qquad-$

$=\qquad\leq$

$>$

$\sigma\ =\qquad-\qquad=\qquad\sigma$

$=\ \sigma\qquad\sigma\ <$

$\sigma\ \geq$

$= \sigma \quad {}^{\gamma}$

$\leq \frac{}{\gamma} \qquad \frac{\sigma}{\sigma} \qquad \sigma \;\geq \qquad\qquad = \frac{}{\gamma} \qquad \frac{\sigma}{\sigma}$

$= \qquad\qquad \begin{matrix} \leq \\ > \end{matrix}$

$\gamma =$

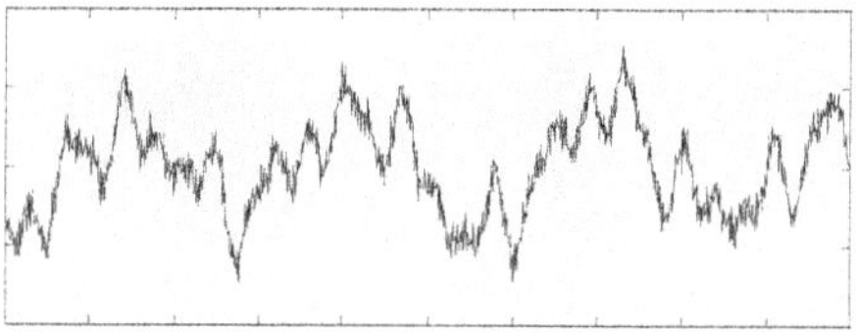

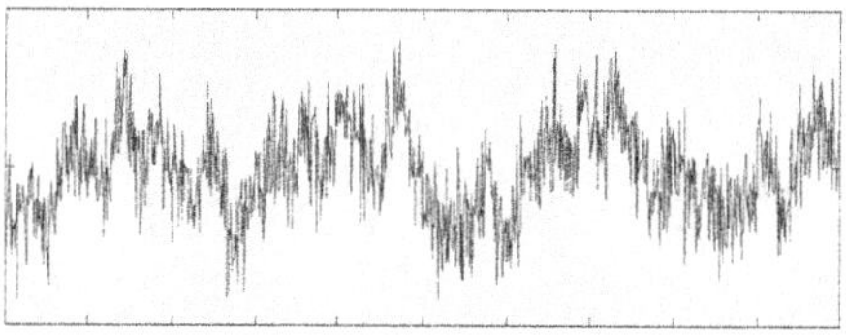

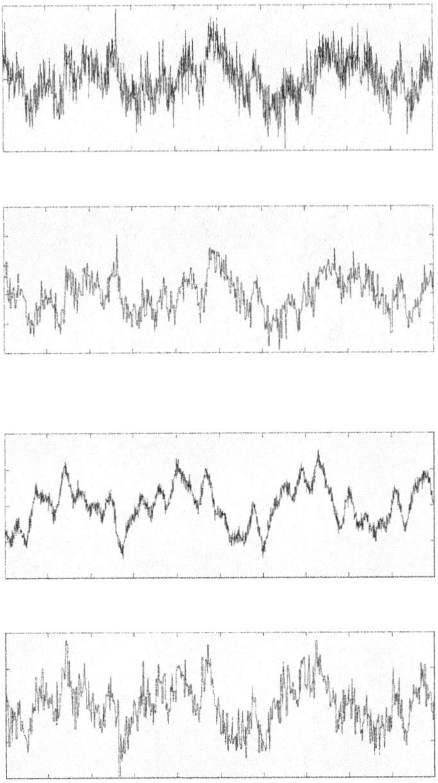

]

ϕ ψ

$$= \quad +$$

$$\sigma$$

$$= \quad + \quad = \quad \cdots$$

$$\in$$

σ $\in$

$$= \phi \quad - \quad + \quad = \phi \quad - \quad +$$

$$\in$$

$$- \quad = \quad - \quad - \quad \cdots \quad - \quad \phi = \phi \quad \phi \quad \cdots \phi$$

$$\phi =$$

$$\sigma = \quad -$$

$$= \quad \cdots$$

$$\_ = \begin{matrix} & & \cdots & - \\ & & \cdots & - \\ \vdots & \vdots & \vdots & \vdots \\ - & - & \cdots & \end{matrix}$$

$$= \quad - \quad +$$

$$= \quad +$$

$$= \begin{matrix} \phi & \phi & \cdots & \phi \\ & & \cdots & \\ \vdots & \vdots & \vdots & \vdots \\ & \cdots & & \end{matrix} \qquad = \cdots \qquad = \cdots$$

$\in$

$$= \quad - \quad \cdots \quad - \quad +$$

$$= \quad - \quad -$$

$$= \cdots \qquad = \phi \quad - \quad + \frac{}{+\sigma} \quad -\phi \quad -$$

$$= \phi \quad - \quad - \qquad = \phi \quad \_\phi \quad + \sigma$$

$\sigma \quad = \quad - \quad = \quad \cdots \quad \cdots$

$-$

$- \quad - \quad >$

$$- = \quad - \quad - \; + \frac{}{+\sigma} \quad -\phi \quad -$$

$$= \; \cdots \; -$$

$$= \; \phi \quad - \; - \quad - \; - \quad -$$

$$= \; \cdots \; \cdots \quad {}_{-}\phi$$

$\cdots \quad \cdots$

$$= \quad - \; + \quad -\phi \quad -$$

$+ \quad -$

$$= \quad + \; - \quad = \; \cdots \quad + \; -$$

$$\sigma \quad = \quad - \quad = \quad \cdots \quad {}_{+ \; -} \quad \cdots$$

$$\sigma \quad - \sigma \quad = \frac{+}{+ \quad + \; + \sigma} > \quad {}_{= \cdots -}$$

$-$

$$_{+} = \quad _{-} + \frac{\qquad}{+\sigma} \quad -\phi \quad -$$

$$- \frac{\qquad}{+\sigma} \quad - \quad - - \quad - \quad \phi \quad - -$$

$$_{+} \quad _{-} \quad +$$

$$= \quad \overset{-}{+} \quad -$$

$$-\phi \quad - \quad = \quad +\sigma$$

$$- \quad - - \quad - \quad \phi \quad - - \quad =$$

$$_{+} = \quad _{-} - \frac{\qquad}{+\sigma} = \cdots -$$

$$-$$

$$= \quad +$$

$$= \qquad = \frac{\qquad}{+\sigma}$$

$$\sigma \quad = \quad - \quad = \frac{\sigma}{+\sigma}$$

$$- \quad = — \quad \sigma \quad + \quad \sigma$$

$$= \qquad -$$

$\cdots \quad -$

$\cdots \quad = \quad -$

σ

$=$

$= \quad =$

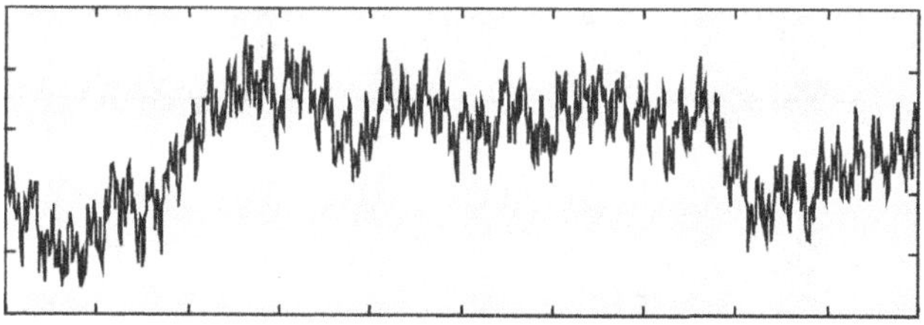

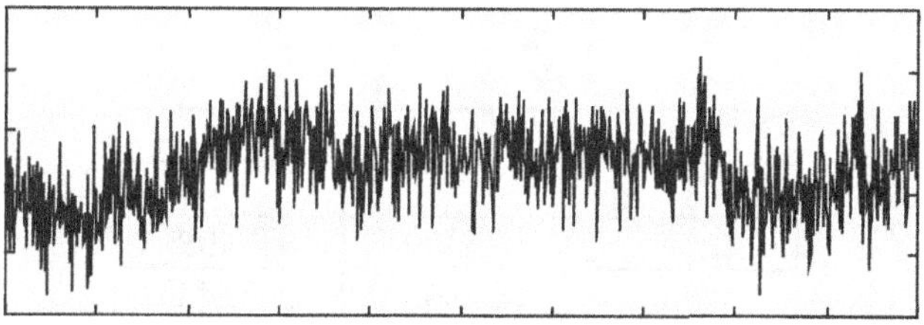

$=$

$\geq$

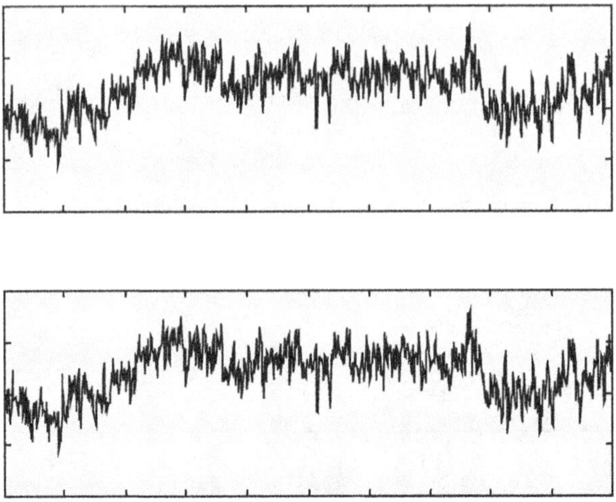

≥

σ

= =

=

H		
0.55	0.5499	0.5498
0.60	0.5551	0.5543
0.65	0.5627	0.5606
0.70	0.5726	0.5683
0.75	0.5849	0.5770
0.80	0.5994	0.5862
0.85	0.6161	0.5954
0.90	0.6345	0.6040
0.95	0.6544	0.6115

H		
0.55	0.4610	0.4550
0.60	0.4520	0.4454
0.65	0.4442	0.4370
0.70	0.4374	0.4297
0.75	0.4315	0.4231
0.80	0.4262	0.4173
0.85	0.4216	0.4122
0.90	0.4176	0.4076
0.95	0.4141	0.4035

p=1 J=11	0.5879	p=2 J=3	0.5805
p=2 J=11	0.5795	p=3 J=3	0.5782
p=3 J=11	0.5772	p=4 J=3	0.5770

jpli2222@sina.com, jpli2222@yahoo.com

yytang@comp.hkbu.edu.hk

$\times$

$-$

$=$

$=$

α

α α

α $\frac{\pi}{}$

α

$\alpha\beta$

α β $\frac{\pi}{}$

α β α β α β α β

α $\frac{\pi}{}$ β $-\frac{\pi}{}$

= −

= −

α β

α β

α β α β α β α β α β α β

α β

α β

α β

α β

α β α β α β α β α β α β

α β γ α β γ

α β γ θ α β γ θ

α β γ α β γ

α β γ θ α β γ θ

α α α α α

α = − 时,

α α α α α α

α α

α α

α α

α α α $=$ $-$

α α α α α α

α α α $=$ $-$

α α α α α α

π

$=$ $-$

$=$ $-$

α β γ α β γ

α β γ α β γ

α β γ α β γ

α β γ α β γ α β γ α β γ α β γ

α β γ

α β γ α β γ

α β γ α β γ α β γ α β γ

α β γ

α β γ α β γ α β γ

α β γ

α β γ

α β γ α β γ

α β γ α β γ

α β γ π

α β γ α γ β α β γ

α β γ α γ β α β γ

α π β π γ π

γ

α α α α

α α α α α α

α α α α α α α

α

α α

α α

$\sum C(N)$ and $\sum S(N)$

$$\sum C(N) = \alpha \quad \alpha \quad \alpha$$

$$\sum S(N) = \alpha \quad \alpha \quad \alpha$$

α α α

α α α α α α

α α α

$\underline{\pi}$ α α α

α α α

A Design of Automatic Speech Playing System Based on Wavelet Transform*

Yishu Liu, Jinyu Cen, Qian Sun, and Lihua Yang

Department of Scientific Computing and Computer Applications
Zhongshan University, Guangzhou 510275, P. R. China

Abstract. This paper introduces a novel approach to store speech words after cutting the signals and decomposing them through Mallat's decompostion algorithm, and generate a speech phrase by connecting such word data and reconstructing it through Mallat's reconstruction algorithm. This way, speech signals of good quality can be produced easily from a small library of compressed speech words.

Keywords: speech signal processing, wavelet basis, Mallat's algorithm.

1 Introduction

Automatic Speech Playing plays an important role in speech signal processing. It is widely applied in many modern business automatic processing, such as electronic business, ATA (automatic time announcing) system, banks' ATM system, IP telephone card service, etc.. An automatic speech playing system consists of a speech library and a speech connecting algorithm. Generally speaking, a speech library must be as small as possible and it must be guaranteed that the library has great ability in generating every kind of phrases and sentences. In the light of different practical application, accordingly we can choose the most basic speech units (in Chinese they are Chinese words' speech data) to build an speech library. In order to reduce the volume of the library, its data can be compressed provided the result is acceptable. In this paper, we extract single Chinese word speech's main part, which is then decomposed twice by wavelet decomposition algorithm. An approximation of the original speech obtained this way is made an element of the basic speech library. The library so built is very small. In phrase/sentence generating, the basic speech units are put together in proper order, then form a playable speech signal through wavelet reconstruction algorithm. Wavelet reconstruction makes the signals smoother, so that the audio effect is good. And this algorithm is potent in phrase/sentence generating.

As a significant breakthrough of Fourier Analysis, wavelet has attached much attention in many fields from applied mathematics to signal processing. The idea of multi-resolution analysis underlying wavelet theory makes it possible to get the signals at different scales whose lengths are reduced by half successively. By

* This work was supported by the Foundation for University Key Teacher by the Ministry of Education of China, NSFC(19871095) and GPNSFC(990227).

Y. Y. Tang et al. (Eds.): WAA 2001, LNCS 2251, pp. 321–325, 2001.

neglecting the details, we can get an approximation of the signal, whose length is greatly shortened but whose main characteristics remain. Some basic facts on wavelets will be stated in the next section.

A textbook-like reference on speech signal processing can be found in [4].

2 Some Basic Facts on Wavelets

Wavelet analysis has become an effective mathematical tool to process signals locally on time and frequency, which was developed more than ten years ago. With a wavelet basis, the space $L^2(R)$ can be decomposed into the orthogonal sum of a sequence of closed spaces. That means, any signal with finite energy can be expressed as the sum of a series of local frequency structures. Such a decomposition is convenient for signal compression and smoothing. In this section, some basic facts on wavelet theory used in this paper are stated without proofs. A textbook-like introduction on wavelet theory can be found in [1,2,3].

Let $L^2(R)$ be the space of all the finite energy signals, i.e.,

$$L^2(R) = \left\{ f(t) \ : \ \int_{-\infty}^{\infty} |f(t)|^2 dt < \infty \right\}.$$

The well-known Mutliresolution Analysis (MRA) is defined as follows.

Definition 1 *Let $\phi(x) \in L^2(R)$. The sequence of closed subspaces of $L^2(R)$ which are defined by*

$$V_j = \{\phi_{j,k}(x) = 2^{j/2}\phi(2^j x - k), \ k \in Z\}, \quad (j \in Z),$$

is called an orthonormal Mutliresolution Analysis (MRA) of $L^2(R)$ if the following there conditions are satisfied:

1) $V_j \subseteq V_{j+1}, \quad (\forall j \in Z)$;

2) $\overline{\bigcup_{j\in Z} V_j} = L^2(R), \quad \bigcap_{j\in Z} V_j = \{0\}$;

3) $\{\phi(t-k)\}_{k\in Z}$ is an orthonormal basis of V_0.

Then $\phi(t)$ is said to be the corresponding scaling funtion of the MRA.

It can be easily derived that $\{\phi_{j,k}(t) = 2^{j/2}\phi(2^j t - k)\}_{k\in Z}$ is an orthonormal basis of V_j.

For any $j \in Z$, we let $V_j = V_{j-1} \oplus W_{j-1}$, where W_{j-1} is the orthogonal complement of V_{j-1} in V_j. We can find a wavelet function $\psi(t) \in W_0$ such that $\{\psi_{j,k}(t) = 2^{j/2}\psi(2^j t - k)\}_{k\in Z}$ is an orthonormal basis of W_j.

Therefore, for any $j \in Z$, we have $L^2(R) = V_j \oplus W_j \oplus W_{j+1} \oplus W_{j+2} \oplus \ldots\ldots$ It can thus be inferred that $\forall f(t) \in L^2(R), \ f(t) = A_j + \sum_{n>j} D_n$, where

$$\begin{cases} A_j = \sum_n a_n(j) \cdot \phi_{j,n} \,, & a_n(j) = \langle f, \phi_{j,n} \rangle \\ D_j = \sum_n d_n(j) \cdot \psi_{j,n} \,, & d_n(j) = \langle f, \psi_{j,n} \rangle \end{cases}$$

A_j is said to be the approximation of f(t) at 2^j scale, it reflects the main information of f(x). And D_j is said to be the details of f(t) between scales 2^{j+1} and 2^j.

From $V_j = V_{j-1} \oplus W_{j-1}$, we know that $\phi_{j-1,0}$ and $\psi_{j-1,0}$ can be expanded in $\{\varphi_{j,n}\}_{n\in Z}$ to obtain scale equations as follows:

$$\phi_{j-1,0}(t) = \sum_n h(n)\phi_{j,n}(t) \ , \quad \psi_{j-1,0}(t) = \sum_n g(n)\phi_{j,n}(t) \ ,$$

where

$$h(n) = \langle 2^{-\frac{1}{2}}\phi(\frac{t}{2}), \phi(t-n)\rangle \ , \quad g(n) = \langle 2^{-\frac{1}{2}}\psi(\frac{t}{2}), \phi(t-n)\rangle = (-1)^{1-n}h(1-n).$$

And therefore, S. Mallat's decomposition and reconstruction algorithms are obtained:

Algorithm 2

Decomposition:

$$\begin{cases} a_k(j-1) = \sum\limits_n h(n-2k)a_n(j) \\ d_k(j-1) = \sum\limits_n g(n-2k)a_n(j) \end{cases}, \quad (k \in Z).$$

Reconstruction:

$$a_k^j = \sum_n h(k-2n)a_n(j-1) + \sum_n g(k-2n)d_n(j-1), \quad (k \in Z).$$

3 The Construction of Basic Speech Library

The authors of the paper recorded some speech words. After our surveying and analyzing their wave forms in MatLab, a conclusion was arrived at: the waves are all composed of smooth parts and steep ones. Such an example is shown in the left of Fig.1. It is the speech data of Chinese pronunciation "jiu" of "nine".

In order to extract the signal's main information (the steep part), an algorithm is presented as follows:

1. For each speech signal $f_{[i]}^{(l)}$, where l=0,1,2,......and i=0, 1, 2,..., $N_l - 1$ (N_l is an integer), do
   (a) Divide the interval $[0, N_l - 1]$ equally into M parts;
   (b) Calculate num[j], j=0, 1, 2,..., M-1, where num[j] denotes the number of $f_{[i]}^{(l)}$ which satisfies: i belongs to the j'th part and $f_{[i]}^{(l)} \geq \varepsilon$;
   (c) Find the maximum among num[j], j=0, 1,..., M-1. Suppose it is num[maxj]. To select j's around maxj such that $num[j] \geq \delta \times num[maxj]$ and all the selected j's are adjoining. Suppose they are $j_0, j_1, ..., j_k$.
   (d) Let=$g_{[i]}^{(l)} = f_{[i+\frac{N}{M}j_0]}^{(l)}$, i=0, 1, 2,..., $\frac{N}{M}(j_k - j_0 + 1) - 1$.
2. Put all $g^{(l)}$ together in proper order to form a new signal.

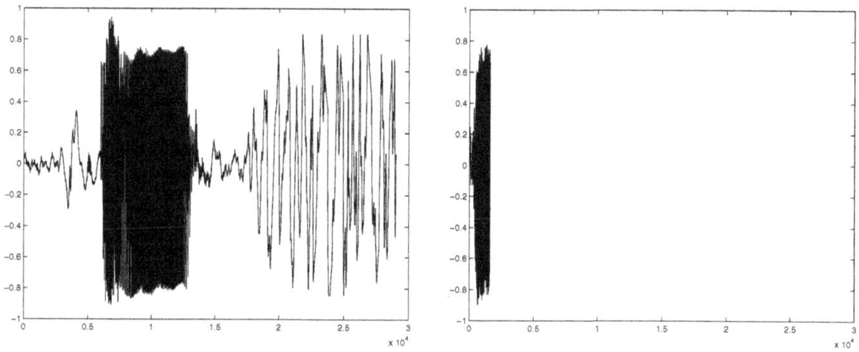

Fig. 1. Left: speech signal of "jiu", Right: the extracted and decomposed signal

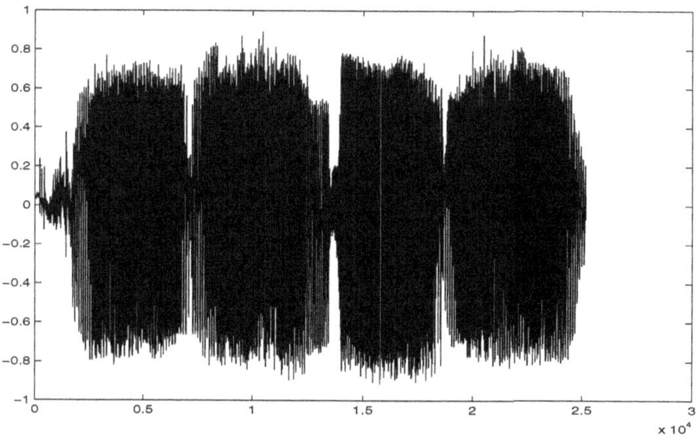

Fig. 2. The connected speech signal

The signal extracted still has some redundance. To save speech data efficiently, we further decompose it twice with Mallat's decomposition algorithm with respect to Daubechies 10. Then the result is stored into the library as a basic element. The right of Fig.1 is an example for Chinese word "jiu".

4 Connection and Reconstruction of Speech Signals

When the system is assigned to play a phrase such as the Chinese phrase of nine dollars and two cents: "jiu yuan er jiao", four words will be chosen to connect the phrase which is then reconstructed by Mallat's algorithm with respect to Daubechies 10. It can be shown in mathematics that such a reconstruction can smooth the connection points of a signal to obtain a better audio effect.

The experiment is shown in Fig. 2. It has been played and shown to be audio acceptable.

References

1. C. K. Chui. *An Introduction to Wavelets.* Academic Press, Boston, 1992. 322
2. I. Daubechies. *Ten Lectures on Wavelets.* Society for Industrial and Applied Mathemathics, Philadelphia, 1992. 322
3. S. Mallat. *Wavelet Tour of Signal Processing.* Academic Press, San Diego, USA, 2nd edition, 1998. 322
4. L. Rabiner and B. H. Juang. *Fundamentals of Speech Recognition.* Prentice-Hall International, Inc., 1993. 322

General Design of Wavelet High-Pass Filters from Reconstructional Symbol*

Lihua Yang[1], Qiuhui Chen[1], and Yuan Y. Tang[2]

[1] Department of Scientific Computing and Computer Applications
Zhongshan University, Guangzhou 510275, P. R. China
[2] Department of Computer Science, Hong Kong Baptist University

Abstract. For given reconstructional low-pass filters, the general solutions of matrix equation $\tilde{M}(\xi)M^*(\xi) = I$ for the construction of orthogonal or biorthogonal wavelet filter banks are presented.
Keywords: matrix equation, MRA, wavelets, filter.

1 Introduction

In the theory of wavelet Analysis, it is well-known that the key to construct an orthogonal wavelet base or a pair of biorthogonal wavelet bases from MRA (Multiresolution Analysis) is to design the filter banks $\{\tilde{m}_\mu(\xi), m_\mu(\xi) \ |\mu \in E_d\}$,with E_d being the set of all the vertices of $[0,1]^d$, such that

$$\tilde{M}(\xi) := (\tilde{m}_\mu(\xi + \pi\nu))_{\mu,\nu \in E_d}, \quad M(\xi) := (m_\mu(\xi + \pi\nu))_{\mu,\nu \in E_d}, \tag{1}$$

satisfy:

$$M^*(\xi)\tilde{M}(\xi) = \tilde{M}(\xi)M^*(\xi) = I_{2^d} \quad a.e.\ \xi \in T^d. \tag{2}$$

Usually, the question on the solutions of (2) can be described as follows:

Question 1 Assume $m_0(\xi)$, the filter function of a MRA, is given. We are needed to construct $\tilde{m}_0(\xi), m_\mu(\xi)$, $\tilde{m}_\mu(\xi) \in L^\infty(T)$ $(\mu \in E_d\backslash\{0\})$ such that (2) holds.

Question 2 Assume $m_0(\xi)$ and $\tilde{m}_0(\xi)$, the filter functions of a pair of biorthogonal MRAs, are given. We need to construct $m_\mu(\xi)$, $\tilde{m}_\mu(\xi) \in L^\infty(T)$ $(\mu \in E_d\backslash\{0\})$ such that (2) holds.

It is essentially the problem of matrix extension which can be solved by constructing a or a pair of particular solution(s), or constructing all the possible solutions. Up till now, many results have been developed on the construction of wavelet bases from MRAs mainly by constructing a or a pair of particular solution(s) of (2)(see [1,2,6,3]). In this paper, we present all analytic solutions of (2) based a special solution. We also design an algorithm to get a special solution for the matrix equation (2) and illustrate some examples to verify our results.

---

* This work was supported by the Foundation for University Key Teacher by the Ministry of Education of China, NSFC(19871095) and GPNSFC(990227).

Y. Y. Tang et al. (Eds.): WAA 2001, LNCS 2251, pp. 326–330, 2001.

2 Main Results

2.1 The general solutions of matrix equation (2)

We first restate the **Question 1 and 2** by polyphase factorization of filters. We call respectively $m_{\mu,\tau}(\xi)$, $\tilde{m}_{\mu,\tau}(\xi)$ the polyphase components of $m_\mu(\xi)$, $\tilde{m}_\mu(\xi)$ if

$$m_\mu(\xi) = \sum_{\tau\in E_d} m_{\mu,\tau}(2\xi)e^{-i\tau\xi} \quad \text{and} \quad \tilde{m}_\mu(\xi) = \sum_{\tau\in E_d} \tilde{m}_{\mu,\tau}(2\xi)e^{-i\tau\xi}.$$

Let $(\nu_0, \cdots, \nu_s)$ be a permutation of E_d with $s = 2^d - 1$ and $\nu_0 = 0 \in E_d$. For simplicity, we denote $\tilde{\mathbf{m}}(\xi) := (\tilde{m}_{\mu,\nu}(\xi))_{\mu,\nu\in E_d}$, $\mathbf{m}(\xi) := (m_{\mu,\nu}(\xi))_{\mu,\nu\in E_d}$. Then the filters m_μ, $\tilde{m}_\mu$ and their polyphase components have the following relations

$$M(\xi) = \mathbf{m}(2\xi)E(\xi), \quad \tilde{M}(\xi) = \tilde{\mathbf{m}}(2\xi)E(\xi) \tag{3}$$

with Vandemonde matrix $E(\xi)$ defined by $E(\xi) = \left(e^{-i\nu_j\cdot(\xi+\pi\nu_k)}\right)_{j,k=0}^{s}$. It is easy to conclude that (2) leads to

$$2^d\tilde{\mathbf{m}}(\xi)\mathbf{m}^*(\xi) = I, \quad a.e.\ \xi \in T^d. \tag{4}$$

It is an well-known that constructing $\tilde{M}(\xi), M(\xi)$ satisfying (2) is equivalent to constructing $2\pi Z^d$-periodic matrices $\tilde{\mathbf{m}}(\xi)$, $\mathbf{m}(\xi)$ satisfying (4). An equivalent discription of **Question 1,2** is stated as follows:

Question 1 Assume the first row of a matrix $\mathbf{m}(\xi)$ is given and satisfies $\sum_{\nu\in E_d}|m_{0\nu}(\xi)|^2 \neq 0$, $a.e.\ \xi \in T^d$. We are needed to construct the other rows and the $2\pi Z^d-$ periodic matix $\tilde{\mathbf{m}}(\xi)$ such that (4) holds.

Question 2 Assume the first rows of matrixes $\mathbf{m}(\xi)$ and $\tilde{\mathbf{m}}(\xi)$ are given and satisfy:

$$2^d \sum_{\nu\in E_d} \tilde{m}_{0,\nu}(\xi)\bar{m}_{0,\nu}(\xi) = 1, \quad a.e.\ \xi \in T^d.$$

We are needed to construct the other rows such that (4) holds.

Let e_0 be the column vector whose first element equals to 1 and others being 0s. The following notations will be used in the following theorem.

$$\Delta_d(\xi) := \sum_{\nu\in E^d} |m_0(\xi/2 + \pi\nu)|^2 = 2^d \sum_{\nu\in E^d} |m_{0,\nu}(\xi)|^2\ ; \tag{5}$$

$$\tilde{\Delta}_d(\xi) := \sum_{\nu\in E^d} |\tilde{m}_0(\xi/2 + \pi\nu)|^2 = 2^d \sum_{\nu\in E^d} |\tilde{m}_{0,\nu}(\xi)|^2\ . \tag{6}$$

$$U(\xi)e_0 = \frac{2^{d/2}}{\sqrt{\Delta_d(\xi)}}\mathbf{m}^t(\xi)e_0 \qquad a.e.\ \xi \in T^d. \tag{7}$$

$$\tilde{U}(\xi)e_0 = \frac{2^{d/2}}{\sqrt{\tilde{\Delta}_d(\xi)}}\tilde{\mathbf{m}}^t(\xi)e_0 \qquad a.e. \xi \in T^d \tag{8}$$

Theorem 1. *Suppose that $2\pi Z^d$-periodic functions $\{m_{0,\nu}(\xi)\}_{\nu\in E^d}$ satisfy $\sum_{\nu\in E_d}|m_{0,\nu}(\xi)|^2\neq 0$ a.e. $\xi\in T^d$ and $U(\xi)$ is a $2\pi Z^d$-periodic unitary matrix satisfying (7). Then the $2\pi Z^d$ -periodic solutions $\mathbf{m}(\xi)$ and $\tilde{\mathbf{m}}(\xi)$ of (4) can be expressed as:*

$$\begin{cases} \mathbf{m}(\xi)=\begin{pmatrix} m_{0,0}(\xi) & \cdots & m_{0,\nu_s}(\xi) \\ (c(\xi), & A^t(\xi))U^t(\xi) & \end{pmatrix}, & a.e.\ \xi\in T^d; \\ \tilde{\mathbf{m}}(\xi)=\frac{2^{-d/2}}{\sqrt{\Delta_d(\xi)}}\begin{pmatrix} 1 & -\overline{c^t(\xi)A^{-1}(\xi)} \\ 0_{(2^d-1)\times 1} & 2^{-d/2}\sqrt{\Delta_d(\xi)}\ \overline{A^{-1}}(\xi) \end{pmatrix}U^t(\xi), & a.e.\ \xi\in T^d. \end{cases} \tag{9}$$

If there exist $2\pi Z^d$-periodic functions $\{\tilde{m}_{0,\nu}(\xi)\}_{\nu\in E^d}$ satisfying $2^d\sum_{\nu\in E_d}\tilde{m}_{0,\nu}(\xi)\bar{m}_{0,\nu}(\xi)=1$ a.e. $\xi\in T^d$, then the solution of (4) with the first rows of $\mathbf{m}(\xi)$ and $\tilde{\mathbf{m}}(\xi)$ being $(m_{0,\nu}(\xi))_{\nu\in E_d}$ and $(\tilde{m}_{0,\nu}(\xi))_{\nu\in E_d}$ respectively can be expressed as follows:

$$\begin{cases} \mathbf{m}(\xi)=\begin{pmatrix} m_{0,0}(\xi) & \cdots & m_{0,\nu_s}(\xi) \\ (c(\xi), & A^t(\xi))\,U^t(\xi) & \end{pmatrix}, & a.e.\ \xi\in T^d; \\ \tilde{\mathbf{m}}(\xi)=\begin{pmatrix} \tilde{m}_{0,0}(\xi) & \cdots & \tilde{m}_{0,\nu_s}(\xi) \\ (0_{(2^d-1)\times 1}, & 2^{-d}\overline{A^{-1}}(\xi))U^t(\xi) & \end{pmatrix} & a.e.\ \xi\in T^d, \end{cases} \tag{10}$$

where $A(\xi)$ is a $2\pi Z^d-$periodic nonsingular matrix of order 2^d-1 and

$$c(\xi)=2^{d/2}\sqrt{\Delta_d(\xi)}\ [0,\ A^t(\xi)]U^t(\xi)\begin{pmatrix} \bar{\tilde{m}}_{0,0}(\xi) \\ \vdots \\ \bar{\tilde{m}}_{0,\nu_s}(\xi) \end{pmatrix}.$$

Furthermore, if $\tilde{U}(\xi)$ is a $2\pi Z^d$-periodic unitary matrix satisfying (8), then $\mathbf{m}(\xi)$ can be rewritten as follows:

$$\mathbf{m}(\xi)=\begin{pmatrix} m_{0,0}(\xi) & \cdots & m_{0,\nu_s}(\xi) \\ (0, & A^t(\xi)L(\xi))\,\tilde{U}^t(\xi) & \end{pmatrix}. \tag{11}$$

where

$$L(\xi)=(0,\ I_{2^d-1})\,U^t(\xi)\left(I_{2^d}-\sqrt{\Delta_d(\xi)\tilde{\Delta}_d(\xi)}\bar{\tilde{U}}(\xi)e_0e_0^tU^t(\xi)\right)\bar{\tilde{U}}(\xi)\begin{pmatrix} 0 \\ I_{2^d-1} \end{pmatrix} \tag{12}$$

is a nonsingular matrix of order 2^d-1. Particularly, if $\tilde{m}_{0,\nu}(\xi)=m_{0,\nu}(\xi)$ a.e. $\xi\in T^d$ $(\forall\nu\in E_d)$, we have $L(\xi)=I_{2^d-1}$.

2.2 The Construction of Unitary Matrix $U(\xi)$

This section focuses on the construction of the $2\pi Z^d$- periodic unitary matrix $U(\xi)$ which satisfies (7) for a given nonzero vector $\mathbf{m}_0(\xi):=(m_{0,0}(\xi),\cdots,m_{0,\nu_s}(\xi))^t$. For simplicity, we denote

$$\|\mathbf{m}_0(\xi)\|:=\sqrt{\sum_{\nu\in E_d}|m_{0\nu}(\xi)|^2}=2^{-d/2}\sqrt{\Delta_d(\xi)}. \tag{13}$$

Theorem 2. *Let* $\|\mathbf{m}_0(\xi)\| \neq 0$ *and* $\overline{m_{00}(\xi)} = |m_{00}(\xi)|e^{-i\theta(\xi)}$ *a.e.* $\xi \in T$. *Then,* $U(\xi)$, *which is defined by*

$$U(\xi) = \frac{1}{\|\mathbf{m}_0(\xi)\|}\begin{pmatrix} m_{00}(\xi) & \overline{m_{0\nu_1}(\xi)} & \cdots & \overline{m_{0\nu_s}(\xi)} \\ m_{0\nu_1}(\xi) & & & \\ \vdots & & M_1(\xi) & \\ m_{0\nu_s}(\xi) & & & \end{pmatrix}, \tag{14}$$

with

$$M_1(\xi) = \frac{e^{-i\theta(\xi)}}{\|\mathbf{m}_0(\xi)\| + |m_{00}(\xi)|}\begin{pmatrix} m_{0\nu_1}(\xi)\overline{m_{0\nu_1}(\xi)} & \cdots & m_{0\nu_1}(\xi)\overline{m_{0\nu_s}(\xi)} \\ \vdots & & \vdots \\ m_{0\nu_s}(\xi)\overline{m_{0\nu_1}(\xi)} & \cdots & m_{0\nu_s}(\xi)\overline{m_{0\nu_s}(\xi)} \end{pmatrix}$$
$$-\|\mathbf{m}_0(\xi)\|e^{-i\theta(\xi)}I_{2^d-1},$$

is a unitary matrix satisfying (7). Furthermore, If there exists constant $c > 0$ *such that* $\|\mathbf{m}_0(\xi)\| > c$ $(\forall \xi \in T^d)$, *then* $U(\xi)$ *is smooth as* $\mathbf{m}_0(\xi)$ *and* $\theta(\xi)$.

2.3 Examples

For simplicity, we denote $x := e^{-i\xi_1}, y := e^{-i\xi_2}$. It can be verified that the following polynomials

$$m_0(\xi_1, \xi_2) = \frac{1}{8}(1, x, x^2, x^3)\begin{pmatrix} 1 & 1 & 1 & -1 \\ 1 & -1 & 1 & 1 \\ 1 & 1 & -1 & 1 \\ -1 & 1 & 1 & 1 \end{pmatrix}\begin{pmatrix} 1 \\ y \\ y^2 \\ y^3 \end{pmatrix}$$

satisfies

$$|m_0(\xi_1, \xi_2)|^2 + |m_0(\xi_1 + \pi, \xi_2)|^2 + |m_0(\xi_1, \xi_2 + \pi)|^2 + |m_0(\xi_1 + \pi, \xi_2 + \pi)|^2 = 1.$$

The unitary matrix in Theorem 2.2 is

$$U(\xi_1, \xi_2) :=$$
$$\frac{1}{4}\begin{pmatrix} 1+x+y-xy & 1-x+y+xy & 1+x-y+xy & -1+x+y+xy \\ 1-x+y+xy & 1+x+y-xy & 1-x-y-xy & -1-x+y-xy \\ 1+x-y+xy & 1-x-y-xy & 1+x+y-xy & -1+x-y-xy \\ -1+x+y+xy & -1-x+y-xy & -1+x-y-xy & 1+x+y-xy \end{pmatrix}$$

which leads to the following three high-pass filters

$$\begin{cases} m_1(\xi_1,\xi_2) = \frac{1}{8}(1,x,x^2,x^3) \begin{pmatrix} 1 & 1 & 1 & -1 \\ 1 & -1 & 1 & 1 \\ -1 & -1 & 1 & -1 \\ 1 & -1 & -1 & -1 \end{pmatrix} \begin{pmatrix} 1 \\ y \\ y^2 \\ y^3 \end{pmatrix} \\ m_2(\xi_1,\xi_2) = \frac{1}{8}(1,x,x^2,x^3) \begin{pmatrix} 1 & 1 & -1 & 1 \\ 1 & -1 & -1 & -1 \\ 1 & 1 & 1 & -1 \\ -1 & 1 & -1 & -1 \end{pmatrix} \begin{pmatrix} 1 \\ y \\ y^2 \\ y^3 \end{pmatrix} \\ m_3(\xi_1,\xi_2) = \frac{1}{8}(1,x,x^2,x^3) \begin{pmatrix} -1 & -1 & 1 & -1 \\ -1 & 1 & 1 & 1 \\ 1 & 1 & 1 & -1 \\ -1 & 1 & -1 & -1 \end{pmatrix} \begin{pmatrix} 1 \\ y \\ y^2 \\ y^3 \end{pmatrix} \end{cases}$$

Since there are different choices for the unitary matrix $A(x,y)$, we can also construct other high-pass filter for the given low-pass filter $m_0(\xi_1,\xi_2)$. For example, for the unitary matrix

$$A(x,y) = \begin{pmatrix} \frac{x}{\sqrt{2}} & 0 & \frac{y}{\sqrt{2}} \\ 0 & xy & 0 \\ \frac{x}{\sqrt{2}} & 0 & -\frac{y}{\sqrt{2}} \end{pmatrix},$$

we can get another three high-pass filters corresponding to the low-pass filter $m_0(\xi_1,\xi_2)$. We omit the details here.

References

1. A. COHEN, I. DAUBECHIES, AND J. C. FEAUVEAU, *Biorthogonal bases of compactly supported wavelets,* Comm. Pure Appl. Math., 45:485-560,1992. 326
2. K. GROCHNING, *Analyse multi-echelle et bases d'ondelettes,* Acad. Sci. Paris., Serie 1,305:13-17,1987. 326
3. R. Q. JIA AND C. A. MICCHELLI, *Using the refinement equation for the construction of pre-wavelets V:extensibility of trigonometric polynomials,* Computing, Vol. 48, 61-72,1992. 326
4. D. X. Zhou, Construction of real-valued wavelets by symmetry, preprint.
5. S. Mallat, Review of Multifrequency Channel Decomposition of Images and Wavelet Models, Technical report 412, Robotics Report 178, New York Univ., (1988).
6. C. A. MICCHELLI AND YUESHENG XU, *Reconstruction and decomposition algorithms for biorthogonal multiwavelets,* Multidimensional Systems and Signal Processing 8, 31-69,1997. 326

Realization of Perfect Reconstruction Non-uniform Filter Banks via a Tree Structure

Wing-kuen Ling and Peter Kwong-Shun Tam

Department of Electronic and Information Engineering
The Hong Kong Polytechnic University
Hung Hom, Kowloon, Hong Kong
Hong Kong Special Administrative Region, China
Tel: (852) 2766-6238, Fax: (852) 2362-8439
Email: bingo@encserver.eie.polyu.edu.hk

Abstract. It is well known that a tree structure filter bank can be realized via a non-uniform filter bank, and perfect reconstruction is achieved if and only if each branch of the tree structure can provide perfect reconstruction. In this paper, the converse of this problem is studied. We show that a perfect reconstruction non-uniform filter bank with decimation ratio $\{2,4,4\}$ can be realized via a tree structure and each branch of the tree structure achieves perfect reconstruction.

1 Introduction

It is well known that the tree structure filter bank shown in figure 1b can be realized via a non-uniform filter bank shown in figure 1a, and perfect reconstruction can be achieved if and only if each branch of the tree structure can provide perfect reconstruction [1-4]. However, is the converse true? That is, given any perfect reconstruction non-uniform filter shown in figure 1a, can it be realized via a tree structure shown in figure 1b? In general, a perfect reconstruction non-uniform filter bank cannot be realized by a tree structure [11]. This paper works on this problem.

There are some advantages of realizing a non-uniform filter bank via a tree structure, such as reducing the filter length in the filters [5], and improving the computation complexity and implementation speed [5]. In section II, we show how a perfect reconstruction non-uniform filter bank can be converted to a tree structure filter bank. Some illustrative examples are demonstrated in section III. Finally, a conclusion is given in section IV.

2 Realization of Non-uniform Filter Bank Via a Tree Structure

Theorem 1

A non-uniform filter bank with decimation ratio $\{2,4,4\}$ achieves perfect reconstruction if and only if it can be realized via a tree structure and each branch of the tree structure achieves perfect reconstruction.

Proof:

Since the if part was well known [1-4], we only prove the only if part.

A non-uniform filter bank shown in figure 1a achieves perfect reconstruction if and only if $\exists c \in \mathbf{C}$ and $\exists m \in \mathbf{Z}$ such that:

Y. Y. Tang et al. (Eds.): WAA 2001, LNCS 2251, pp. 331-335, 2001.

$$\left(\begin{bmatrix} \frac{1}{2}\cdot H_0(z) & \frac{1}{4}\cdot H_1(z) & \frac{1}{4}\cdot H_2(z) \\ 0 & \frac{1}{4}\cdot H_1(z\cdot W) & \frac{1}{4}\cdot H_2(z\cdot W) \\ \frac{1}{2}\cdot H_0(z\cdot W^2) & \frac{1}{4}\cdot H_1(z\cdot W^2) & \frac{1}{4}\cdot H_2(z\cdot W^2) \\ 0 & \frac{1}{4}\cdot H_1(z\cdot W^3) & \frac{1}{4}\cdot H_2(z\cdot W^3) \end{bmatrix}\right)\cdot\begin{pmatrix} G_0(z) \\ G_1(z) \\ G_2(z) \end{pmatrix} = \begin{pmatrix} c\cdot z^m \\ 0 \\ 0 \\ 0 \end{pmatrix}. \tag{1}$$

This directly implies that:

$$\det\left(\begin{bmatrix} H_1(z\cdot W) & H_2(z\cdot W) \\ H_1(z\cdot W^3) & H_2(z\cdot W^3) \end{bmatrix}\right)=0. \tag{2}$$

Since

$$\det\left(\begin{bmatrix} H_1(z\cdot W) & H_2(z\cdot W) \\ H_1(z\cdot W^3) & H_2(z\cdot W^3) \end{bmatrix}\right)=0 \Rightarrow \det\left(\begin{bmatrix} H_1(z) & H_2(z) \\ H_1(z\cdot W^2) & H_2(z\cdot W^2) \end{bmatrix}\right)=0, \tag{3}$$

hence

$$\left(\begin{bmatrix} \frac{1}{2}\cdot H_0(z) & \frac{1}{4}\cdot H_1(z) & \frac{1}{4}\cdot H_2(z) \\ 0 & \frac{1}{4}\cdot H_1(z\cdot W) & \frac{1}{4}\cdot H_2(z\cdot W) \\ \frac{1}{2}\cdot H_0(z\cdot W^2) & \frac{1}{4}\cdot H_1(z\cdot W^2) & \frac{1}{4}\cdot H_2(z\cdot W^2) \\ 0 & \frac{1}{4}\cdot H_1(z\cdot W^3) & \frac{1}{4}\cdot H_2(z\cdot W^3) \end{bmatrix}\right)\cdot\begin{pmatrix} G_0(z) \\ G_1(z) \\ G_2(z) \end{pmatrix} = \begin{pmatrix} c\cdot z^m \\ 0 \\ 0 \\ 0 \end{pmatrix}$$

$$\Rightarrow \det\left(\begin{bmatrix} H_1(z) & H_2(z) \\ H_1(z\cdot W) & H_2(z\cdot W) \end{bmatrix}\right)\neq 0 \text{ and } \det\left(\begin{bmatrix} H_0(z) & H_2(z) \\ H_0(z\cdot W^2) & H_2(z\cdot W^2) \end{bmatrix}\right)\neq 0. \tag{4}$$

The converse is also true. That is if:

$$\det\left(\begin{bmatrix} H_1(z) & H_2(z) \\ H_1(z\cdot W) & H_2(z\cdot W) \end{bmatrix}\right)\neq 0,\ \det\left(\begin{bmatrix} H_0(z) & H_2(z) \\ H_0(z\cdot W^2) & H_2(z\cdot W^2) \end{bmatrix}\right)\neq 0 \text{ and } \det\left(\begin{bmatrix} H_1(z\cdot W) & H_2(z\cdot W) \\ H_1(z\cdot W^3) & H_2(z\cdot W^3) \end{bmatrix}\right)=0,$$

then there exist $G_0(z)$, $G_1(z)$ and $G_2(z)$ such that:

$$\left(\begin{bmatrix} \frac{1}{2}\cdot H_0(z) & \frac{1}{4}\cdot H_1(z) & \frac{1}{4}\cdot H_2(z) \\ 0 & \frac{1}{4}\cdot H_1(z\cdot W) & \frac{1}{4}\cdot H_2(z\cdot W) \\ \frac{1}{2}\cdot H_0(z\cdot W^2) & \frac{1}{4}\cdot H_1(z\cdot W^2) & \frac{1}{4}\cdot H_2(z\cdot W^2) \\ 0 & \frac{1}{4}\cdot H_1(z\cdot W^3) & \frac{1}{4}\cdot H_2(z\cdot W^3) \end{bmatrix}\right)\cdot\begin{pmatrix} G_0(z) \\ G_1(z) \\ G_2(z) \end{pmatrix} = \begin{pmatrix} c\cdot z^m \\ 0 \\ 0 \\ 0 \end{pmatrix}. \tag{5}$$

Let $H_i(z)=\sum_{l=0}^{3} z^{-l}\cdot E_{i,l}(z^4)$, for $i=0,1,2$, then from equation (2), there exist $R(z)$, $R'(z)$ and $R''(z)$ such that:

$$E_{1,1}(z^4)=R(z^4)\cdot E_{1,0}(z^4) \text{ and } E_{2,1}(z^4)=R(z^4)\cdot E_{2,0}(z^4) \text{ and} \tag{6}$$

$$E_{1,3}(z^4)=R'(z^4)\cdot E_{1,2}(z^4) \text{ and } E_{2,3}(z^4)=R'(z^4)\cdot E_{2,2}(z^4) \text{ and} \tag{7}$$

$$\{R(z^4)=R'(z^4) \text{ or } \{E_{1,0}(z^4)=R''(z^4)\cdot E_{1,2}(z^4) \text{ and } E_{2,0}(z^4)=R''(z^4)\cdot E_{2,2}(z^4)\}\}. \tag{8}$$

But $E_{1,0}(z^4)=R''(z^4)\cdot E_{1,2}(z^4)$ and $E_{2,0}(z^4)=R''(z^4)\cdot E_{2,2}(z^4)$ contradict equation (4). Hence, we have $R(z^4)=R'(z^4)$. $R(z^4)=R'(z^4)$, which implies:

$$\frac{H_2(z)}{H_1(z)}=\frac{E_{2,0}(z^4)+z^{-2}\cdot E_{2,2}(z^4)}{E_{1,0}(z^4)+z^{-2}\cdot E_{1,2}(z^4)}. \tag{9}$$

Hence, there exist $F'_1(z)$, $F_0(z)$, and $F_1(z)$ such that $H_1(z)=F'_1(z)\cdot F_0(z^2)$ and $H_2(z)=F'_1(z)\cdot F_1(z^2)$, respectively. And the non-uniform filter bank shown in figure 1a can be realized via a tree structure shown in figure 1b.

From equation (4), we have:

$$E_{1,2}(z^4)\cdot E_{2,0}(z^4)-E_{1,0}(z^4)\cdot E_{2,2}(z^4)\neq 0 \text{ and} \tag{10}$$

$$\{E_{0,1}(z^4)\cdot E_{2,0}(z^4)-E_{0,0}(z^4)\cdot E_{2,1}(z^4)\neq 0 \text{ or} \tag{11}$$

$$E_{0,3}(z^4)\cdot E_{2,2}(z^4)-E_{0,2}(z^4)\cdot E_{2,3}(z^4)\neq 0 \text{ or} \tag{12}$$

$$E_{0,1}(z^4)\cdot E_{2,2}(z^4)+E_{0,3}(z^4)\cdot E_{2,0}(z^4)-E_{0,0}(z^4)\cdot E_{2,3}(z^4)-E_{0,2}(z^4)\cdot E_{2,1}(z^4)\neq 0\}. \tag{13}$$

Let $F_1(z^2)$ be the numerator of $H_2(z)/H_1(z)$, and $F_0(z^2)$ be the denominator of $H_2(z)/H_1(z)$, respectively. We have:

$$\det\left(\begin{bmatrix} F_0(z) & F_1(z) \\ F_0(-z) & F_1(-z) \end{bmatrix}\right)=2\cdot z^{-1}\cdot\left(E_{1,2}(z^2)\cdot E_{2,0}(z^2)-E_{1,0}(z^2)\cdot E_{2,2}(z^2)\right)\neq 0, \text{ and} \tag{14}$$

$$\begin{aligned}\det\left(\begin{bmatrix} H_0(z) & F_1'(z) \\ H_0(-z) & F_1'(-z) \end{bmatrix}\right)&=\frac{2\cdot z^{-1}\cdot\left(E_{0,1}(z^4)\cdot E_{2,0}(z^4)-E_{0,0}(z^4)\cdot E_{2,1}(z^4)\right)+2\cdot z^{-5}\cdot\left(E_{0,3}(z^4)\cdot E_{2,2}(z^4)-E_{0,2}(z^4)\cdot E_{2,3}(z^4)\right)}{E_{2,0}(z^4)+z^{-2}\cdot E_{2,2}(z^4)}\\ &+\frac{2\cdot z^{-3}\cdot\left(E_{0,1}(z^4)\cdot E_{2,2}(z^4)+E_{0,3}(z^4)\cdot E_{2,0}(z^4)-E_{0,0}(z^4)\cdot E_{2,3}(z^4)-E_{0,2}(z^4)\cdot E_{2,1}(z^4)\right)}{E_{2,0}(z^4)+z^{-2}\cdot E_{2,2}(z^4)}\\ &\neq 0.\end{aligned} \tag{15}$$

Hence, each branch of the tree structure achieves perfect reconstruction. ■

3 Illustrative Examples

3.1 Non-tree Structure Filter Bank

Consider an example of $H_0(z)=1+z^{-1}$, $H_1(z)=1-z^{-1}$, and $H_2(z)=z^{-2}$, respectively. Since $H_1(z)/H_2(z)=z^2\cdot(1-z^{-1})$, there does not exist $F'_1(z)$, $F_0(z)$, and $F_1(z)$ such that $H_1(z)=F'_1(z)\cdot F_0(z^2)$ and $H_2(z)=F'_1(z)\cdot F_1(z^2)$, respectively. Hence, this non-uniform filter bank cannot be realized via a tree structure. By theorem 1, this non-uniform filter bank does not achieve perfect reconstruction.

It is worth to note that by converting the non-uniform filter bank to a uniform filter bank shown in figure 2 [6-10], perfect reconstruction can be achieved. However, $G'_1(z)\neq z^{-2}G'_0(z)$, this implies that the corresponding synthesis filter $G_0(z)$ shown in figure 1a is time varying.

3.2 Tree Structure Filter Bank

Consider another example with $H_0(z)=2\cdot(1+z^{-1}+z^{-2}+z^{-3})$, $H_1(z)=4\cdot(2+6z^{-1}+4z^{-2}+12z^{-3})$, and $H_2(z)=4\cdot(5+15z^{-1}+7z^{-2}+21z^{-3})$, respectively. Since $H_1(z)/H_2(z)=(2+4z^{-2})/(5+7z^{-2})$, there exists $F'_1(z)$, $F_0(z)$, and $F_1(z)$ such that $H_1(z)=F'_1(z)\cdot F_0(z^2)$ and $H_2(z)=F'_1(z)\cdot F_1(z^2)$, respectively. Hence, this non-uniform filter bank can be realized via a tree structure. It can be checked easily that each branch in the tree structure achieves perfect reconstruction. Hence, this non-uniform filter bank achieves perfect reconstruction.

4 Conclusion

In this paper, we show that a non-uniform filter bank with decimation ratio {*2,4,4*} achieves perfect reconstruction if and only if it can be realized via a tree structure and each branch of the tree structure achieves perfect reconstruction. The advantage of realizing a non-uniform filter bank via a tree structure is to reduce the computation complexity and provide a fast implementation for a non-uniform filter bank [5].

Acknowledgement

The work described in this paper was substantially supported by a grant from the Hong Kong Polytechnic University with account number G-V968.

References

1. Vaidyanathan P. P.: Lossless Systems in Wavelet Transforms. IEEE International Symposium on Circuits and Systems, ISCAS, Vol. 1. (1991) 116-119.
2. Soman K. and Vaidyanathan P. P.: Paraunitary Filter Banks and Wavelet Packets. IEEE International Conference on Acoustics, Speech, and Signal Processing, ICASSP, Vol. 4. (1992) 397-400.
3. Sodagar I., Nayebi K. and Barnwell T. P.: A Class of Time-Varying Wavelet Transforms. IEEE International Conference on Acoustics, Speech, and Signal Processing, ICASSP, Vol. 3. (1993) 201-204.
4. Soman A. K. and Vaidyanathan P. P.: On Orthonormal Wavelets and Paraunitary Filter Banks. IEEE Transactions on Signal Processing, Vol. 41, No. 3. (1993) 1170-1183.
5. Vaidyanathan P. P.: Multirate Systems and Filter Banks. Englewood Cliffs, NJ: Prentice Hall, 1993.
6. Hoang P. Q. and Vaidyanathan P. P.: Non-Uniform Multirate Filter Banks: Theory and Design. IEEE International Symposium on Circuits and Systems, ISCAS, Vol. 1. (1989) 371-374.
7. Li J., Nguyen T. Q. and Tantaratana S.: A Simple Design Method for Nonuniform Multirate Filter Banks. Conference Record of the Twenty-Eight Asilomar Conference on Signals, Systems and Computers, Vol. 2. (1995) 1015-1019.
8. Makur A.: BOT s Based on Nonuniform Filter Banks. IEEE Transactions on Signal Processing, Vol. 44, No. 8. (1996) 1971-1981.
9. Li J., Nguyen T. Q. and Tantaratana S.: A Simple Design Method for Near-Perfect-Reconstruction Nonuniform Filter Banks. IEEE Transactions on Signal Processing, Vol. 45, No. 8. (1997) 2105-2109.
10. Omiya N., Nagai T., Ikehara M. and Takahashi S. I.: Organization of Optimal Nonuniform Lapped Biorthogonal Transforms Based on Coding Efficiency. IEEE International Conference on Image Processing, ICIP, Vol. 1. (1999) 624-627.
11. Akkarakaran S. and Vaidyanathan P. P.: New Results and Open Problems on Nonuniform Filter-Banks. IEEE International Conference on Acoustics, Speech, and Signal Processing, ICASSP, Vol. 3. (1999) 1501-1504.

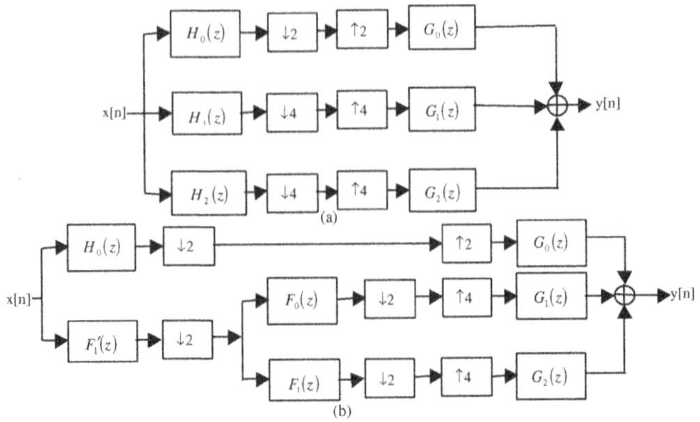

Fig. 1. (a) Non-uniform filter bank (b) Tree structure filter bank

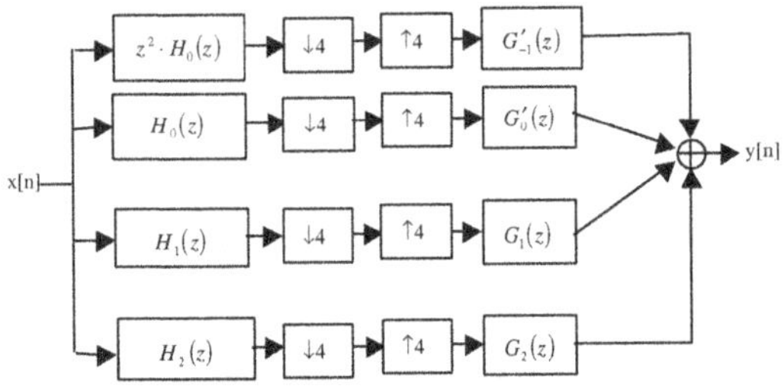

Fig. 2. Realization of non-uniform filter bank via a uniform filter bank

Set of Decimators for Tree Structure Filter Banks

Wing-kuen Ling and Peter Kwong-Shun Tam

Department of Electronic and Information Engineering
The Hong Kong Polytechnic University
Hung Hom, Kowloon, Hong Kong
Hong Kong Special Administrative Region, China
Tel: (852) 2766-6238, Fax: (852) 2362-8439
Email: bingo@encserver.eie.polyu.edu.hk

Abstract. In this paper, we propose a novel method to test if a set of decimators can be generated by a tree structure filter bank. The decimation ratio is first sorted in an ascending order. Then we group the largest decimators with the same decimation ratio together and form a new set of decimators. A set of decimators can be generated by a tree structure filter bank if and only if by repeating the above procedure, all the decimators can be grouped together. Some examples are illustrated to show that the proposed method is simple and easy to implement.

1 Introduction

Non-uniform filter banks have taken an important role in this decade and they are widely applied in the area of digital image compression [3, 6, 7, 9, 13]. By realizing a non-uniform filter bank in a tree structure [1, 2, 4, 5, 10-12], the filter lengths in the filters can be reduced, improving the computation complexity and the implementation speed [15]. However, not all the non-uniform filter banks can be realized via a tree structure [5, 8, 10-12]. This paper is to propose a method to test if a set of decimators can be generated by a tree structure filter bank.

In order to tackle this problem, a method to compute the number of combinations of sub-trees is proposed [8]. However, if the number of decimators is large, it is very complicated to compute the number of combinations of sub-trees. Also, this method is order dependent, which will give a wrong result by changing the order of the decimators in the set [8].

2 Proposed Algorithm

Theorem 1

Let the ordered set of decimators $\{n_0,\ldots,n_0,n_1,\ldots,n_1,\ldots,n_{N-1},\ldots,n_{N-1}\}$ be **D**, where $n_i > n_j$ for $i > j$, and the multiplicity of n_i in **D** be p_i. By grouping the largest decimators with the same decimation ratio together and forming a new set of decimators, a set of decimators can be generated by a tree structure filter bank if and only if by repeating the above procedures, all the decimators can be grouped together.

Proof:

Y. Y. Tang et al. (Eds.): WAA 2001, LNCS 2251, pp. 336-340, 2001.

Consider the only if part first. If **D** can be generated by a tree structure filter bank, then there should be no branch coming out from the decimators n_{N-1}. Otherwise, n_{N-1} is not the greatest number in **D** . Hence, by grouping some or all of the decimators with decimation ratio n_{N-1} together, the branch corresponding to the grouped decimators is removed. Suppose k_{N-1} decimators are grouped together, where $2 \le k_{N-1} \le p_{N-1}$, then the effective decimation ratio corresponding to the grouped decimators is n_{N-1}/k_{N-1}. And the new set of the decimators become $\{n_0,\ldots,n_0,n_1,\ldots,n_1,\ldots,n_{N-2},\ldots,n_{N-2},n_{N-1}/k_{N-1},n_{N-1},\ldots,n_{N-1}\}$.

If **D** can be generated by a tree structure filter bank, then $\exists n_i \in$ **D** such that $n_i = n_{N-1}/k_{N-1}$. Hence, by repeating the above procedure, all the branches are removed and eventually there is only one decimator left in **D** , which is 1. And this proves the only if part.

For the if part, since $\exists n_i \in$ **D** such that $n_i = n_{N-1}/k_{N-1}$, we can construct a sub-tree corresponding to those k_{N-1} channels. By repeating the above procedure, the non-uniform filter bank can be realized in the form of a tree structure. Hence, this proves the if part and the theorem. ■

Some problems are: When should we group all of the decimators with decimation ratio n_{N-1} together, that is $k_{N-1} = p_{N-1}$? When should we group part of them together, that is $k_{N-1} < p_{N-1}$? If we group part of them together, how many decimators should we group?

If n_{N-1}/k_{N-1} is an integer, then we group all the decimators together. Otherwise, we group part of them. If we can divide those p_{N-1} decimators into m sessions, with each session containing k^i_{N-1} decimators and n_{N-1}/k^i_{N-1} being an integer, then we group in this way.

There are two possible cases for **D** that cannot be generated by a tree structure filter bank. One is the failure of dividing those p_{N-1} decimators into m sessions, in which each session contains k^i_{N-1} decimators and n_{N-1}/k^i_{N-1} is an integer. The other case is by iterating the above procedure q times, we have **D** $=\{n_0,\ldots,n_0,n_1,\ldots,n_1,\ldots,n_j,\ldots,n_j,n_{j+1}\}$, and $p_{j+1}=1$. Both cases lead to the conclusion that this non-uniform filter bank cannot be realized via a tree structure.

3 Illustrative Examples

It is well known that $\{2,4,4\}$, $\{2,6,6,6\}$ can be generated by a tree structure [14]. By applying our proposed algorithm, it gives the same conclusion as below:

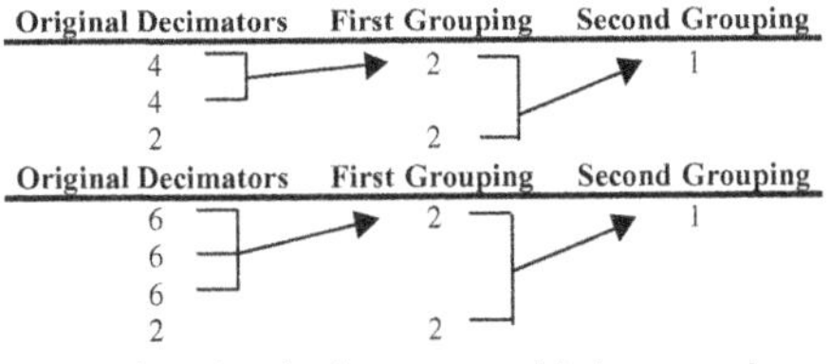

Since all the decimators are grouped together, the above two sets of decimators can be generated by a tree structure.

Consider another example as follows:

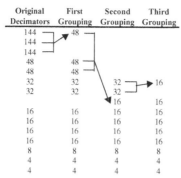

Original Decimators	First Grouping	Second Grouping	Third Grouping
144	48		
144			
144			
48	48		
48	48		
32	32	32	16
32	32	32	
		16	16
16	16	16	16
16	16	16	16
16	16	16	16
16	16	16	16
8	8	8	8
4	4	4	4
4	4	4	4

After the third grouping, we have six decimators with the decimation ratio *16* in the set. Since *16/6* is not an integer, we split those six decimators into two sessions. The first session contains four decimators and the other session contains two decimator. The proposed algorithm then gives:

Third Grouping	Fourth Grouping	Fifth Grouping	Sixth Grouping	Seventh Grouping
16				
16				
16				
16				
16	16	8	4	1
16	16			
8	8	8		
4	4	4	4	
4	4	4	4	
	4	4	4	

Since all the decimators are grouped together, this set of decimators can be generated by a tree structured filter bank. There is another way to split those six decimators with decimation ratio *16* into different sessions, which is to split into three sessions, in which each session contains two decimators. And it will give the same conclusion as above.

There are some sets of decimators which cannot be generated by a tree structure, such as *{2,3,6}*, *{2,6,10,12,12,30,30}* [14]. By applying our proposed algorithm, it gives the same conclusion as below:

Original Decimators	First Grouping
30	15
30	
12	12
12	12
10	10
6	6
2	2

After the first grouping, there is only one decimator with decimation ratio *15* in the set, and we cannot proceed with the proposed algorithm further. Hence, this set of decimators cannot be generated by a tree structure filter bank.

Consider the last example with the set of decimators *{5,5,5,7,7,35,35,35,35}*. Although $p_{N-1}\neq 1$, there does not exist k_{N-1}, such that $2\leq k_{N-1}\leq p_{N-1}$. Hence, this set of decimators cannot be generated by a tree structure filter bank.

4 Conclusion

In this paper, we have proposed a novel method to test if a set of decimators can be generated by a tree structure filter bank. The proposed method is order independent, simple and easy to implement because there is no need to consider the number of combinations of sub-trees [8]. If all the decimators can be grouped together, thenthe set of decimators can be generated by a tree structure filter bank.

Acknowledgement

The work described in this paper was substantially supported by a grant from the Hong Kong Polytechnic University with account number G-V968.

References

1. Vaidyanathan P. P.: Lossless Systems in Wavelet Transforms. IEEE International Symposium on Circuits and Systems, ISCAS, Vol. 1. (1991) 116-119.
2. Soman A. K. and Vaidyanathan P. P.: Paraunitary Filter Banks and Wavelet Packets. IEEE International Conference on Acoustics, Speech, and Signal Processing, ICASSP, Vol. 4. (1992) 397-400.
3. Bamberger R. H., Eddins S. L. and Nuri V.: Generalizing Symmetric Extension: Multiple Nonuniform Channels and Multidimensional Nonseparable IIR Filter Banks. IEEE International Symposium on Circuits and Systems, ISCAS, Vol. 2. (1992) 991-994.
4. Sodagar I., Nayebi and Barnwell T. P.: A Class of Time-Varying Wavelet Transforms. IEEE International Conference on Acoustics, Speech, and Signal Processing, ICASSP, Vol. 3. (1993) 201-204.
5. Soman A. K. and Vaidyanathan P. P.: On Orthonormal Wavelets and Paraunitary Filter Banks. IEEE Transactions on Signal Processing, Vol. 41, No. 3. (1993) 1170-1183.
6. Vaidyanathan P. P.: Orthonormal and Biorthonormal Filter Banks as Convolvers, and Convolutional Coding Gain. IEEE Transactions on Signal Processing, Vol. 41, No. 6. (1993) 2110-2130.
7. Soman A. K. and Vaidyanathan P. P.: Coding Gain in Paraunitary Analysis/Synthesis Systems. IEEE Transactions on Signal Processing, Vol. 41, No. 5. (1993) 1824-1835.
8. Kovaèeviæ J. and Vetterli M.: Perfect Reconstruction Filter Banks with Rational Sampling Factors. IEEE Transactions on Signal Processing, Vol. 41, No. 6. (1993) 2047-2066.
9. Bamberger R. H., Eddins S. L. and Nuri V.: Generalized Symmetric Extension for Size-Limited Multirate Filter Banks. IEEE Transactions on Image Processing, Vol. 3, No. 1. (1994) 82-87.
10. Makur A.: BOT s Based on Nonuniform Filter Banks. IEEE Transactions on Signal Processing, Vol. 44, No. 8. (1996) 1971-1981.
11. Li J., Nguyen T. Q. andTantaratana S.: A Simple Design Method for Near-Perfect-Reconstruction Nonuniform Filter Banks. IEEE Transactions on Signal Processing, Vol. 45, No. 8. (1997) 2105-2109.
12. Akkarakaran S. and Vaidyanathan P. P.: New Results and Open Problems on Nonuniform Filter-Banks. IEEE International

Conference on Acoustics, Speech, and Signal Processing, ICASSP, Vol. 3. (1999) 1501-1504.

13. Omiya N., Nagai T., Ikehara M. and Takahashi S. I.: Organization of Optimal Nonuniform Lapped Biorthogonal Transforms Based on Coding Efficiency. IEEE International Conference on Image Processing, ICIP, Vol. 1. (1999) 624-627.
14. Hoang P. Q. and Vaidyanathan P. P.: Non-uniform Multirate Filter Banks: Theory and Design. IEEE International Symposium on Circuits and Systems, ISCAS, Vol. 1. (1989) 371-372.
15. Vaidyanathan P. P.: Multirate Systems and Filter Banks. Englewood Cliffs, NJ: Prentice Hall, 1993.

Set of Perfect Reconstruction Non-uniform Filter Banks via a Tree Structure

Wing-kuen Ling and Peter Kwong-Shun Tam

Department of Electronic and Information Engineering
The Hong Kong Polytechnic University
Hung Hom, Kowloon, Hong Kong
Hong Kong Special Administrative Region, China
Tel: (852) 2766-6238, Fax: (852) 2362-8439
Email: bingo@encserver.eie.polyu.edu.hk

Abstract. In this paper, we propose a novel method to test if a non-uniform filter bank can achieve perfect reconstruction via a tree structure. The set of decimators is first sorted in an ascending order. A non-uniform filter bank can achieve perfect reconstruction via a tree structure if and only if some or all of the channels corresponding to the maximum decimation ratio can be grouped into one channel, and the procedure can be repeated until all the channels are grouped together.

1 Introduction

Non-uniform filter banks play an important role in this decade and they are widely applied to digital image compression [3, 6, 7, 9, 13]. By realizing a non-uniform filter bank via a tree structure [1, 2, 4, 5, 10-12], the filter length in the filters is reduced, improving the computation complexity and the implementation speed [14]. However, not all the non-uniform filter banks can be realized via a tree structure [5, 8, 10-12]. A method to compute the number of combinations of sub-trees is proposed [8] to test if the decimators in the non-uniform filter bank can be generated by a tree structure. However, even though the decimators can be generated by a tree structure, this does not imply that the non-uniform filter bank can be generated by a tree structure. This is because the analysis filters are ignored in the consideration. In this paper, the necessary and sufficient conditions for a non-uniform filter bank to be realized by a tree structure are addressed.

The necessary and sufficient conditions are discussed in section II and illustrative examples are presented in section III. Finally, a conclusion is given in section IV.

2 Necessary and Sufficient Conditions for Realizing a Non-uniform Filter Bank via a Tree Structure

Let the ordered set of decimators $\{n_0, \ldots, n_0, \ldots, n_{N-1}, \ldots, n_{N-1}\}$ be $\mathbf{D}$, where $n_i > n_j$ for $i > j$, and the multiplicity of n_i in $\mathbf{D}$ be p_i. Let the corresponding analysis filters and synthesis filters be $\{H_{0,0}(z), \cdots, H_{0,p_0-1}(z), \cdots, H_{N-1,0}(z), \cdots, H_{N-1,p_{N-1}-1}(z)\}$ and $\{G_{0,0}(z), \cdots, G_{0,p_0-1}(z), \cdots, G_{N-1,0}(z), \cdots, G_{N-1,p_{N-1}-1}(z)\}$, respectively.

If there exists a set of filters $\{H'_{N-1}(z), H'_{N-1,k_0}(z), \cdots, H'_{N-1,k_{K_{N-1}-1}}(z)\}$, where $k_i \in [0\ p_{N-1}-1]$ for $i = 0, 1, \ldots, K_{N-1}-1$ and $K_{N-1} \in [2\ p_{N-1}]$,

Y. Y. Tang et al. (Eds.): WAA 2001, LNCS 2251, pp. 341-346, 2001.

such that:

- $n_{N-1}/K_{N-1} \in \mathbb{Z}$, (1)
- $H'_{N-1}(z)\cdot H'_{N-1,k_i}\left(z^{\frac{n_{N-1}}{K_{N-1}}}\right) = H_{N-1,k_i}(z)$ and, (2)
- $\det\left(\begin{bmatrix} H'_{N-1,k_0}(z) & H'_{N-1,k_1}(z) & \cdots & H'_{N-1,k_{K_{N-1}-1}}(z) \\ H'_{N-1,k_0}(z\cdot W_{N-1}) & H'_{N-1,k_1}(z\cdot W_{N-1}) & \cdots & H'_{N-1,k_{K_{N-1}-1}}(z\cdot W_{N-1}) \\ \vdots & \vdots & \ddots & \vdots \\ H'_{N-1,k_0}\left(z\cdot W_{N-1}{}^{K_{N-1}-1}\right) & H'_{N-1,k_1}\left(z\cdot W_{N-1}{}^{K_{N-1}-1}\right) & \cdots & H'_{N-1,k_{K_{N-1}-1}}\left(z\cdot W_{N-1}{}^{K_{N-1}-1}\right) \end{bmatrix}\right) \neq 0$, (3)

where $W_{N-1} = e^{-\frac{j2\pi}{K_{N-1}}}$, then by a proper design of the synthesis filters, those K_{N-1} channels can be grouped together into one channel with the analysis filter $H'_{N-1}(z)$ and the decimator $\downarrow n_{N-1}/K_{N-1}$.

Now, we have a new set of decimators and analysis/synthesis filters. Let the new set of decimators $\{n'_0, \ldots, n'_i, \ldots, n'_{N'-1}\}$ be $\mathbf{D}'$ and the multiplicity of n'_i in $\mathbf{D}'$ be p'_i. Let the corresponding analysis/synthesis filters be $\left\{H^{new}{}_{0,0}(z), \cdots, H^{new}{}_{0,p'_0-1}(z), \cdots, H^{new}{}_{N'-1,0}(z), \cdots, H^{new}{}_{N'-1,p'_{N'-1}-1}(z)\right\}$ and $\left\{G^{new}{}_{0,0}(z), \cdots, G^{new}{}_{0,p'_0-1}(z), \cdots, G^{new}{}_{N'-1,0}(z), \cdots, G^{new}{}_{N'-1,p'_{N'-1}-1}(z)\right\}$, respectively.

By repeating the above grouping procedure, if all the channels can be grouped together, and eventually only one channel is left, then the non-uniform filter bank can achieve perfect reconstruction via a tree structure.

Theorem 1

A non-uniform filter bank can achieve perfect reconstruction via a tree structure if and only if all the channels can be grouped together by the above grouping procedure.

Proof.

The if part is proved in the above. Now, let's consider the only if part. Since the non-uniform filter bank can be realized by a tree structure, $\exists n_i \in \mathbf{D}$ such that $n_i = n_{N-1}/K_{N-1}$, and a set of filters $\left\{H'_{N-1}(z), H'_{N-1,k_0}(z), \cdots, H'_{N-1,k_{K_{N-1}-1}}(z)\right\}$ such that $H'_{N-1}(z)\cdot H'_{N-1,k_i}\left(z^{\frac{n_{N-1}}{K_{N-1}}}\right) = H_{N-1,k_i}(z)$. But do those filters satisfy equation (3)? Or in other words, if some of the analysis filters in a sub-tree are linearly dependent, does there exist a set of synthesis filters such that the whole system still achieves perfect reconstruction?

Assume $\det\left(\begin{bmatrix} H'_{N-1,k_0}(z) & H'_{N-1,k_1}(z) & \cdots & H'_{N-1,k_{K_{N-1}-1}}(z) \\ H'_{N-1,k_0}(z\cdot W_{N-1}) & H'_{N-1,k_1}(z\cdot W_{N-1}) & \cdots & H'_{N-1,k_{K_{N-1}-1}}(z\cdot W_{N-1}) \\ \vdots & \vdots & \ddots & \vdots \\ H'_{N-1,k_0}\left(z\cdot W_{N-1}{}^{K_{N-1}-1}\right) & H'_{N-1,k_1}\left(z\cdot W_{N-1}{}^{K_{N-1}-1}\right) & \cdots & H'_{N-1,k_{K_{N-1}-1}}\left(z\cdot W_{N-1}{}^{K_{N-1}-1}\right) \end{bmatrix}\right) = 0$, $\exists G_{N-1,k_0}(z), \cdots, G_{N-1,k_{K_{N-1}-1}}(z)$ and a non-zero transfer function $T(z)$ such that:

$$\begin{bmatrix} H'_{N-1}(z)\cdot H'_{N-1,k_0}\left(z^{\frac{n_{N-1}}{K_{N-1}}}\right) & H'_{N-1}(z)\cdot H'_{N-1,k_1}\left(z^{\frac{n_{N-1}}{K_{N-1}}}\right) & \cdots & H'_{N-1}(z)\cdot H'_{N-1,k_{K_{N-1}-1}}\left(z^{\frac{n_{N-1}}{K_{N-1}}}\right) \\ H'_{N-1}(z\cdot W)\cdot H'_{N-1,k_0}\left(z^{\frac{n_{N-1}}{K_{N-1}}}\cdot W^{\frac{n_{N-1}}{K_{N-1}}}\right) & H'_{N-1}(z\cdot W)\cdot H'_{N-1,k_1}\left(z^{\frac{n_{N-1}}{K_{N-1}}}\cdot W^{\frac{n_{N-1}}{K_{N-1}}}\right) & \cdots & H'_{N-1}(z\cdot W)\cdot H'_{N-1,k_{K_{N-1}-1}}\left(z^{\frac{n_{N-1}}{K_{N-1}}}\cdot W^{\frac{n_{N-1}}{K_{N-1}}}\right) \\ \vdots & \vdots & \ddots & \vdots \\ H'_{N-1}\left(z\cdot W^{K_{N-1}-1}\right)\cdot H'_{N-1,k_0}\left(z^{\frac{n_{N-1}}{K_{N-1}}}\cdot W^{\frac{n_{N-1}}{K_{N-1}}(K_{N-1}-1)}\right) & H'_{N-1}\left(z\cdot W^{K_{N-1}-1}\right)\cdot H'_{N-1,k_1}\left(z^{\frac{n_{N-1}}{K_{N-1}}}\cdot W^{\frac{n_{N-1}}{K_{N-1}}(K_{N-1}-1)}\right) & \cdots & H'_{N-1}\left(z\cdot W^{K_{N-1}-1}\right)\cdot H'_{N-1,k_{K_{N-1}-1}}\left(z^{\frac{n_{N-1}}{K_{N-1}}}\cdot W^{\frac{n_{N-1}}{K_{N-1}}(K_{N-1}-1)}\right) \end{bmatrix}$$

$$\begin{bmatrix} G_{N-1,k_0}(z) \\ G_{N-1,k_1}(z) \\ \vdots \\ G_{N-1,k_{K_{N-1}-1}}(z) \end{bmatrix} = \begin{bmatrix} T(z) \\ 0 \\ \vdots \\ 0 \end{bmatrix}, \tag{4}$$

where $W = e^{-\frac{j 2\pi}{n_{N-1}}}$.

Since $\det\left(\begin{bmatrix} H'_{N-1,k_0}(z) & H'_{N-1,k_1}(z) & \cdots & H'_{N-1,k_{K_{N-1}-1}}(z) \\ H'_{N-1,k_0}(z \cdot W_{N-1}) & H'_{N-1,k_1}(z \cdot W_{N-1}) & \cdots & H'_{N-1,k_{K_{N-1}-1}}(z \cdot W_{N-1}) \\ \vdots & \vdots & \ddots & \vdots \\ H'_{N-1,k_0}\left(z \cdot W_{N-1}{}^{K_{N-1}-1}\right) & H'_{N-1,k_1}\left(z \cdot W_{N-1}{}^{K_{N-1}-1}\right) & \cdots & H'_{N-1,k_{K_{N-1}-1}}\left(z \cdot W_{N-1}{}^{K_{N-1}-1}\right) \end{bmatrix}\right) = 0$, by letting $z = z^{\frac{n_{N-1}}{K_{N-1}}}$, we have:

$$\det\left(\begin{bmatrix} H'_{N-1}(z) \cdot H'_{N-1,k_0}\left(z^{\frac{n_{N-1}}{K_{N-1}}}\right) & H'_{N-1}(z) \cdot H'_{N-1,k_1}\left(z^{\frac{n_{N-1}}{K_{N-1}}}\right) & \cdots & H'_{N-1}(z) \cdot H'_{N-1,k_{K_{N-1}-1}}\left(z^{\frac{n_{N-1}}{K_{N-1}}}\right) \\ H'_{N-1}(z \cdot W) \cdot H'_{N-1,k_0}\left(z^{\frac{n_{N-1}}{K_{N-1}}} \cdot W^{\frac{n_{N-1}}{K_{N-1}}}\right) & H'_{N-1}(z \cdot W) \cdot H'_{N-1,k_1}\left(z^{\frac{n_{N-1}}{K_{N-1}}} \cdot W^{\frac{n_{N-1}}{K_{N-1}}}\right) & \cdots & H'_{N-1}(z \cdot W) \cdot H'_{N-1,k_{K_{N-1}-1}}\left(z^{\frac{n_{N-1}}{K_{N-1}}} \cdot W^{\frac{n_{N-1}}{K_{N-1}}}\right) \\ \vdots & \vdots & \ddots & \vdots \\ H'_{N-1}\left(z \cdot W^{K_{N-1}-1}\right) \cdot H'_{N-1,k_0}\left(z^{\frac{n_{N-1}}{K_{N-1}}} \cdot W^{\frac{n_{N-1}}{K_{N-1}}(K_{N-1}-1)}\right) & H'_{N-1}\left(z \cdot W^{K_{N-1}-1}\right) \cdot H'_{N-1,k_1}\left(z^{\frac{n_{N-1}}{K_{N-1}}} \cdot W^{\frac{n_{N-1}}{K_{N-1}}(K_{N-1}-1)}\right) & \cdots & H'_{N-1}\left(z \cdot W^{K_{N-1}-1}\right) \cdot H'_{N-1,k_{K_{N-1}-1}}\left(z^{\frac{n_{N-1}}{K_{N-1}}} \cdot W^{\frac{n_{N-1}}{K_{N-1}}(K_{N-1}-1)}\right) \end{bmatrix}\right) = 0 . \tag{5}$$

Let the matrix in equation (5) be **H**. By examining equation (5) and applying Cramer' s rule to equation (4), we find that the determinants of the matrices by deleting the first row and any columns are zero. By the modulation principle, we find that the determinants of the matrices by deleting the last row and any columns are zero. Let the rank of the matrix by deleting the first row of **H** be r, and that of the matrix by keeping the first $r+1$ rows of **H** be $H' = \begin{bmatrix} h'_0 & \cdots & h'_{K_{N-1}-1} \end{bmatrix} = \begin{bmatrix} h_{0,0} & h_{0,1} \\ h_{S,0} & h_{S,1} \end{bmatrix}$, where $\mathbf{h}'_i$ is the ith column of **H'** and $\mathbf{h}_{0,0}$ are the first r elements of the first row of **H'**. Since $H' \cdot \begin{bmatrix} g_a \\ g_b \end{bmatrix} = \begin{bmatrix} T(z) \\ 0 \\ \vdots \\ 0 \end{bmatrix}$, where $\mathbf{g}_a$ is a vector containing the first r synthesis filters, we have $\mathbf{h}_{0,0} \cdot \mathbf{g}_a + \mathbf{h}_{0,1} \cdot \mathbf{g}_b = T(z)$ and $\mathbf{h}_{S,0} \cdot \mathbf{g}_a + \mathbf{h}_{S,1} \cdot \mathbf{g}_b = \mathbf{0}$. This implies that $(\mathbf{h}_{0,1} - \mathbf{h}_{0,0} \cdot \mathbf{h}_{S,0}{}^{-1} \cdot \mathbf{h}_{S,1}) \cdot \mathbf{g}_b = T(z)$, and $\left[\det\left(\begin{bmatrix} h'_0 & \cdots & h'_{r-1} & h'_r \end{bmatrix}\right) \; \det\left(\begin{bmatrix} h'_0 & \cdots & h'_{r-1} & h'_{r+1} \end{bmatrix}\right) \; \cdots \; \det\left(\begin{bmatrix} h'_0 & \cdots & h'_{r-1} & h'_{K_{N-1}-1} \end{bmatrix}\right)\right] \cdot g_b = T(z) \cdot \det(h_{S,0}) = 0$, which contradicts the assumption. Hence, if some of the analysis filters in a sub-tree are linearly dependent, there does not exist a set of synthesis filters such that the whole system achieves perfect reconstruction. This proves the only if part and the theorem. ■

3 Illustrative Examples

3.1 Uniform Filter Bank

Consider an M-channel uniform filter bank with analysis filters $\{H_0(z), H_1(z), \ldots, H_{M-1}(z)\}$. In this case, $N=1$ and $n_0=p_0=K_0=M$. By selecting $H'_0(z)=1$, $H'_{0,i}(z)=H_i(z)$, for $i=0,1,\ldots,M-1$, this M-channel uniform filter bank can achieve perfect reconstruction via a tree structure if and only if:

$$\det\left(\begin{bmatrix} H'_{0,0}(z) & H'_{0,1}(z) & \cdots & H'_{0,M-1}(z) \\ H'_{0,0}(z \cdot W) & H'_{0,1}(z \cdot W) & \cdots & H'_{0,M-1}(z \cdot W) \\ \vdots & \vdots & \ddots & \vdots \\ H'_{0,0}\left(z \cdot W^{M-1}\right) & H'_{0,1}\left(z \cdot W^{M-1}\right) & \cdots & H'_{0,M-1}\left(z \cdot W^{M-1}\right) \end{bmatrix}\right) \neq 0, \tag{6}$$

where $W = e^{-\frac{j2\pi}{M}}$ [14].

3.2 Perfect Reconstruction Dyadic Tree Structure Filter Bank

Consider the non-uniform filter bank shown in figure 1 [1, 2, 5]:

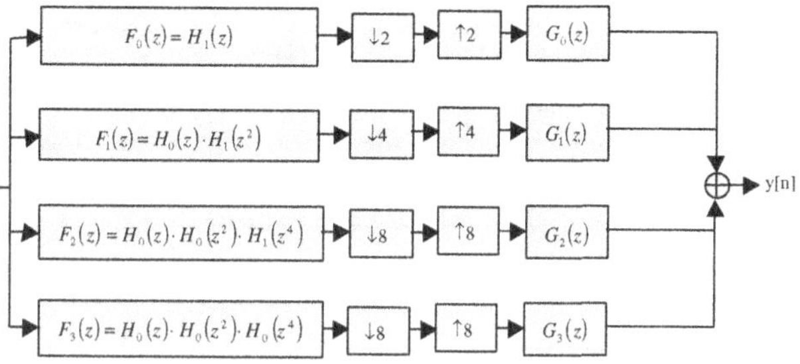

Fig. 1. Perfect reconstruction dyadic tree structure filter bank

In this case, $n_0=2$, $n_1=4$, $n_2=8$, $p_0=1$, $p_1=1$, $p_2=2$ and $N=3$. By selecting $K_j=2$, $H'_0(z)=1$, $H'_1(z)=H_0(z)$, $H'_2(z)=H_0(z)\cdot H_0(z^2)$, $H'_{j,0}(z)=H_1(z)$, and $H'_{j,1}(z)=H_0(z)$, for $j=0,1,2$, this non-uniform filter bank can achieve perfect reconstruction via a tree structure if and only if $\det\left(\begin{bmatrix} H_0(z) & H_1(z) \\ H_0(-z) & H_1(-z)\end{bmatrix}\right) \neq 0$ [1, 2, 5].

3.3 Perfect Reconstruction Tree Structure Filter Bank

Consider the non-uniform filter bank shown in figure 2:

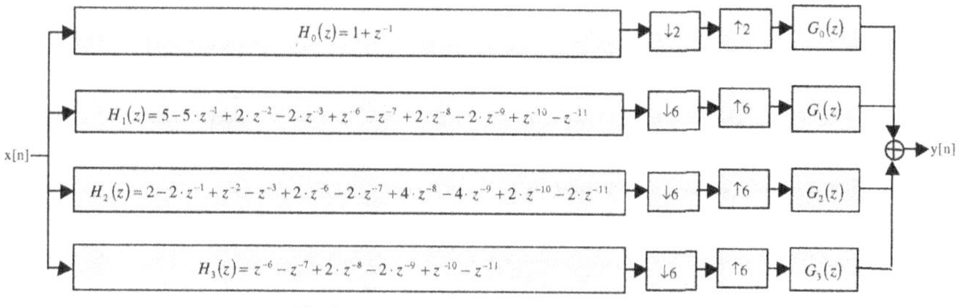

Fig. 2. Perfect reconstruction tree structure filter bank

In this case, $n_0=2$, $n_1=6$, $p_0=1$, $p_1=3$ and $N=2$. By selecting $K_1=3$, $H'_1(z)=1-z^{-1}$, $H'_{1,0}(z)=5+2\cdot z^{-1}+z^{-3}+2\cdot z^{-4}+z^{-5}$, $H'_{1,1}(z)=2+z^{-1}+2\cdot z^{-3}+4\cdot z^{-4}+2\cdot z^{-5}$, and $H'_{1,2}(z)=z^{-3}+2\cdot z^{-4}+z^{-5}$ [14], we can group the last three channels together into one channel with the new analysis filter $H'_1(z)=1-z^{-1}$ and the decimator $\downarrow 2$. Similarly, by selecting $K_0=2$, $H'_0(z)=1$, $H'_{0,0}(z)=1+z^{-1}$, and $H'_{0,1}(z)=1-z^{-1}$, this non-uniform filter bank can achieve perfect reconstruction via a tree structure.

3.4 Not Perfect Reconstruction Tree Structure Filter Bank Due to the Dependent Kernel

Consider the same non-uniform filter bank shown in figure 2 with $H_0(z)$ is changed to $F(z)\cdot(1-z^{-1})$, where $F(z)=F(-z)$. The last three channels are grouped together with the same procedure as above, and we have two channels left with decimator $\downarrow 2$, and the analysis filters are $F(z)\cdot(1-z^{-1})$ and $(1-z^{-1})$, respectively. Since $\det\left(\begin{bmatrix} H_0(z) & 1-z^{-1} \\ H_0(-z) & 1+z^{-1} \end{bmatrix}\right)=0$, we conclude that this non-uniform filter bank cannot achieve perfect reconstruction even through it can be realized via a tree structure.

3.5 Cannot Be Realized Via a Tree Structure Filter Bank Due to Structural Problem

Consider the same non-uniform filter bank shown in figure 2 with $H_1(z)$ changed to $(1-z^{-1})\cdot F_1(z)$, $H_2(z)$ changed to $(1-z^{-1})\cdot F_2(z)$, $H_3(z)$ changed to $(1-z^{-1})\cdot F_3(z)$, where $F_1(z)/F_2(z)$ and $F_2(z)/F_3(z)$ are not rational functions of z^2. In this case, the last three channels cannot be grouped together. Hence, this non-uniform filter bank cannot be realized via a tree structure.

3.6 Incompatible Non-uniform Filter Bank

Consider an incompatible non-uniform filter bank [15] with the set of decimators $\{2,3,6\}$. Since $p_i=1$, $\forall i$, there does not exist $K_j\in[2\ p_j]$. Hence, an incompatible non-uniform filter bank cannot be realized via a tree structure [15].

3.7 Compatible Non-uniform Filter Bank, But Cannot Be Realized Via a Tree Structure

Consider a non-uniform filter bank with the set of decimators $\{5,5,5,7,7,35,35,35,35\}$. In this case, $n_0=5$, $n_1=7$, $n_2=35$, $p_0=3$, $p_1=2$, $p_2=4$, and $N=3$. Since there does not exist $K_2\in[2\ p_2]$ such that $n_2/K_2\in\mathbf{Z}$, this non-uniform filter bank cannot be realized via a tree structure.

4 Conclusion

In this paper, we propose a novel method to test if a non-uniform filter bank can achieve perfect reconstruction via a tree structure. The advantage of realizing a non-uniform filter bank via a tree structure is to reduce the computation complexity and provide fast implementation for non-uniform filter bank [14].

Acknowledgement

The work described in this paper was substantially supported by a grant from the Hong Kong Polytechnic University with account number G-V968.

References

1. Vaidyanathan P. P.: Lossless Systems in Wavelet Transforms. IEEE International Symposium on Circuits and Systems, ISCAS, Vol. 1. (1991) 116-119.
2. Soman A. K. and Vaidyanathan P. P.: Paraunitary Filter Banks and Wavelet Packets. IEEE International Conference on Acoustics, Speech, and Signal Processing, ICASSP, Vol. 4. (1992) 397-400.
3. Bamberger R. H., Eddins S. L. and Nuri V.: Generalizing Symmetric Extension: Multiple Nonuniform Channels and Multidimensional Nonseparable IIR Filter Banks. IEEE International Symposium on Circuits and Systems, ISCAS, Vol. 2. (1992) 991-994.
4. Sodagar I., Nayebi K. and Barnwell T. P.: A Class of Time-Varying Wavelet Transforms. IEEE International Conference on Acoustics, Speech, and Signal Processing, ICASSP, Vol. 3. (1993) 201-204.
5. Soman A. K. and Vaidyanathan P. P.: On Orthonormal Wavelets and Paraunitary Filter Banks. IEEE Transactions on Signal Processing, Vol. 41, No. 3. (1993) 1170-1183.
6. Vaidyanathan P. P.: Orthonormal and Biorthonormal Filter Banks as Convolvers, and Convolutional Coding Gain. IEEE Transactions on Signal Processing, Vol. 41, No. 6. (1993) 2110-2130.
7. Soman A. K. and Vaidyanathan P. P.: Coding Gain in Paraunitary Analysis/Synthesis Systems. IEEE Transactions on Signal Processing, Vol. 41, No. 5. (1993) 1824-1835.
8. Kovaèeviæ J. and Vetterli M.: Perfect Reconstruction Filter Banks with Rational Sampling Factors. IEEE Transactions on Signal Processing, Vol. 41, No. 6. (1993) 2047-2066.
9. Bamberger R. H., Eddins S. L. and Nuri V.: Generalized Symmetric Extension for Size-Limited Multirate Filter Banks. IEEE Transactions on Image Processing, Vol. 3, No. 1. (1994) 82-87.
10. Makur A.: BOT's Based on Nonuniform Filter Banks. IEEE Transactions on Signal Processing, Vol. 44, No. 8. (1996) 1971-1981.
11. Li J., Nguyen T. Q. and Tantaratana S.: A Simple Design Method for Near-Perfect-Reconstruction Nonuniform Filter Banks. IEEE Transactions on Signal Processing, Vol. 45, No. 8. (1997) 2105-2109.
12. Akkarakaran S. and Vaidyanathan P. P.: New Results and Open Problems on Nonuniform Filter-Banks. IEEE International Conference on Acoustics, Speech, and Signal Processing, ICASSP, Vol. 3,. (1999) 1501-1504.
13. Omiya N., Nagai T., Ikehara M. and Takahashi S. I.: Organization of Optimal Nonuniform Lapped Biorthogonal Transforms Based on Coding Efficiency. IEEE International Conference on Image Processing, ICIP, Vol. 1. (1999) 624-627.
14. Vaidyanathan P. P.: Multirate Systems and Filter Banks. Englewood Cliffs, NJ: Prentice Hall, 1993.
15. Hoang P. Q. and Vaidyanathan P. P.: Non-Uniform Multirate Filter Banks: Theory and Design. IEEE International Symposium on Circuits and Systems, ISCAS, Vol. 1. (1989) 371-374.

Joint Time-Frequency Distributions for Business Cycle Analysis*

Sharif Md. Raihan[1], *Yi Wen*[2], *and Bing Zeng*[1]

[1] Department of Electrical and Electronic Engineering
The Hong Kong University of Science and Technology
Clear Water Bay, Hong Kong, China

[2] Department of Economics
Cornell University
Ithaca, NY 14853, USA

Abstract: The joint time-frequency analysis (JTFA) is a signal processing technique in which signals are represented in both the time domain and the frequency domain simultaneously. Recently, this analysis technique has become an extremely powerful tool for analyzing nonstationary time series. One basic problem in business-cycle studies is how to deal with nonstationary time series. The market economy is an evolutionary system. Economic time series therefore contain stochastic components that are necessarily time dependent. Traditional methods of business cycle analysis, such as the correlation analysis and the spectral analysis, cannot capture such historical information because they do not take the time-varying characteristics of the business cycles into consideration. In this paper, we introduce and apply a new technique to the studies of the business cycle: the wavelet-based time-frequency analysis that has recently been developed in the field of signal processing. This new method allows us to characterize and understand not only the timing of shocks that trigger the business cycle, but also situations where the frequency of the business cycle shifts in time. Applying this new method to post war US data, we are able to show that 1973 marks a new era for the evolution of the business cycle since World War II.

Keywords: Wavelets, time-frequency analysis, business cycle, non-stationary time series, scalogram, and spectrum.

* This work has been supported by a grant, HKUST6176/98H, from the Research Grants Council of the Hong Kong Special Administrative Region, China.

Y. Y. Tang et al. (Eds.): WAA 2001, LNCS 2251, pp. 347-358, 2001.

I. Introduction

The analysis of nonstationary signal cannot be accomplished by classical time domain representations such as correlation methods, or by frequency domain representations based on the Fourier transform [2]. To analyze business cycles that evolve over time, we need to develop a concept of time-frequency distribution that takes into account jointly and simultaneously the information of time and frequency.

The business cycle, one of the most puzzling phenomena in capitalistic, free-market economies, has long been the central focus of macroeconomic researches. The biggest challenge to researchers in this field is to capture business cycle patterns that vary in nature across time. Economic time series contain stochastic components that are necessarily time dependent.

Although time-frequency analysis has its origin almost 50 years ago [Gabor, 1946; Ville, 1948], significant advances occurred only in the last 15 years or so. Recently, time-frequency representations have become an extremely powerful tool for analyzing nonstationary signals in many fields: such as engineering, medical sciences, and astronomy, to name just a few. A number of articles have also been published to deal with applications in economics and finance [10].

So far, many alternative transforms have been developed to overcome the problems associated with classical spectral analysis, we introduce in this paper a new technique of time series analysis to business cycle studies: a joint time-frequency distribution based on the wavelet transform. This new technique enables us to capture the evolutionary aspects of the spectral distribution of the business cycle across time.

In this paper, we compare the wavelet-based time-frequency analysis to a traditional approach based on the windowed Fourier transform. We show that the wavelet transform has many advantages over the traditional approach in that the wavelet transform has a beautiful property: its window size adjusts itself optimally to longer basis functions at low frequencies and to shorter basis functions at high frequencies. Consequently, it has sharp frequency resolution for low frequency movements and sharp time resolution for high frequency movements. Thus, the new method is capable of capturing simultaneously the time-varying nature of low frequency cycles and the frequency distribution of sudden and abrupt shocks in the original time series.

The rest of the sections are organized as follows. Section II describes the windowed Fourier transform and spectrogram. Section III describes the wavelet transform and scalogram. Section IV explains the implementation of the wavelet transform when applying to actual data. Section V uses artificial signals to demonstrate the advantages of wavelet transform over the windowed Fourier transform. Section VI applies the wavelet-based time-frequency analysis to economic data. Finally, we conclude the paper in section VII.

II. The Windowed Fourier Transform and Spectrogram

Fourier transform (FT), most widely used classical representation, is a mathematical technique for transforming a signal from the time domain to the frequency domain.

However, in the transformation process, the time information of the signal is completely lost. When we look at the FT of a signal, we observe no information about when a particular event took place. For signals in which the time information is not important but the frequency contents are of primary interest, this limitation is of little consequence. Thus, Fourier analysis is useful for analyzing periodic and stationary signals whose moments do not change much over time. However, many interesting and important signals are not stationary and need to be analyzed in both time and frequency domain simultaneously.

For many years, the representation of a signal in a joint time-frequency space has been of interest in the signal processing area, especially when one is dealing with time-varying nonstationary signals. Performing a mapping of a one-dimensional signal of time into a two-dimensional function of time and frequency is thus needed in order to extract relevant time-frequency information. We refer to several excellent review papers on distributions for the time-frequency (TF) analysis [3, 6].

A classical linear time-frequency representation, called the windowed Fourier transform (WFT), has been extensively used for nonstationary signal analysis since its introduction by Gabor [5]. The basic idea of WFT is to find the spectrum of a signal $x(t)$ at a particular time τ by analyzing a small portion of the signal around this time point. Specifically, the signal is multiplied by a window function $w(t)$ centered at time point τ, and the spectrum of the windowed signal, $x(t)w^*(t-\tau)$, is calculated by

$$WFT_x(\tau,\omega) = \int_{-\infty}^{\infty} x(t)w^*(t-\tau)e^{-j\omega t}dt, \tag{1}$$

where ω is the angular frequency and * denotes the complex conjugation. Because multiplication by a relatively short window $w(t-\tau)$ effectively suppresses the signal outside a neighborhood around the analysis time point $t=\tau$, the WFT is a 'local' spectrum of the signal $x(t)$ around τ.

Spectrogram is the most familiar representation to obtain the energy distribution of the signal. The spectrogram of a signal $x(t)$ is defined as the squared magnitudes of the WFT:

$$SP_x(\tau,\omega) = \left| \int_{-\infty}^{\infty} x(t)w^*(t-\tau)e^{-j\omega t}dt \right|^2. \tag{2}$$

The WFT has many useful properties [9], including a well-developed theory [1]. It is one of the most efficient methods in computation. But a crucial feature inherent in the WFT method is that the length of the window can be selected arbitrarily, but is fixed exogenously once the selection is made. To enhance the time information, therefore, one must choose a short window; and to enhance the frequency resolution, one must choose a long window, which means that the time information (nonstationarities) occurring within the window interval is smeared. The length of the window is therefore the main issue involved in practice.

III. The Wavelet Transform and Scalogram

In recent years, an alternative representation, called the wavelet transform, has been widely adopted in the literature [4, 7, 14, 17]. One major advantage afforded by wavelet transform is that the windows vary endogenously in an optimal way. With this transform one can process data at different resolutions. In order to isolate signal discontinues, for example, one would like to have some very short basis functions. At the same time, in order to obtain detailed frequency analysis, one would like to have some very long basis functions. A way to achieve this is to have short basis functions for high-frequency movements and long ones for low-frequency movements. This is exactly what can be achieved with the wavelet transform. WT have an infinite set of possible basis functions. Thus, wavelet analysis provides immediate access to information that can be obscured by other time-frequency methods such as Fourier analysis.

The wavelet transform is defined as the convolution of a signal $x(t)$ with a wavelet function $\Psi(t)$, called mother wavelet, shifted in time by a translation parameter τ, and dilated by a scale parameter a, as shown by the following equation

$$WT_x(\tau,a) = \frac{1}{\sqrt{|a|}} \int_{-\infty}^{\infty} x(t)\Psi^*\left(\frac{t-\tau}{a}\right)dt, \tag{3}$$

where $\Psi^*(.)$ is the complex conjugate of the basic wavelet function $\Psi(t)$, the parameter a is the scaling factor that controls the length of the analyzing wavelet; and τ is the translation parameter.

The squared modulus of the wavelet transform, called scalogram, is defined as

$$SCAL_x(\tau,a) = \left|\frac{1}{\sqrt{|a|}} \int_{-\infty}^{\infty} x(t)\Psi^*\left(\frac{t-\tau}{a}\right)dt\right|^2 \tag{4}$$

The wavelet transform of a signal depends on two parameters: scale (or frequency) and time. This leads to a so-called time-scale representation that provides a tool for the analysis of nonstationary signals [7, 14].

There is a dozen of wavelet function available, such as Morlet, Mexican hat, Haar, Shannon, Daubechies wavelet function, etc. The choice of the wavelet function depends on the specific application. With respect to time and frequency localization, the Haar and Shannon wavelets take opposite extremes. Having compact support in time, the Haar wavelet has poor decay in frequency, whereas the Shannon wavelet has compact support in frequency with poor decay in time. Other wavelets typically fall in the middle of these two extremes. In fact, having exponential decay in both the time and frequency domain, the Morlet wavelet has optimal joint time-frequency concentration [16]. The wavelet that is used for analysis of economics fluctuations in this paper is Morlet wavelet, which is a modulated Gaussian function with exponential decay property. It is defined as

$$\Psi(t) = e^{-\frac{t^2}{2a^2}} e^{j2\pi ft}, \tag{5}$$

where f is the modulation (frequency) parameter. The scale parameter a and the frequency parameter f are related to each other by the relationship:

$$a = f_0 / f, \tag{6}$$

where f_0 is the central wavelet frequency.

IV. Implementations

In WFT, the signal is divided into small enough segments, where these segments of the signal can be assumed to be stationary. For this purpose, a window function w is chosen. The width of this window must be equal to the segment of the signal where its stationarity is valid. This window function is first placed at the beginning of the signal and the Fourier transform is performed. Then the window is shifted to a new location and another Fourier transform is computed. This procedure continues until the end of the signal is reached. The spectrogram is computed accordingly as the squared modulus of the windowed Fourier transform.

The wavelet transform is done in a similar manner to the WFT. The signal is multiplied by a wavelet function and the wavelet transform is computed according to equation (3) for different values of the scale parameter (a) at different time location (τ). Suppose $x(t)$ is the signal to be analyzed. The mother wavelet is chosen to serve as a prototype for all wavelets in the process. All the wavelets that are used subsequently are the stretched (or compressed) and shifted versions of the mother wavelet. The computation starts with a value of the scaling factor $a = a_1$, and the wavelet is placed at the beginning of the signal. Since the wavelet function has only finite time duration, it serves just like a window in the WFT. The constant $1/\sqrt{a_1}$ is for normalization purpose so that the transformed signal will have the same energy at every scale. Next, with the same scale $a = a_1$, the wavelet function is shifted to the next sample point, and the wavelet transform is computed again. This procedure is repeated until the wavelet reaches the end of the signal. The result is a sequence of numbers corresponding to the scale $a = a_1$.

Next, the scale factor is changed to $a = a_2$, and the whole procedure described above is repeated. When the process is completed for all desired values of a, the result is an energy distribution of the original signal along the two-dimensional time-frequency space.

V. Applications to Test Signals

To show the effects and the advantages of wavelet-based time-frequency analysis over the traditional WFT based time-frequency analysis, we present scalograms and spectrograms of two test signals. The signals are of length 512 points each. The WFT uses a Hanning window, and the scalogram is obtained with the Morlet wavelet. The horizontal axis is time and the vertical axis is frequency in both scalograms and spectrograms respectively.

The first test signal used is composed of two parts: the 1st part is a time-varying low frequency sinusoidal cycle, and the 2nd part is a constant high frequency cycle with some sample points gap in the middle of the signal. The signal is shown in the top window in Figure 1.*a*, and the power spectrum is shown in the left window in Figure 1.*a*.

The central window in Figure 1.*a* shows that the scalogram is able to capture not only the frequency location of the time-varying low frequency cycle, but also the exact timing of the missing signals presented in the constant high frequency cycle. There is no energy distribution in the middle of the scalogram due to the missing data points in the high frequency cycle (notice the sharp breaking edges in the middle of the scalogram).

WFT, on the other hand, is unable to simultaneously capture all the information adequately. With a short window (Figure 1.*b*), the time information with respect to the exact timing of the missing data points is captured, but the frequency location of the low frequency cycle is not localized at all along the frequency axis. With a large window (Figure 1.*c*), on the other hand, the frequency locations of the cycles are well localized along the frequency axis, but the exact location and timing of the missing data points are not very well captured or localized along the time axis.

The second test signal showing in Figure 2.*a* (top window) is composed of sine waves whose frequency shifts periodically across time in the low frequency region. Along the sample, however, there are three sharp transitory impulses. The power spectrum of the test signal is shown in the left window of Figure 2.*a*. It is seen there that the power spectrum is completely silent about the time-varying nature of the cycle and about the white noise impulses. Instead, it shows that there are simultaneously several major cycles contained in the low frequency region.

The central window in Figure 2.*a*, however, shows how remarkably the scalogram captures not only the time-varying nature of the low frequency cycle, but also the exact timing of the white noise impulses. Notice that the frequency of the shifting cycle is highly localized along the frequency dimension on one hand, and the timing of the frequency shift is also highly localized along the time dimension on the other hand.

As a comparison, the spectrogram based on WFT is shown in Figure 2.*b* and Figure 2.*c*. We see there that the spectrogram either gives an imprecise frequency localization of the time-varying low frequency cycle when the window size is small enough to adequately capture the timing of the high frequency impulses (Figure 2.*b*), or misses the impulses entirely when the window size is large enough to capture adequately the frequency location of the time-varying low frequency cycle in the original signal (Figure 2.*c*). This is so because both the time and the frequency resolutions of WFT are fixed once the window length is fixed. In contrast, scalogram allows good frequency resolution at low frequencies and good time resolution at high frequencies.

VI. Application to Economic Data

Since Second World War, the US economy has experienced several important institutional changes. These institutional changes have likely had important impact on the structure of the US economy. The US economy has also experienced several unprecedented shocks that may also have brought deep structural adjustment to the economy. The oil price shock during the early 70s, for example, could have resulted in a fundamental reorganization of the input-output structure in the economy, especially with regard to the energy-intensive industries.

It is then of great interest to investigate whether these changes have also brought fundamental changes to the nature of the US business cycle. In particular, it is of great interest to know whether the old business cycles observed by economists almost half century ago are still alive, and whether new business cycles have emerged during those years of social changes and economic development.

Applying the wavelet-based time-frequency transform to the growth rate of real GDP (1960:1 - 1996:3), we find that the US business cycle has the following defining features:

1) Business cycles through out the sample period are concentrated mostly in the frequency region below 10 quarters per cycle. They are triggered mostly by external shocks.

2) Business cycles become far more active during the 70s and 80s after the oil price shocks in the early 70s. The two most active business cycles occurred around 1974 and 1983, both are triggered apparently by external impulses. The periodicity of the two cycles is about 6 years per cycle.

3) There exist business cycles that are not triggered by any external shocks to GDP, such as the 1991 business cycle. On the other hand, strong external shocks to GDP do not necessarily trigger business cycles, such as the shocks during 1977-1978.

Figure 3 shows the contour of energy distribution of the US GDP growth across time and frequency. The time series (top window) reveals very little about the frequency location of the cycles, while the spectrum (left window) reveals nothing about the timing of the different cycles. The scalogram (center), however, shows that there have been three major business cycles since 1960. The first occurred in 1961, triggered by a sharp external impulse during that year. The 1961 cycle has a frequency of 0.1 cycles per quarter (or 10 quarters per cycle) and is short lived (it lasted about one year). The second major cycle took place in 1973, apparently triggered by two impulses during 1972 and 1973, and was greatly intensified by another impulse near 1975. This business cycle lasted about 3-4 years and peaked at the frequency of about 0.04 cycles per quarter (or 25 quarters per cycle). The third major cycle occurred during 1982-1984, apparently triggered by a shock in 1982. This cycle lasted about 3 years and peaked also at a frequency similar to the 1973 cycle. The 1973 cycle and the 1982 cycle dominated all other business cycles since 1960. Notice that the cycle in 1991 is very mild compared to the three major cycles mentioned above. It is apparently not triggered by any external shocks to GDP. The scalogram also reveals that a major shock around 1977-1978 did not trigger any business cycle around that time. In addition, there is a short-lived business cycle in 1966 triggered by an external impulse that is not obvious or noticeable, however, in the original time series (see top window in Figure 3).

We think that these findings are of great importance to the business cycle theory. They not only help us identify the important historical shocks that triggered the business cycle, but also provide important information regarding the evolution of the business cycle across time. If the business cycle is unstable over time, for example, then there is the need for finding a common propagation mechanism to explain that instability. Without exception, existing real business cycle models all predict a stable business cycle with the same characteristic frequencies. But the scalogram shows otherwise: business cycles come and go; they emerge at different frequencies and at different times; they are not at all alike.

VII. Conclusions

A new technique of nonstationary time series analysis based on joint time-frequency representation was proposed. Two popular time-frequency distributions, the wavelet transform and the windowed Fourier transform were compared for this purpose. Our analyses showed that the wavelet-based time-frequency analysis is superior to the Fourier transform based time-frequency analysis. Applying the wavelet-based analysis to economic data, we found that business cycles in the US have not been stable over time. In particular, business cycles became far more active since the oil price crisis in the early 70s.

References

[1] Allen J. B., Rabiner L. R., "A unified approach to short-time Fourier analysis and synthesis," Proceedings of the IEEE, vol. 65, no. 11, 1977, pp. 1558-64.

[2] Boashash B., "Theory, implementation and application of time-frequency signal analysis using the Wigner-Ville distribution," Journal of Electrical and Electronics Engineering, vol. 7, no. 3, 1987, pp. 166-177.

[3] Cohen L., "Time-frequency distributions – a review," Proceedings of the IEEE, vol. 77, no. 7, 1989, pp. 941-981.

[4] Daubechies I., "The wavelet transform, time-frequency localization and signal analysis," IEEE Transactions on Information Theory, vol. 36, no. 5, 1990, pp. 961-1005.

[5] Gabor D., "Theory of communication," J. Inst. Elec. Eng., vol. 93, 1946, pp. 429-457.

[6] Hlawatsch F., Boudreaux-Bartels G. F., "Linear and quadratic time-frequency signal representation," IEEE Signal Processing Magazine, 1992, pp. 21-67.

[7] Kaiser G., "A friendly guide to wavelets," Birkhauser, Boston, 1994.

[8] Lin Z., "An introduction to time-frequency signal analysis," Sensor Review, vol. 17, no. 1, 1997, pp. 46-53.

[9] Nawab S. N., Quatieri T. F., "Short-time Fourier transform," In Lim, J. S. and Oppenheim, A. V. (Eds), Advanced Topics in Signal Processing, Prentice-Hall, Englewood Cliffs, NJ, 1988.

[10] Ramsey J., "The contribution of wavelets to the analysis of economic and financial data," Phil. Trans. R. Soc. Lond. A (forthcoming), 1996.

[11] Ramsey J., Zhang Z., "The analysis of foreign exchange rates using waveform dictionaries," Journal of Empirical Finance, 4, 1997, pp. 341-372.

[12] Ramsey J., Usikov D., Zaslavsky G., "An analysis of US stock price behavior using wavelets," Fractals, vol. 3, no. 2, 1995, pp. 377-389.

[13] Rioul O., Flandrin P., "Time-scale energy distributions: a general class extending wavelet transforms," IEEE Transactions on Signal Processing, vol. 40, no. 7, 1992, pp. 1746-57.
[14] Rioul O., Vetterli M., "Wavelets and signal processing," IEEE Signal Processing Magazine, 1991, pp. 14-38.
[15] Stankovic L., "An analysis of some time-frequency and time-scale distributions," Annals of Telecommunications, vol. 49, no. 9-10, 1994, pp. 505-517.
[16] Teolis A., "Computational signal processing with wavelets," 1964.
[17] Vetterli M., Harley C., "Wavelets and filter banks: theory and design," IEEE Transactions on Signal Processing, vol. 40, no. 9, 1992, pp. 2207-2232.
[18] Wen Y., Zeng B., "A simple nonlinear filter for economic time series analysis," Economics Letters, 64, 1999, pp. 151-160.

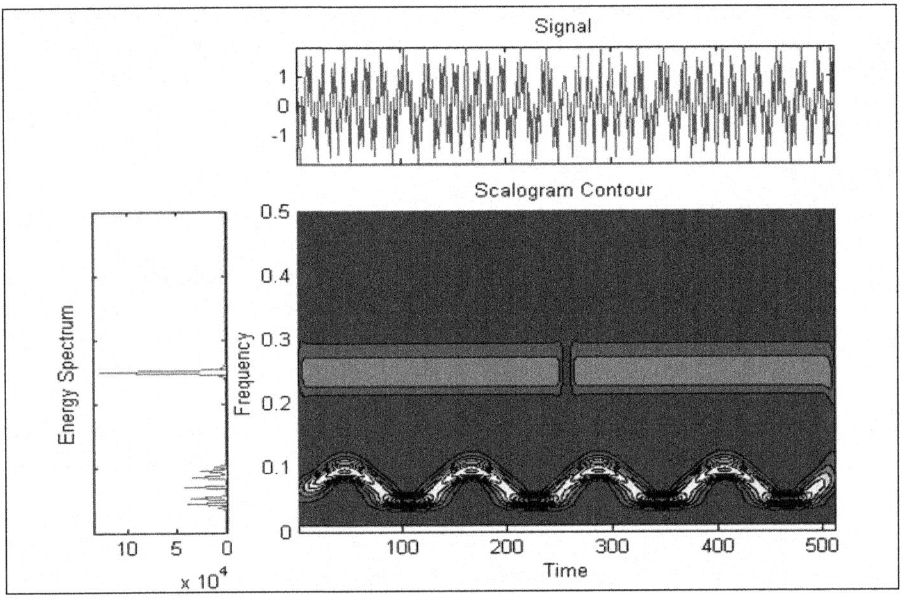

Figure 1.*a*: Scalogram contour with signal (top) and spectrum (left).

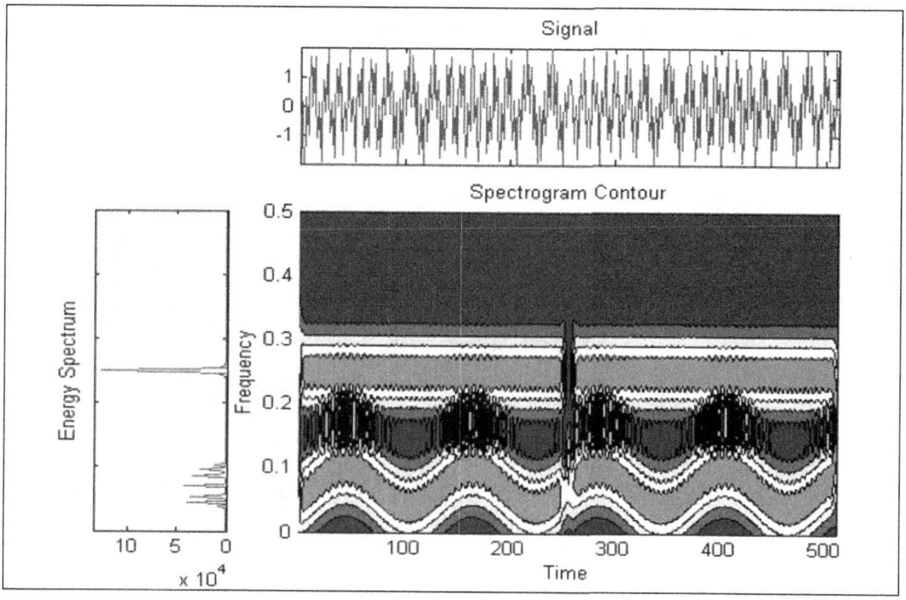

Figure 1.*b*: Spectrogram contour with signal (top) and spectrum (left) (window = 13).

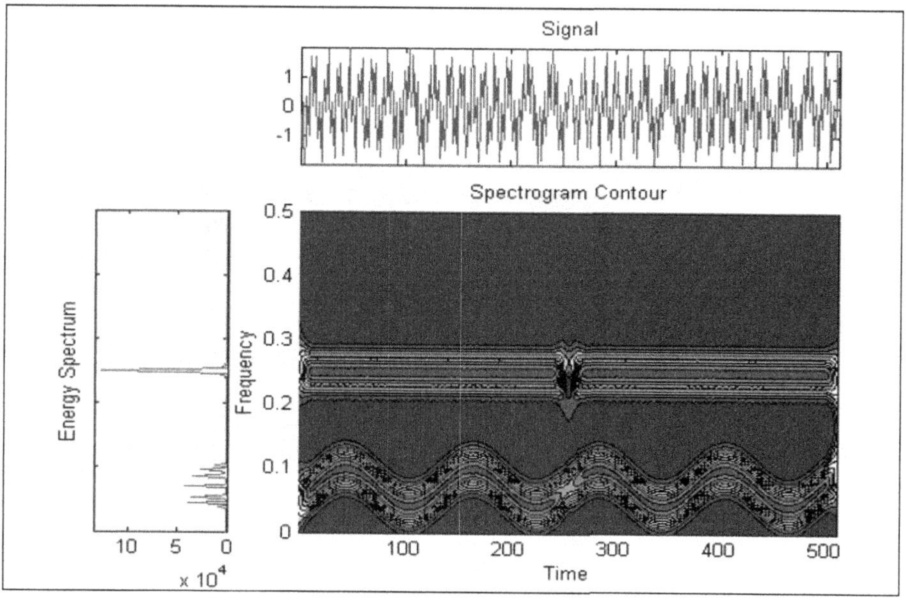

Figure 1.*c*: Spectrogram contour with signal (top) and spectrum (left) (window = 27).

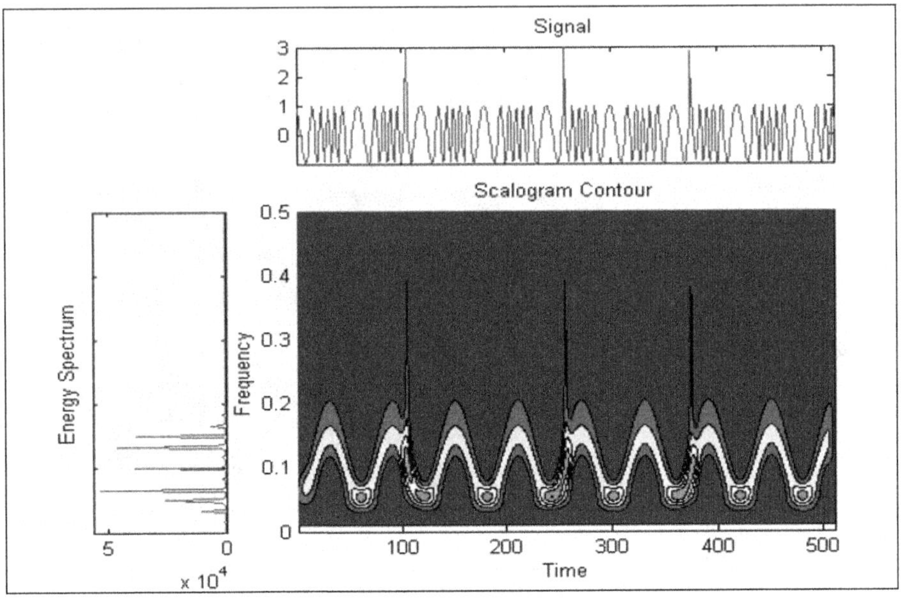

Figure 2.*a*: Scalogram contour with signal (top) and spectrum (left).

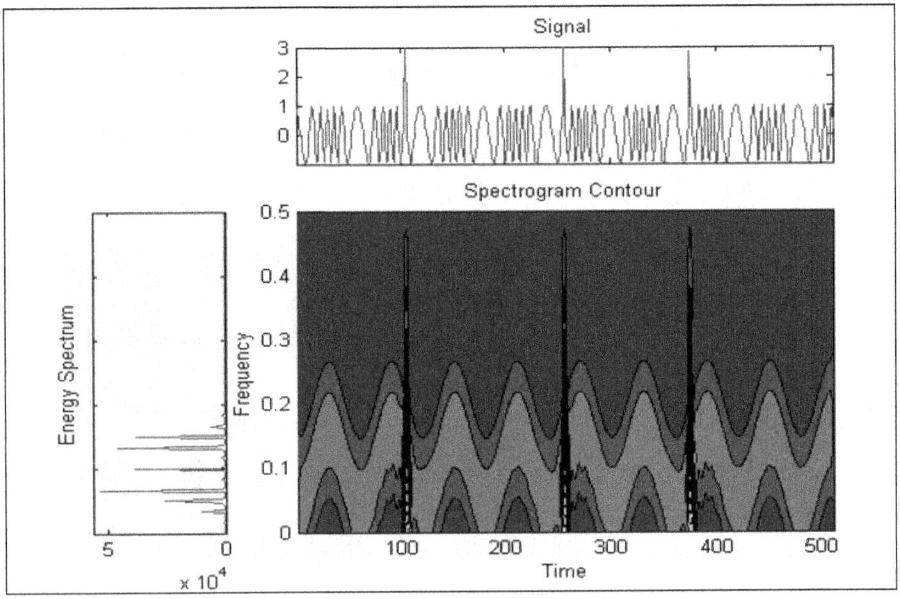

Figure 2.*b*: Spectrogram contour with signal (top) and spectrum (left) (window = 7).

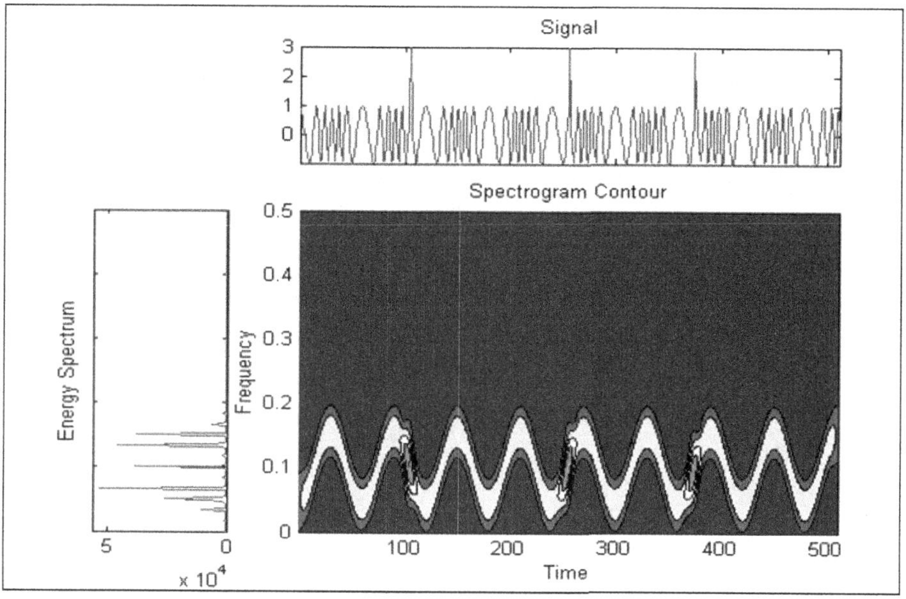

Figure 2.*c*: Spectrogram contour with signal (top) and spectrum (left) (window = 21).

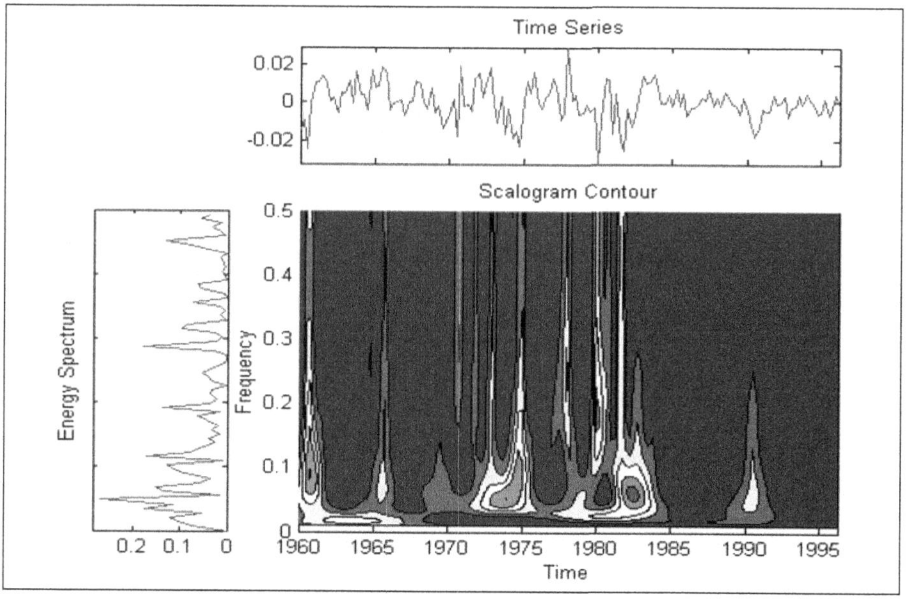

Figure 3: Scalogram contour with time series (top) and spectrum (left).
U.S. GDP growth rate (1960:1 - 1996:3).

$$\Phi \quad = \quad \Phi \quad -$$

$$\phi \quad = \quad \phi \quad -$$

$$= \quad - \quad {}_{+}\Phi \quad +$$

$$\Phi$$

$$=$$

$$_{+} \quad = \quad \delta$$

$$\delta$$

$$\phi \quad = \quad \phi \quad -$$

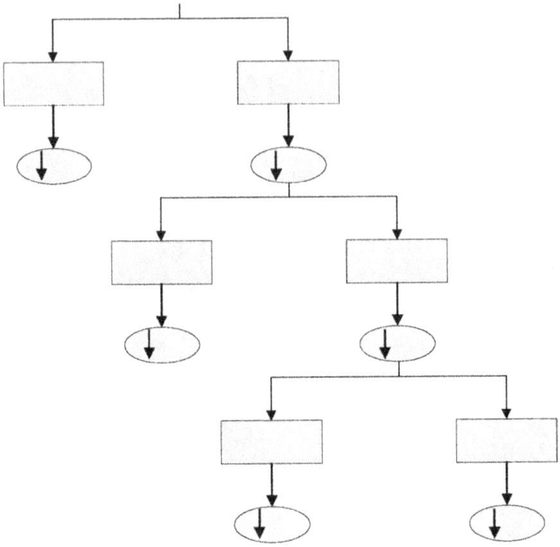

$$= - \quad - \; + \quad -$$
$$=$$

$$= - \quad - \quad + \quad + \; - \quad -$$
$$=$$

```
1. counter1 = 2 * (i - 1)
2. counter2 = mod (counter1, data size)
3. counter3 = mod (counter1 + min( data size, length(1)) -
            1), data size)
4.  for  n from counter2 until counter3 do
   4.1. calculate index to get highpass value, mod(n -
        counter1, data size)
   4.2. calculate multiplication between highpass value
  with subscript index + 1 and data with subscript
n + 1
   4.3. store result in variable,  b
5. end for
6. Result  tally  in 4.3
```

```
1.  flag = 0
2.  call function get_h to get highpass value from lowpass
    value
3.  while  data size >= 2
    3.1 value of highpass and lowpass will be set
    3.2 if  flag = 0 then
        3.2.1 for I := 0 to ( data size /2) do
        3.2.1.1 call  lowpass, result copy to array d.
        3.2.1.2 call  highpass, result copy to array h.
        3.2.1.3 copy value in d  to array  temp
        3.2.1.4 Set flag = 1
        3.2.2  end for.
        3.2.3 copy value in array h to final array
       result w in descending way.
    3.3 else
        3.3.1 repeat process 3.2.1.1 and 3.2.1.2 but
              inputs for lowpass and highpass are the
              elements in array temp
        3.3.2 intializing array temp.
        3.3.3 repeat process 3.2.1.3.
        3.3.4 repeat process 3.2.3 to get the next result
              of highpass
     3.4 end if.
     3.5 Data size = ( Data size /2)
4.  end while.
5.  end.
```

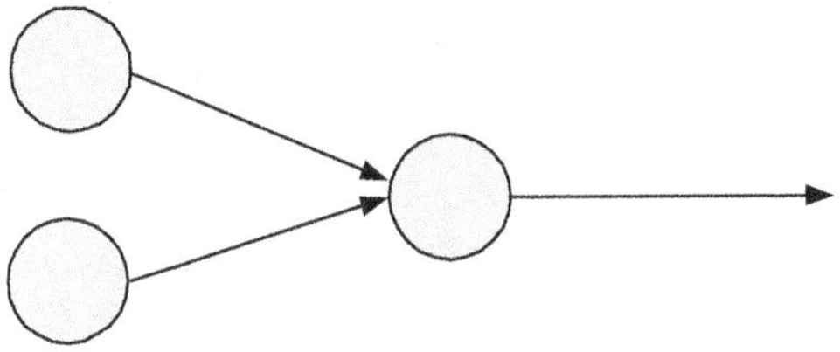

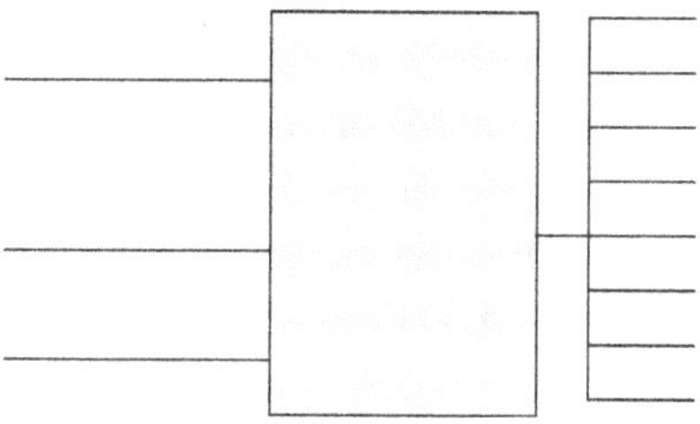

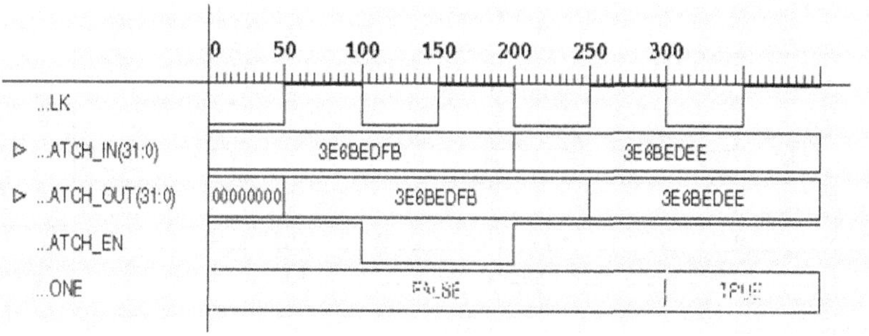

0
50
100
150
200
250
300
...LK
...ATCH_IN(31:0)
3E6BEDFB
3E6BEDEE
...ATCH_OUT(31:0)
00000000
3E6BEDFB
3E6BEDEE
...ATCH_EN
...ONE

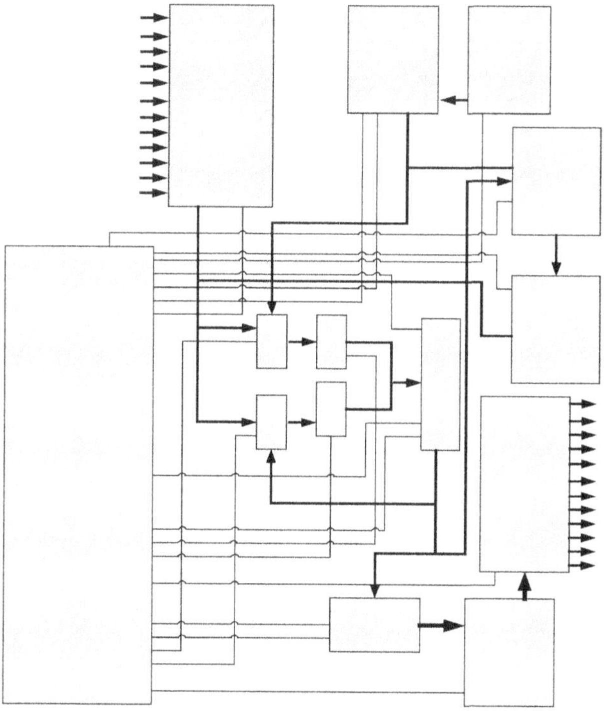

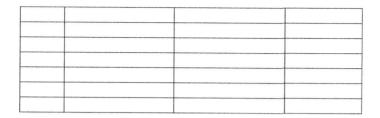

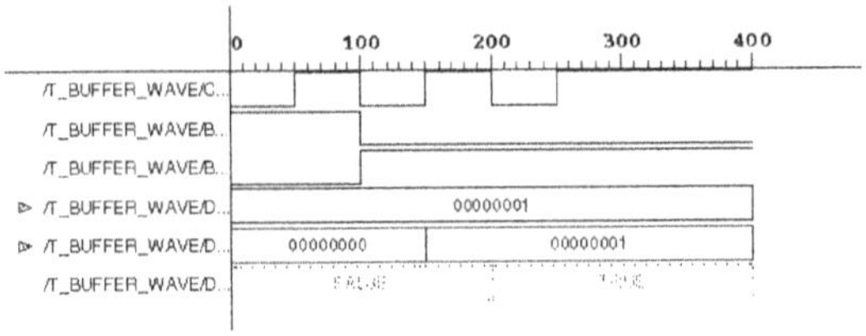
0
100
200
300
400
/T_BUFFER_WAVE/C...
/T_BUFFER_WAVE/B...
/T_BUFFER_WAVE/B...
/T_BUFFER_WAVE/D...
00000001
/T_BUFFER_WAVE/D...
00000000
00000001
/T_BUFFER_WAVE/D...

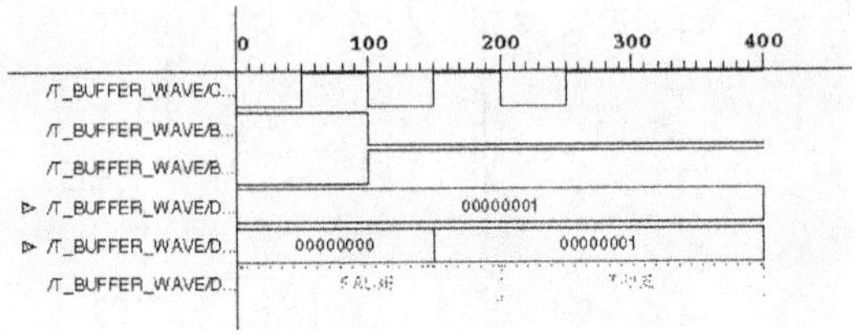
0
100
200
300
400
/T_BUFFER_WAVE/C...
/T_BUFFER_WAVE/B...
/T_BUFFER_WAVE/B...
/T_BUFFER_WAVE/D...
00000001
/T_BUFFER_WAVE/D...
00000000
00000001
/T_BUFFER_WAVE/D...

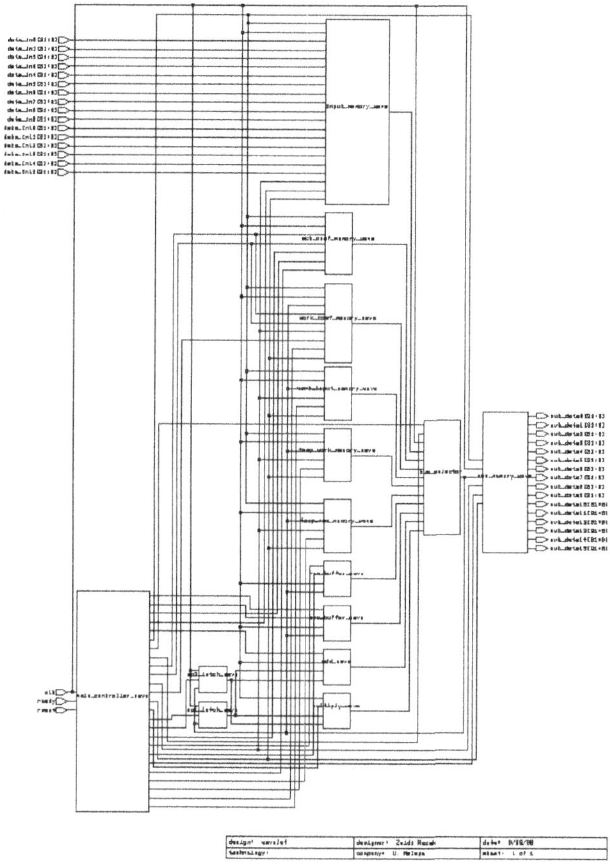
design: wavelet
designer: Zaidi Razak
technology:
company: U. Malaya
sheet: 1 of 1

{ahmadi,guyd}@ppc.ubc.ca

{tafreshi,sassani}@mech.ubc.ca

$\omega_\gamma = \quad - \quad =$

ω

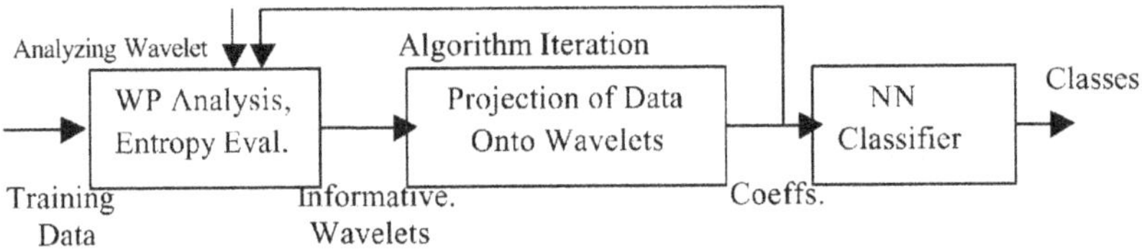

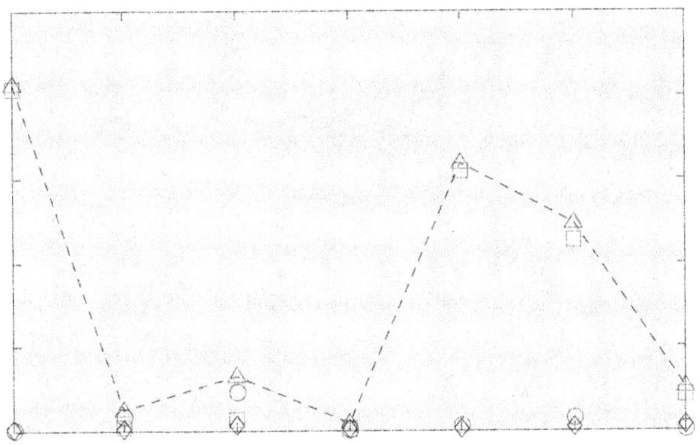

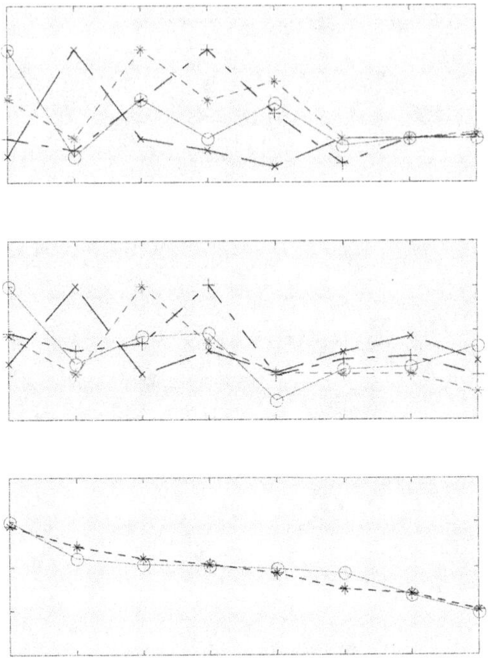

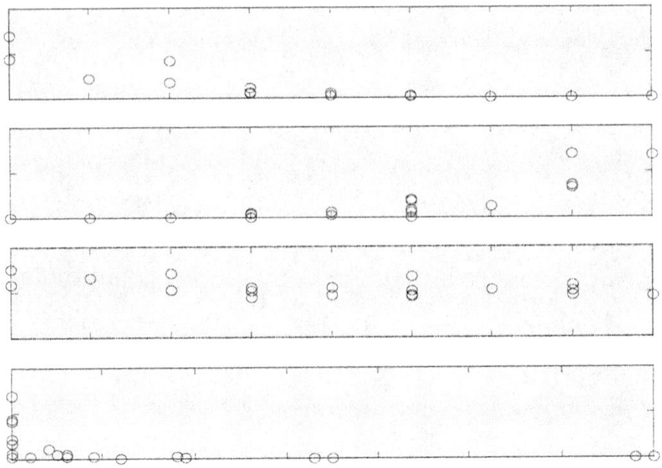

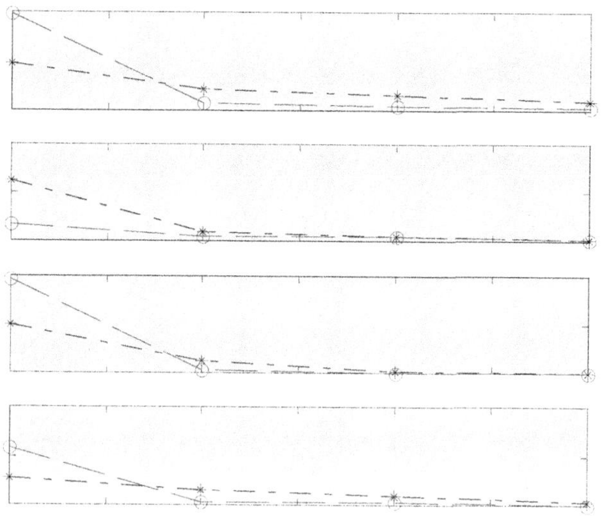

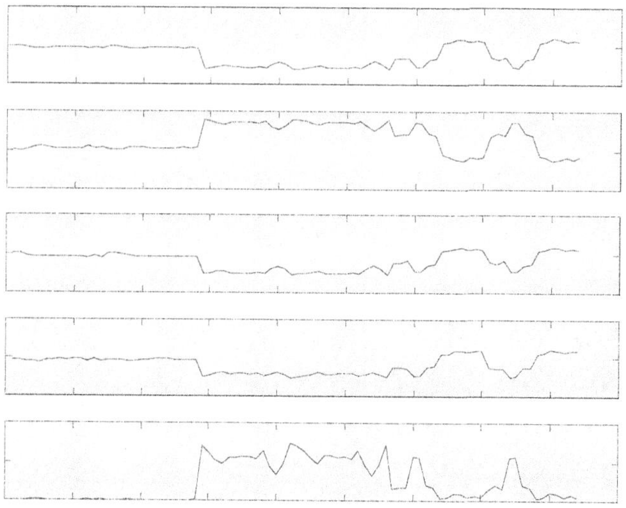

$$(\tau) = \ (\tau)\ (\tau - \ )$$

τ

$$(\omega) = \frac{}{\sqrt{\ \pi}} \quad {}^{-\ \omega\tau} \ (\tau)\ \ \tau$$

$$= \frac{}{\sqrt{\ \pi}} \quad {}^{-\ \omega\tau} \ (\tau)\ (\tau - \ )\ \ \tau$$

$$(\ \ \omega) = |\ \ (\omega)| \ = \left| \frac{}{\sqrt{\ \pi}} \quad {}^{-\ \omega\tau} \ (\tau)\ (\tau - \ )\ \ \tau \right|$$

ω ω

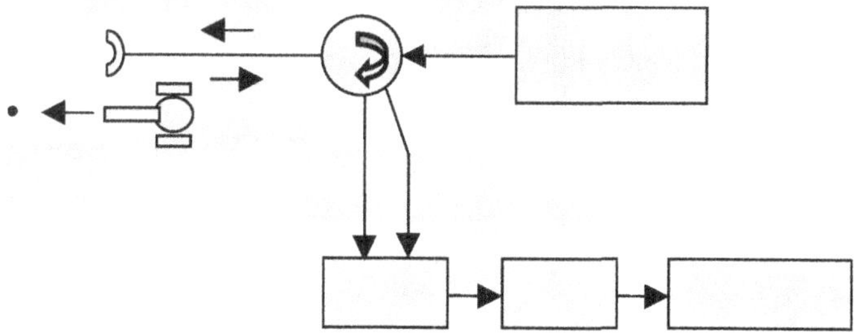

$$= \quad \times \quad \times \lambda$$

λ λ

$$\lambda = -$$

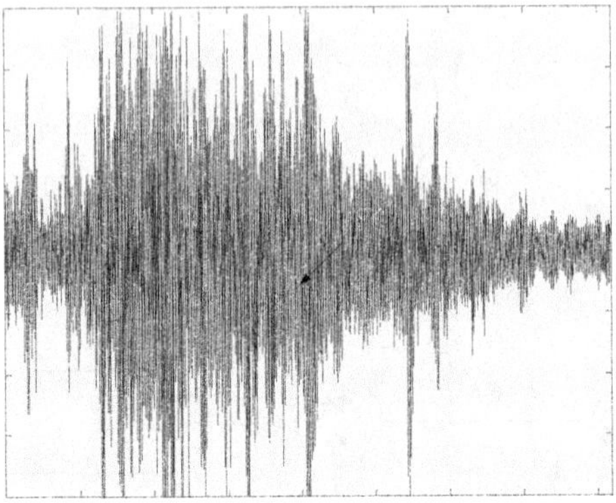

$$= \quad \times \quad \times \lambda \times$$

λ

$$= \frac{\lambda}{\lambda} = \frac{}{\sqrt{- \quad \frac{\times}{\pi \times \quad \times}}}$$

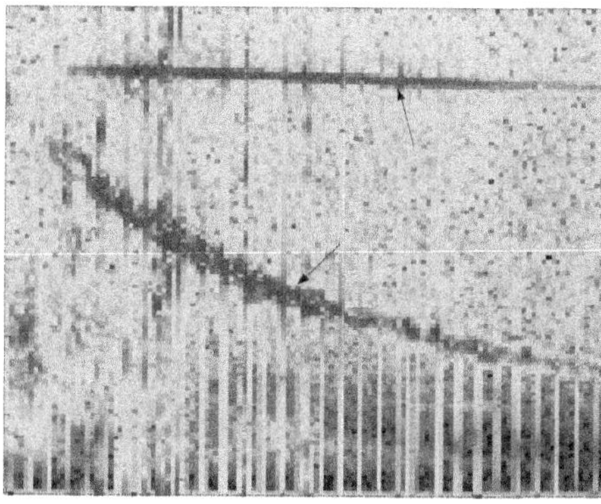

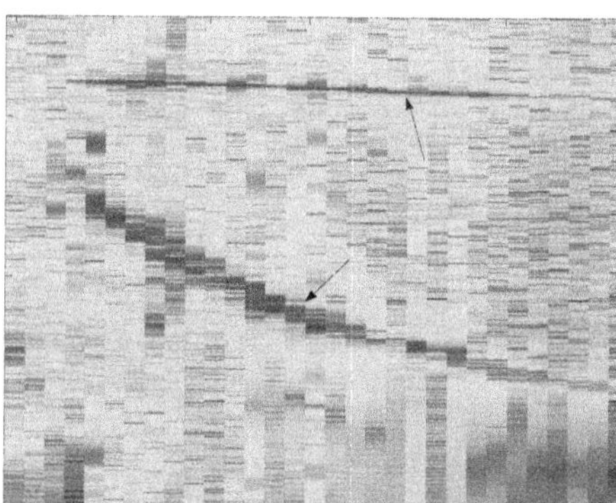

± ×

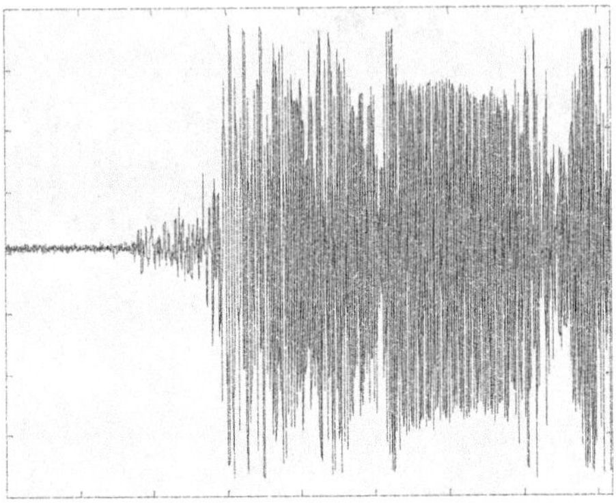

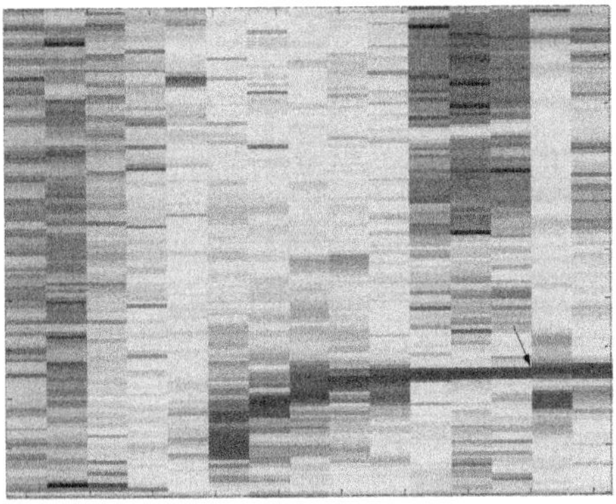

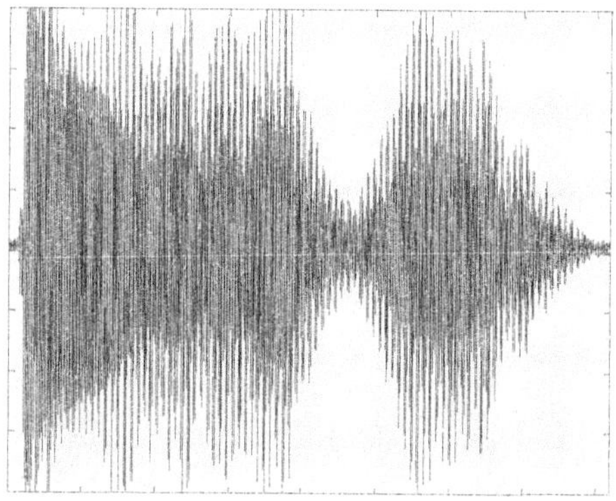

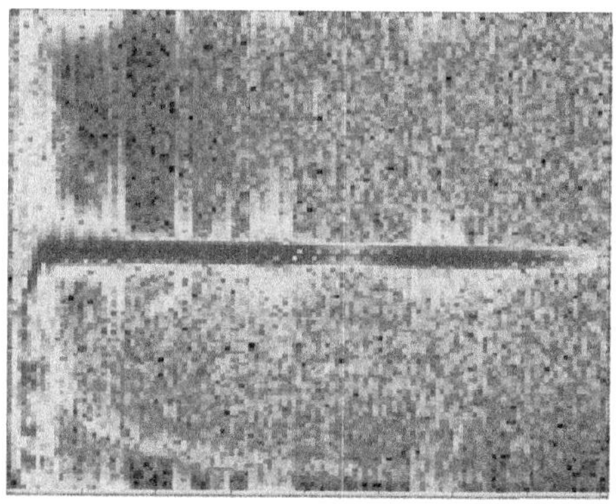

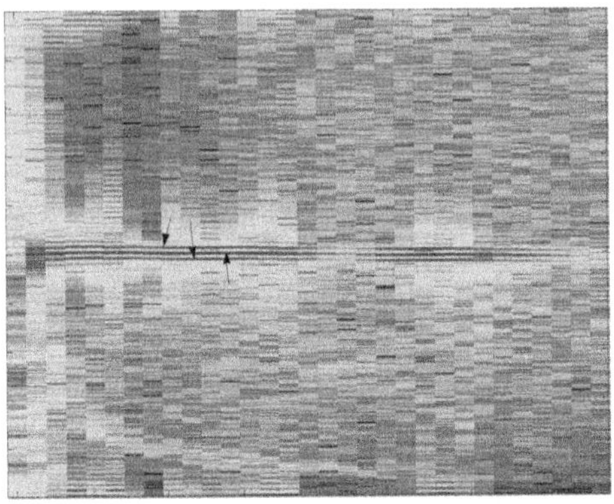

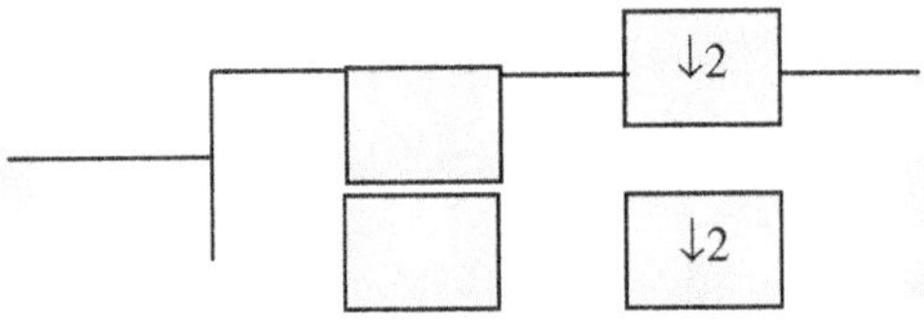

↓2
↓2

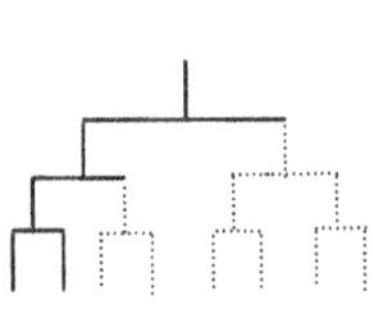

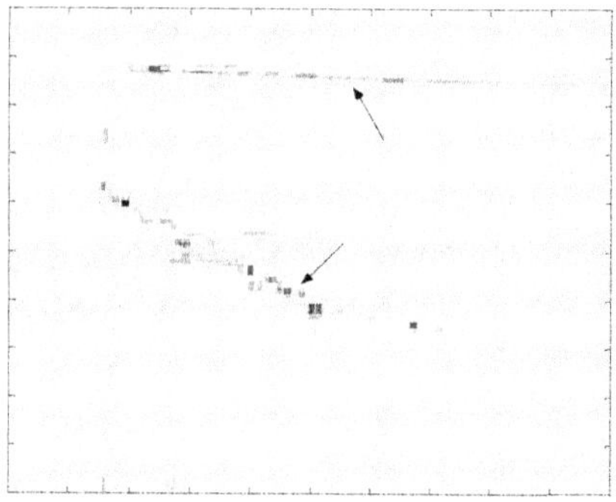

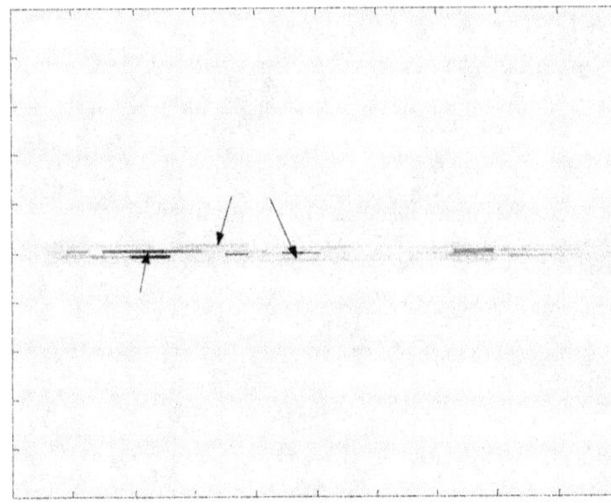

jpli2222@sina.com

yytang@comp.hkbu.edu.hk

$\psi = | \; |^{-}$ ψ ——

ψ

ψ

ψ

ψ

Piecewise Periodized Wavelet Transform and Its Realization, Properties and Applications

Wing-kuen Ling and P. K. S. Tam

Department of Electronic and Information Engineering
The Hong Kong Polytechnic University
Hung Hom, Kowloon, Hong Kong
Hong Kong Special Administrative Region, China
Tel: (852) 2766-6238, Fax: (852) 2362-8439
Email: bingo@encserver.eie.polyu.edu.hk

Abstract. In this paper, the realization of piecewise periodized wavelet transform (PPWT) is introduced and some interesting properties are discussed. The signals in both the time and frequency domains are of finite length. The inverse of the transform exists if the inverse of all the N wavelet kernels exist. Under certain conditions, a time shift of the input leads to a shift in the output of another non-uniform filter bank sub-system. The frequency response of each non-uniform filter bank sub-system consists of a set of frequency components with fundamental frequency different from the input frequency. Some applications, such as adaptive regional image processing techniques, analysis of aperiodic switching systems, and design of frequency to voltage converters, are readily facilitated by the transform.

1 Introduction

Wavelet transform provides an effective means to decompose and analyze a signal in a multi-resolution scale in the wavelet domain [1].

However, there are some applications that require the dividing of a signal into several segments and applying different transforms to different segments. For example, the Karhunen-Loève transform (KLT) for image processing involves the transforming of different image segments with different kernels [2, 3]. In these applications, the signals cannot be so readily represented by the traditional wavelet transform.

In order to represent the signals in those systems by a wavelet transform, a piecewise wavelet transform is introduced [10]. However, since the input signal is segmented in the time domain, each segment is of finite duration. By the Heisenberg uncertainty principle, the corresponding frequency spectrum is of infinite bandwidth.

To work on this problem, a piecewise periodized wavelet transform (PPWT) is introduced in this paper. The domain of a signal **U** is first partitioned into N different segments $\wp_i$, where $i=0,1,\ldots,N-1$,

$$\wp_i \cap \wp_j = \Phi \text{, for } i \neq j,$$

$$\bigcup_{\forall i} \wp_i = U, \tag{1}$$

and Φ is the null set. Then each segment of the signal is truncated by passing it through a window, and by convoluting it with an impulse train. For the continuous-time case, the periodized segmented signal is:

$$\tilde{x}_i(t) = [x(t) \cdot rect_i(t)] * \sum_{k \to -\infty}^{\infty} \delta(t - k \cdot \Im_i),$$

where

$$rect_i(t) = \begin{cases} 1 & ;\text{if } t \in \wp_i, \\ 0 & ;\text{otherwise,} \end{cases}$$

$$\Im_i = \max_{t \in \wp_i}(t) - \min_{t \in \wp_i}(t), \tag{2}$$

$\delta(t)$ and $*$ denote the continuous-time unit impulse function and the convolution operator, respectively. Similarly, for the discrete-time case, the periodized segmented signal is:

$$\tilde{x}_i[n] = (x[n] \cdot rect_i[n]) * \sum_{k \to -\infty}^{\infty} \delta[n - k \cdot \Im_i],$$

where

Y. Y. Tang et al. (Eds.): WAA 2001, LNCS 2251, pp. 398-403, 2001.

$$rect_i[n] = \begin{cases} 1 & ;\text{if } n \in \wp_i, \text{ and} \\ 0 & ;\text{otherwise,} \end{cases}$$

$$\Im_i = \max_{n \in \wp_i}[n] - \min_{n \in \wp_i}[n] \cdot \tag{3}$$

The PPWT is defined as the wavelet transform of the periodized segmented signal. It can be further classified as follows: The continuous-time piecewise periodized wavelet transform (CTPPWT) is defined as:

$$\forall a,b \in \Re,\ i=0,1,\ldots,N-1,\ CTPPWT(i,a,b) = \frac{1}{\sqrt{|b|}} \cdot \int_{t \to -\infty}^{+\infty} \tilde{x}_i(t) \cdot \psi^*_i\left(\frac{t-a}{b}\right) dt \cdot \tag{4}$$

The discrete piecewise periodized wavelet transform (DPPWT) is defined as:

$$\forall k,n \in \mathbf{Z},\ i=0,1,\ldots,N-1,\ DPPWT[i,k,n] = a_i^{-\frac{k}{2}} \cdot \int_{t \to -\infty}^{+\infty} \tilde{x}_i(t) \cdot h_i\left(n \cdot T_i - a_i^{-k} \cdot t\right) dt, \tag{5}$$

where T_i and a_i are the fundamental sampling period and the dilation constant of the ith wavelet kernel, respectively.

The discrete-time piecewise periodized wavelet transform (DTPPWT) is defined as:

$$\forall j,n \in \mathbf{Z},\ i=0,1,\ldots,N-1,\ DTPPWT[i,j,n] = \sum_{k \to -\infty}^{+\infty} \tilde{x}_i[k] \cdot h_{i,j}\left[n \cdot M_{i,j} - k \cdot L_{i,j}\right], \tag{6}$$

where $M_{i,j}$ and $L_{i,j}$ are the downsampling and upsampling ratios of the jth channel of the ith wavelet kernel, respectively.

Based on the above definitions, the realization of CTPPWT, DPPWT and DTPPWT is suggested in section II. A comparison on the properties of the proposed transform and the traditional wavelet transform is discussed in section III. In section IV, the advantages and applications of the proposed transform are investigated. Finally, a conclusion is given in section V.

2 Realization of Piecewise Periodized Wavelet Transform

Since the traditional discrete-time wavelet transform (DTWT) can be realized by a non-uniform filter bank system [4, 5], the DTPPWT can also be realized by a set of N modified non-uniform filter bank sub-systems, as shown in figures 1a and 1b. For the realization of DPPWT, the window functions and the impulse trains are changed to the corresponding continuous-time signals, and the non-uniform filter bank sub-systems become banks of continuous-time filters followed by A/D converters as shown in figures 1c and 1d. The realization of CTPPWT is indicated in figure 1e.

3 Comparing the Properties of Piecewise Periodic Wavelet Transform and Traditional Wavelet Transform

3.1 Time-frequency Localization

For the traditional wavelet transform, if the signal is of finite length in the time domain, then the signal is of infinite length in the frequency domain, and vice versa [6]. However, for PPWT, since the segmented signal is periodized in the time domain, it is of finite length in the frequency domain. For example, if:

$$\wp_0 = \left\{ t \text{ such that } \frac{2 \cdot \pi}{m \cdot \omega_0} \geq t \geq 0 \right\},$$

$$\wp_1 = \left\{ t \text{ such that } 0 > t \geq -\frac{2 \cdot \pi}{n \cdot \omega_0} \right\},$$

$$\wp_2 = \left\{ t \text{ such that } t > \frac{2 \cdot \pi}{m \cdot \omega_0} \right\},$$

$$\wp_3 = \left\{ t \text{ such that } t < -\frac{2 \cdot \pi}{n \cdot \omega_0} \right\},$$

$$x(t)=\begin{cases} e^{j\cdot m\cdot\omega_0\cdot t} & ;\text{for } \frac{2\cdot\pi}{m\cdot\omega_0}\geq t\geq 0, \\ e^{j\cdot n\cdot\omega_0\cdot t} & ;\text{for } 0>t\geq -\frac{2\cdot\pi}{n\cdot\omega_0} \text{ and } m\neq n, \\ 0 & ;\text{otherwise,} \end{cases}$$

$$\psi^*_i(t)=e^{jt} \text{ for } \forall t\in\Re \text{ and } i=0,1,2,3, \tag{7}$$

then we have:

$$CTPPWT(0,a,b)=2\cdot\pi\cdot\sqrt{m\cdot\omega_0}\cdot e^{j\cdot a\cdot m\cdot\omega_0}\cdot\delta\left(b+\frac{1}{m\cdot\omega_0}\right),$$

$$CTPPWT(1,a,b)=2\cdot\pi\cdot\sqrt{n\cdot\omega_0}\cdot e^{j\cdot a\cdot n\cdot\omega_0}\cdot\delta\left(b+\frac{1}{n\cdot\omega_0}\right),$$

$$CTPPWT(2,a,b)=CTPPWT(3,a,b)=0\,. \tag{8}$$

From this example, it can be seen that the signal in the frequency domain is localized within a very small region. In the time domain, the signal is segmented into the regions $\wp_0$ and $\wp_1$ and localized within windows with sizes of $(2\cdot\pi)/(m\omega_0)$ and $(2\cdot\pi)/(n\omega_0)$, respectively.

3.2 Inverse Piecewise Periodized Wavelet Transform

Since the PPWT is made up of N different wavelet kernels, if all these wavelet kernels are invertible, then the periodized segmented signal can be determined from the transformed signal. The periodized segmented signal can be further processed by first truncating it to segments of one period duration. Then the overall output can be readily constructed from summing up the processed segments. The formulae for the inverse piecewise periodized wavelet transform (IPPWT) are shown below:

Inverse continuous-time piecewise periodized wavelet transform (ICTPPWT):

$$\forall t\in\Re,\ x(t)=\sum_{i=0}^{N-1}rect_i(t)\cdot\left[\frac{1}{K_i}\cdot\int_{a\to-\infty}^{+\infty}\int_{b\to-\infty}^{+\infty}CTPPWT(i,a,b)\cdot\psi_i\left(\frac{t-a}{b}\right)\frac{dadb}{b^2}\right],$$

where

$$K_i=\int_{f\to-\infty}^{+\infty}\frac{|\Psi_i(f)|^2}{f}df \text{ and } \Psi_i(f)=\int_{t\to-\infty}^{+\infty}\psi_i(t)\cdot e^{-j\cdot 2\cdot\pi\cdot f\cdot t}dt\,. \tag{9}$$

Inverse discrete piecewise periodized wavelet transform (IDPPWT):

$$\forall t\in\Re,\ x(t)=\sum_{i=0}^{N-1}rect_i(t)\cdot\left[\sum_{k\to-\infty}^{+\infty}\sum_{n\to-\infty}^{+\infty}a_i^{-\frac{k}{2}}\cdot DPPWT[i,k,n]\cdot f_i\left(a_i^{-k}\cdot t-n\cdot T_i\right)\right],$$

where

$$f_i(t)=h_i(-t)\,. \tag{10}$$

Inverse discrete-time piecewise periodized wavelet transform (IDTPPWT):

$$\forall n\in\mathbf{Z},\ x[n]=\sum_{i=0}^{N-1}rect_i[n]\cdot\left[\sum_{j\to-\infty}^{+\infty}\sum_{m\to-\infty}^{+\infty}DTPPWT[i,j,m]\cdot f_{i,j}\left[L_{i,j}\cdot n-M_{i,j}\cdot m\right]\right], \tag{11}$$

where $f_{i,j}[n]$ and $h_{i,j}[n]$ are the filters in the ith perfect reconstruction non-uniform filter bank sub-systems [7].

3.3 Shift Invariance of Discrete-time Piecewise Periodized Wavelet Transform

If the outputs of all the channels of a traditional non-uniform filter bank system are summed up together and all the decimators and expanders have the same downsampling ratio M and upsampling ratio L, respectively, then the non-uniform filter bank system has the (L,M) shift property [8]. Hence, a system has the (L,M) shift characteristics if the input-output relationship of the system is governed by a DTWT as follows:

$$\forall n\in\mathbf{Z},\ y[n]=\sum_{j\to-\infty}^{+\infty}\sum_{k\to-\infty}^{+\infty}x[k]\cdot h_j[n\cdot M-k\cdot L]\,. \tag{12}$$

For DTPPWT, since the input signal is multiplied by a window function $rect_i[n]$, the whole system becomes time varying and does not have the shift invariant property in general. However, if we partition the domain of a signal uniformly and assume that all the wavelet kernels are the same, that is, $\Im_i=\Im$, $rect_i[n]=rect_{i-1}[n-\Im]$, $M_{i,j}=M_j$, $L_{i,j}=L_j$ and $h_{i,j}[n]=h_j[n]$ for $i=0,1,\ldots,N-1$, by summing up all the channels in the ith non-uniform filter bank sub-system together, we have:

$$\forall n\in \mathbf{Z},\ i=0,1,\ldots,N-1,\quad y[i,n]=\sum_{j\to-\infty}^{\infty}\sum_{k\to-\infty}^{\infty}\tilde{x}_i[k]\cdot h_j[n\cdot M_j-k\cdot L_j], \tag{13}$$

the output will shift from the output $y[i,n]$ of the ith non-uniform filter bank sub-system to that of the i-1th non-uniform filter bank sub-system $y[i-1,n]$ if the input is shifted by $\Im$.

3.4 Frequency Response of Discrete-time Piecewise Periodized Wavelet Transform

The frequency response of a traditional non-uniform filter bank system consists of a set of complex exponential functions with frequencies $M_j{\cdot}k_{j,p}{\cdot}\omega_0/L_j$, where $k_{j,p}=0,1,\ldots,L_j-1$ [9]. However, it is not the case for the DTPPWT because the fundamental period of a periodized segmented signal is equal to the window size $\Im_i$, but not that of the input signal. Hence, the fundamental frequency of the output signal of each sub-system is also different from that of the input signal.

In general, if the window size $\Im_i$ is not equal to a multiple of the fundamental period of the input signal $2{\cdot}\pi/\omega_0$, that is, $\Im_i\neq 2{\cdot}\pi{\cdot}k/\omega_0$, where k is a natural number, then the output of the system consists of a set of complex exponential components at frequencies $2{\cdot}\pi{\cdot}M_j{\cdot}k_{j,p}/(\Im_i{\cdot}L_j)$. On the other hand, if $\Im_i=2{\cdot}\pi{\cdot}k/\omega_0$, then the frequency response of the DTPPWT will give a result similar to that of the DTWT. That is, it will consist of a set of complex exponential components at frequencies $M_j{\cdot}k_{j,p}{\cdot}\omega_0/L_j$.

This property can be applied to the analysis and design of frequency to voltage converter. The details are discussed in section IV.

4 Advantages and Applications of Piecewise Periodized Wavelet Transform

4.1 Adaptive Regional Image Processing Techniques

Adaptive regional image processing techniques, such as local histogram equalization, adaptive wavelet denoising and local singularity detection, can be modeled as DTPPWT. For an example in the adaptive wavelet denoising, we may apply a suitable lowpass filter to the quasi-stationary region of an image to suppress the high frequency noise, and a bandpass filter to the singularity region to enhance the image features. This can be modeled as DTPPWT. The equations of section I can be applied with $N=2$, $\wp_0$ the quasi-stationary region, $\wp_1$ the singularity region, $h_0[n]$ a lowpass filter, and $h_1[n]$ a bandpass filter.

Adaptive regional image processing techniques usually give better results than global image processing techniques because the local characteristics of an image are captured and different wavelet kernels are tailor-made to optimize the results. Hence, we gain some advantages on applying DTPPWT instead of the traditional wavelet transform.

4.2 Aperiodic Switching System

Since aperiodic switching systems, such as asynchronous power electronic switching circuits, do not have an (L,M) shift invariant property, those systems cannot be modeled by the traditional wavelet transform. Instead, they can be modeled by PPWT. Since several linear time-invariant (LTI) systems are connected in parallel and triggered only at certain time instants, the wavelet kernels of DTPPWT are the impulse responses of those LTI systems. Then the equations in Section I can be applied with N being the number of the LTI systems and $\wp_i$ being the time segment when the ith LTI system is triggered.

4.3 Frequency to Voltage Converter

Although the fundamental frequency of the periodized segmented signals depends on the window size $\Im_i$, their Fourier coefficients depend on the input frequency. The periodized segmented signals consist of harmonics and one of the harmonics can be picked out using a narrow bandwidth wavelet kernel. Then, the output is a single frequency signal where the frequency depends on the window size $\Im_i$ and the magnitude of this frequency component depends on the input signal and the wavelet kernels. Based on this phenomenon, a frequency to voltage converter can be designed.

One of the advantages of employing the proposed PPWT to design a frequency to voltage converter is to avoid the occurrence of limit cycles because the whole system is linear. Also, the instability problem can be avoided easily because the stability of the system depends only on the wavelet kernels. If all the wavelet kernels are stable, then the output is stable.

5 Conclusion

In this paper, a novel transform is proposed, analyzed and compared to the traditional wavelet transform. A modified non-uniform filter bank realization is given. Some properties of PPWT, such as time and frequency localization, invertibility, shift invariance and frequency response, are discussed. Some applications of PPWT, such as adaptive regional image processing techniques, analysis of aperiodic switching systems, and design of frequency to voltage converters, are discussed.

Acknowledgement

The work described in this paper was substantially supported by a grant from the Hong Kong Polytechnic University with account number G-V968.

References

1. Mallat S. G.: A Theory for Multiresolution Signal Decomposition: The Wavelet Representation. IEEE Transactions on Pattern Analysis and Machine Intelligence, Vol. 11, No. 7. (1989) 674-693.
2. Lee J.: Optimized Quadtree for Karhunen-Loeve Transform in Multispectral Image Coding. IEEE Transactions on Image Processing, Vol. 8, No. 4. (1999) 453-461.
3. Diamantaras K. I. and Strintzis M. G.: Optimal Transform Coding in the Presence of Quantization Noise. IEEE Transactions on Image Processing, Vol. 8, No. 11. (1999) 1508-1515.
4. Vetterli M. and Herley C.: Wavelets and Filter Banks: Theory and Design. IEEE Transactions on Signal Processing, Vol. 40, No. 9. (1992) 2207-2232.
5. Vaidyanathan P. P.: Multirate Systems and Filter Banks. Englewood Cliffs: Prentice-Hall, 1993.
6. Daubechies I.: The Wavelet Transform, Time-Frequency Localization and Signal Analysis. IEEE Transactions on Information Theory, Vol. 36, No. 5. (1990) 961-1005.
7. Nayebi K., Barnwell T. P. and Smith M. J. T.: Nonuniform Filter Banks: A Reconstruction and Design Theory. IEEE Transactions on Signal Processing, Vol. 41, No. 3. (1993) 1114-1127.
8. Shenoy R. G.: Multirate Specifications Via Alias-Component Matrices. IEEE Transactions on Circuits and Systems-II, Vol. 45, No. 3. (1998) 314-320.
9. Shenoy R. G., Burnside D. and Parks T. W.: Linear Periodic Systems and Multirate Filter Design. IEEE Transactions on Signal Processing, Vol. 42, No. 9. (1994) 2242-2256.
10. Stephanakis I. M., Stamou G. and Kollias S.: Piecewise Wiener Filter Model Based on Fuzzy Partition of Local Wavelet Features for Image Restoration. International Joint Conference on Neural Networks, IJCNN, Vol. 4. (1999) 2690-2693.

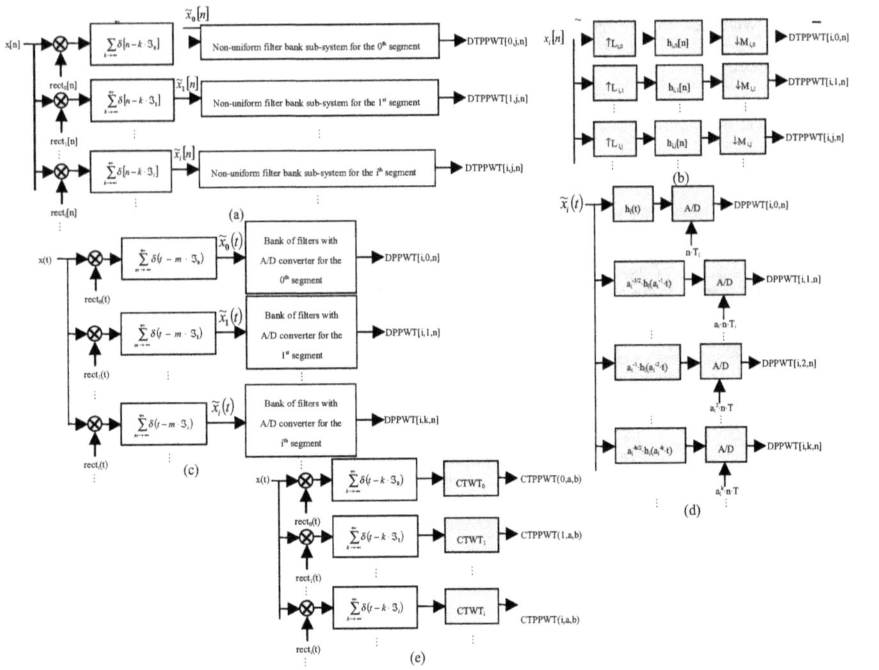

Fig. 1. (a) Realization of DTPPWT (b) Realization of the non-uniform filter bank sub-system for the ith segment (c) Realization of DPPWT (d) Realization of the bank of filters with A/D converters for the ith segment (e) Realization of CTPPWT

$\in$

$$= < \quad \psi \quad > = \frac{}{\left|\sqrt{\ }\right|} \int_{-\infty}^{\infty} \overline{\Psi \frac{-}{}}$$

$\psi \quad \in$

$$\psi \quad = \frac{}{\left|\sqrt{\ }\right|} \psi \frac{-}{} \quad \in \quad \neq$$

ψ

$$\int_{-\infty}^{\infty} \psi \quad =$$

$\in \quad \varphi$

$\subset \quad _{-} \quad \forall \in$

$$\bigcap_{\in} \quad = \quad \bar{\bigcup_{\in}} \quad =$$

$\varphi \quad \in \quad \varphi \quad - \quad _{\in}$

$\in \quad \Leftrightarrow \quad \in \quad _{-}$

$\in \quad \Leftrightarrow \quad ^{-} \quad \in$

$\varphi \quad - \quad \in$

φ

ϕ

φ

$\in \qquad \varphi \qquad \phi$

$\in$

$\in$

$= \qquad = \qquad \varphi$

$<\varphi \quad \varphi_{+} \quad >= \overline{}_{-}$

$<\phi \quad \phi_{+} \quad >= \overline{}_{-}$

$=$

$= \quad {}_{+} \quad + \quad {}_{+}$

$= \quad {}_{+} \quad + \quad {}_{+} \quad + \quad {}_{+}$

$\cdots\cdots$

$= \qquad +$

$= +$

$$= \sum_{=-\infty}^{\infty} \varphi$$

$$= \sum_{=-\infty}^{\infty} \phi$$

$$= \sum_{=-\infty}^{\infty} \overline{}_{-} \quad -$$

$$= \quad {}^{\infty} \quad \overline{\qquad} \quad - \quad - \quad {}_{=-\infty}$$

$\{ \quad \}_{\in}$

$\varphi \qquad \psi \qquad = \quad \otimes \qquad \{ \quad \}_{\in}$

$\Phi \quad \varphi \quad \varphi$

$\Psi \quad \varphi \quad \psi$

$\Psi \quad \psi \quad \varphi$

$\Psi \quad \psi \quad \psi$

$\in$

$$= \qquad = \quad {}_{+} \quad + \quad {}_{\varepsilon=} \quad {}_{+}{}^{\varepsilon}$$

$${}_{+} \quad = \quad {}_{\in} \quad {}_{+} \quad \Phi \quad {}_{+}$$

$${}^{\varepsilon}{}_{+} \quad = \quad {}_{\in} \quad {}^{\varepsilon}{}_{+} \quad \Psi^{\varepsilon} \quad {}_{+}$$

$$c_{J_1+1,m_1,m_2} = \sum_{k_1,k_2\in Z} h_{k_1-2m_1} h_{k_2-2m_2} c_{J_1,k_1,k_2}$$

$${}_{+} \quad = \quad {}_{\in} \quad - \quad -$$

$+ \quad = \quad \in \quad - \quad -$

$+ \quad = \quad \in \quad - \quad -$

$$h_k = \frac{1}{\sqrt{2}} \int_{-\infty}^{+\infty} \phi\left(\frac{x}{2}\right) \overline{\phi(x-k)}\, dx$$

$$g_k = (-1)^{k-1} \overline{h}_{1-k}$$

$= \quad + \quad \varepsilon$

$= + \varepsilon =$

$= \quad \Phi \quad \in$

$\varepsilon \quad = \quad \varepsilon \quad \Psi^{\varepsilon} \quad \in$

Δ

Δ

6

−2

shao_hl@sohu.com

ρ

ρ

ρ ρ

ρ λ

$$\rho \quad = \rho + \rho \quad {}^{+\infty} \quad \lambda -- \quad -\lambda \quad \lambda \ \lambda$$

$\psi \in \quad \cap \qquad \psi \quad =$

ψ

$$\psi \qquad = | \ |^{-} \psi \frac{-}{\quad} \quad \in \quad \in \ -$$

ψ

$\psi \quad =$

$\psi \qquad \psi$

$\in$

$$= < \quad \psi \quad > = | \ |^{-} \qquad \overline{\psi \frac{-}{\quad}}$$

$\psi \in \quad \cap$

$$_{\psi} = \ \pi \ \frac{|\psi \ \omega \ |}{|\omega|} \ \omega < \infty$$

ψ

$\psi \ \omega \ = \qquad \psi \ \omega \ =$

$|\psi \ \omega \ | \leq \quad + |\omega|^{- - \alpha} \qquad \psi$

$\in \quad + \infty \qquad \psi$

$$\psi \quad = \quad - \quad \geq \quad <$$

ψ

$$\psi \quad = \quad \psi \frac{-}{} \quad \in \quad \in \quad +\infty$$

ψ

ψ $\psi =$

ψ ψ

$\alpha >$ ψ

$$|\psi \quad | \leq \quad + | \quad |^{- - \alpha}$$

$\subseteq$ $+\infty$ ψ

$$= < \quad \psi \quad > = \quad \psi \frac{-}{} \quad \alpha >$$

$\in$ $+\infty$

$$\int_{-\infty}^{+\infty} \int^{+\infty} \overline{\quad} \frac{}{} = {}_{\psi} < \quad >$$

$$= \frac{}{\psi} \int^{+\infty} \int_{-\infty}^{+\infty} \psi \frac{}{}$$

$$_{\psi} = \pi \int^{+\infty} \frac{|\psi \ \omega|}{\omega} \omega$$

ψ

$$\int^{+\infty} - \quad = \frac{-}{+} \quad >$$

$$\left|\psi \ \omega \right| = \frac{}{\sqrt{\ \pi}} \times |\omega| \frac{\omega \ + \ \omega \ + \ \omega \ + \ \ \omega \ + \ \ \omega \ + \ \ ^{-}}{\omega \ +}$$

$$= \frac{}{\sqrt{\ \pi}} \frac{|\omega|\sqrt{\omega \ +}}{\omega \ +}$$

$_{\psi}$ — $\pi +$

ψ $\left|\psi \ \omega \right| \leq \ |\omega|^{\alpha} \ + |\omega|^{\ -\frac{\gamma}{}} \ \alpha > \ \ \gamma > \alpha +$

$\omega \neq \ \psi$

$_{=} \left|\psi^{\ -} \ \omega \right| \ \geq \alpha >$ ψ

$+\infty$

$$\left|\psi \ \omega \right| = \frac{}{\sqrt{\ \pi}} \frac{|\omega|\sqrt{\omega \ +}}{\omega \ +}$$

$> \ \ \alpha > \ \ \ \gamma > \alpha +$

$$\frac{}{\sqrt{\ \pi}} \frac{\omega\sqrt{\omega \ +}}{\omega \ +} \leq \ |\omega|^{\alpha} \ + |\omega|^{\ -\frac{\gamma}{}}$$

$= \ \ \alpha = \ \ \gamma = \ \ \ > \alpha +$

$^{\infty}\rho$ λ

$$ {}^{\infty}\rho \quad \lambda $$

$$ \rho = \sum_{\in} \rho \quad \psi $$

$$ \rho \, \psi $$

$$ {}^{\infty}\rho \quad \lambda $$

$$ = \sum_{\in} \rho^{\;--} \quad {}^{+} \frac{+}{\left[\; + \quad + \quad \lambda \;\right]^{/}} - \quad {}^{+} \frac{}{\left[\; + \quad + \quad \lambda \right.} $$

$$ + \quad {}^{-} \frac{-}{\left[\; - \quad + \quad \lambda \;\right]^{/}} - \quad {}^{-} \frac{}{\left[\; - \quad + \quad \lambda \;\right]^{/}} $$

$$ - \sum_{\geq \; \in} \rho^{\;--} \quad \psi \quad \lambda $$

$$ {}^{\infty}\rho \quad \lambda = \rho \quad {}^{\infty}\psi \quad \lambda $$

$$ {}^{\infty}\psi \quad \lambda $$

$$ {}^{\infty} \; {}^{-} \quad \lambda = \frac{}{\sqrt{\quad + \lambda}} $$

$$ {}^{\infty} \; {}^{-} \quad \lambda = \frac{}{+ \lambda^{\;-}} $$

$= \quad =$

$\rho = \quad \Omega \quad \rho = \quad \Omega \quad \rho = \quad \Omega$

Ω

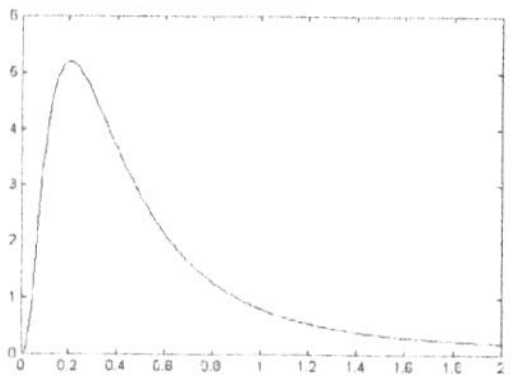

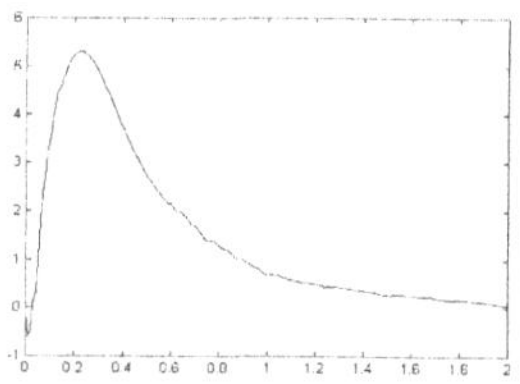

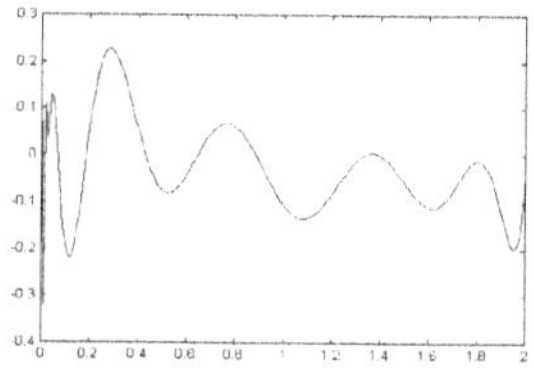

Wavelets Approach in Choosing Adaptive Regularization Parameter*

Feng Lu, Zhaoxia Yang, and Yuesheng Li

Department of Scientific Computing and Computer Applications
Zhongshan University, Guangzhou 510275, P. R. China

Abstract. In noise removal by the approach of regularization, the regularization parameter is global. Constructing the variational model $\min_g \|f - g\|^2_{L_2(R)} + \alpha R(g)$,$g$ is in some wavelets space. Through the wavelets pyramidal decompose and the different time-frequency properties between noise and signal, the regularization parameter is adaptively chosen, the different parameter is chosen in different level for adaptively noise removal.

Keywords: *Sobolev* space, wavelet, noise, adaptive.

1 Wavelets and Discrete Equivalent Norm of Sobolev Space

The model of noisy image is:

$$f = f_0 + \eta \tag{1}$$

where f_0 is original clean image,η is Guassian noise. Our task is to restore the original image f_0 as possible. The regularization approach is always adopted to solve these problems, we consider the variational problems of the form:

$$\min_g \|f - g\|^2_{L_2(R^2)} + \alpha R(g) \tag{2}$$

where $g \in X$; $X \subset L_2(R^2)$

X can be chosen as *Sobolev* space, *Besov* space ,*Lipschitz* space and so on, the *sobolev* space is chosen as X in this paper. α is regularization parameter that determines the trade-off between goodness the fit to the measured data, and the amount of regularization done to the measured image.

In (2), the parameter is global, that the regularization parameter is the same number everywhere. In reference [4], the regularization parameter is chosen as a changeable number with the different gradient in some image. In [2] and [3], to choose the proper parameter, the *Besov* spaces of minimal smoothness can be embedded in $L_2(R)$, and can get the discrete wavelets equivalent norm.

* This work is supported by Natural Science Foundation of Guangdong (9902275), Foundation of Zhongshan University Advanced Research Centre.

Y. Y. Tang et al. (Eds.): WAA 2001, LNCS 2251, pp. 418–423, 2001.

we can easily construct the two dimensional wavelets from one dimensional wavelets ψ and scale function ϕ by setting for $x := (x_1, x_2) \in R$,

$$\psi^{(1)}(x_1, x_2) := \psi(x_1)\phi(x_2);$$
$$\psi^{(2)}(x_1, x_2) := \phi(x_1)\psi(x_2);$$
$$\psi^{(3)}(x_1, x_2) := \psi(x_1)\psi(x_2);$$

If we let $\Psi := \{\psi^{(1)}, \psi^{(2)}, \psi^{(3)}\}$, then the set of functions $\psi_{j,k}(x) := 2^k \psi(2^k x - j)_{\psi\in\Psi, k\in Z, j\in Z^2}$ forms an orthonormal basis for $L_2(R^2)$, that is, for every $f \in L_2(R^2)$, there are coefficients $c_{j,k,\psi} := \int_{R^2} f(x)\psi_{j,k}(x)dx$ such that

$$f = \sum_{j\in Z^2, k\in Z, \psi\in\Psi} c_{j,k,\psi}\psi_{j,k}$$

$$\|f\|^2_{L_2(R^2)} = \sum_{j\in Z^2, k\in Z, \psi\in\Psi} c^2_{i,k,\psi} \tag{3}$$

In reference [2], the discrete equivalent norm of *Sobolev* Space is:

$$\|f\|^2_{W^\beta(L_2(R^2))} \approx \sum_{k\geq 0}\sum_{j\in Z^2}\sum_{\psi\in\Psi} 2^{2\beta k}|c_{j,k,\psi}|^2 \tag{4}$$

where β is the smoothness order of the *Sobolev* Space.

It is an excellent property that a Space Norm can be expressed by the discrete sequence, especially the wavelets coefficient sequence, it makes many problems easier largely.

2 Variational Model and Its Wavelets Solution

From previous work of regularization approach, we can choose the model as follow:

$$\min_g\{\|f - g\|^2_{L_2(R)} + \alpha\|g\|^2_{W^2(L_2(D))}\} \tag{5}$$

where $\alpha > 0$,$W^2(L_2(D)$ represents *SobolevSpace* with two-order smoothness.

Let: $f = \sum_{j,k,\psi} c_{j,k,\psi}\Psi_{j,k}$, $g = \sum_{j,k,\psi} d_{j,k,\psi}\Psi_{j,k}$,

From (4),(5) can be expanded as:

$$\sum_{j,k,\psi}(|c_{j,k,\psi} - d_{j,k,\psi}|^2 + \alpha \cdot 2^{4k}|d_{j,k,\psi}|^2)) \tag{6}$$

In reference [6],*Donoho* points out that for the spectrum analysis of a noisy real image, the spectrum corresponding with the noise is quite small, while the spectrum corresponding with the original image is quite large. (See Figure 1) It means that the "energy" of the noisy image is always "concentrate" on the original image. Because of the wavelets' better property of *Locality* in both time and frequency domain, the wavelets can concentrate the energy, that is, in wavelets transform domain, the energy of original image concentrate on some highlight lines, while almost zeros else where. But for the noise, it is quite different. The wavelets coefficients corresponding with noise is always small, even almost zeros, in every level in wavelets transform domain, and its distribution is quite uniform in all levels. So it is a new way to choose the regularization parameter not as a constant, but changeable with the wavelets coefficients.

Fig. 1. Left:Original Image, Right:Wavelets Coefficients

We can construct the new variational model with changeable parameter:

$$\sum_{j,k,\psi} (|c_{j,k,\psi} - d_{j,k,\psi}|^2 + \alpha(c_{j,k,\psi}) \cdot 2^{4k} |d_{j,k,\psi}|^2)) \tag{7}$$

where $\alpha(t) > 0, t \in \{c_{j,k,\psi}\}$ is the wavelets coefficient, $W^2(L_2(D)$ represents *SobolevSpace* with two-order smoothness. Here, the regularization is not a constant, but a changeable variable with wavelets coefficients.

From this model, we can handle different level with wavelets decomposition with different regularization, when the wavelets coefficient is large, choosing the regularization parameter small for containing more original image , when the wavelets coefficient is small, choosing the parameter large for removing the noise much. So, we can get the regularization image adaptively which containing the information of original image more and removing the noise as well.

Hence, two conditions must be satisfied for choosing regularization parameter:

(1) $\lim_{t \to \infty} \alpha(t) = 0$
(2) $\lim_{t \to 0} \alpha(t) = 1$

In practice, because the wavelets coefficients corresponding with the noise is quite small, we choose function $\alpha(t)$ with decaying rapidly. For example:

$$\alpha(t) := e^{-t^2}, \alpha(t) := \frac{1}{(1+t^2)}$$

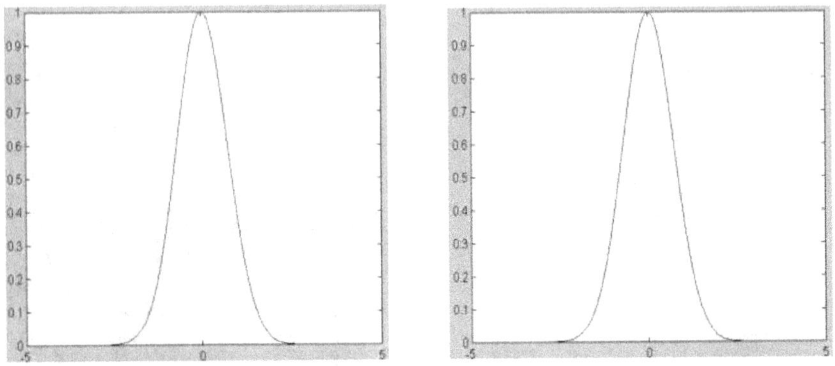

Fig. 2. Left:$\alpha(t) := e^{-t^2}$ Right:$\alpha(t) := \frac{1}{(1+t^2)}$

In reference [7], the formula of window size of decaying function is:

$$\triangle_\alpha := \frac{1}{\|w\|_2}\{\int_{-\infty}^{\infty} x^2|\alpha(x)|^2 dx\}$$

Let $\alpha(t; m, s) := m\alpha(\frac{t}{s})$, to meet the practices, we can change the *Support Set* and *Amplitude* through choosing the proper m, s.

For every j, k, ψ, each term of (7)

$$|c_{j,k,\psi} - d_{j,k,\psi}|^2 + \alpha(c_{j,k,\psi}) \cdot 2^{4k}|d_{j,k,\psi}|^2 \geq 0 \tag{8}$$

Hence, one minimizes (7) just by minimizing separately over $d_{j,k,\psi}$:

$$|c_{j,k,\psi} - d_{j,k,\psi}|^2 + \alpha(c_{j,k,\psi}) \cdot 2^{4k}|d_{j,k,\psi}|^2$$

for each j, k and ψ.

Let:$s := c_{j,k,\psi}, v := d_{j,k,\psi}$,and supposing $v \leq s$,

(8) can be reduced to:

$$F(v) := |s - v|^2 + \alpha(s) \cdot 2^{4k}v^2 \tag{9}$$

Calculating the *derivation* of $F(v)$ for v, we can get the minimizer of (9):

$$v = \frac{s}{1 + \alpha(s) \cdot 2^{4k}} \tag{10}$$

After calculating (10) for all the wavelets coefficients of all levels, we can get the new wavelets coefficients from *regularization* processing. Hence, we can get the restored image by wavelets reconstruction.

3 Experiments

An image *Bird.bmp* is adopted in the experiment, we choose the *Haar* wavelets and $\alpha(t; m, s) := m\alpha(\frac{t}{s}) = me^{-(t/s)^2}$.

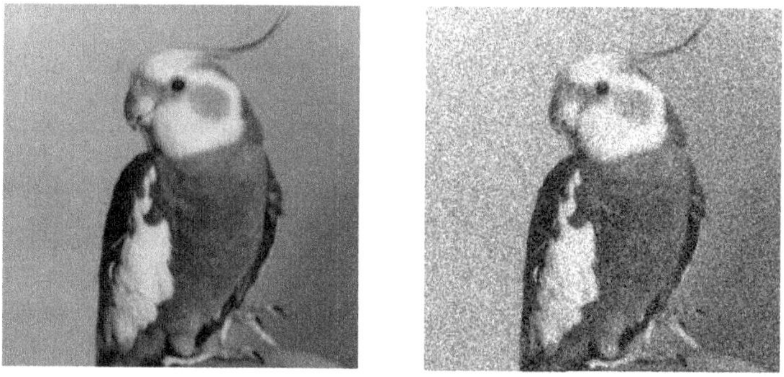

Fig. 3. Left:Original Image of Bird.bmp, Right:Nosiy image with Gaussian white noise, variance $\delta^2 = 18$

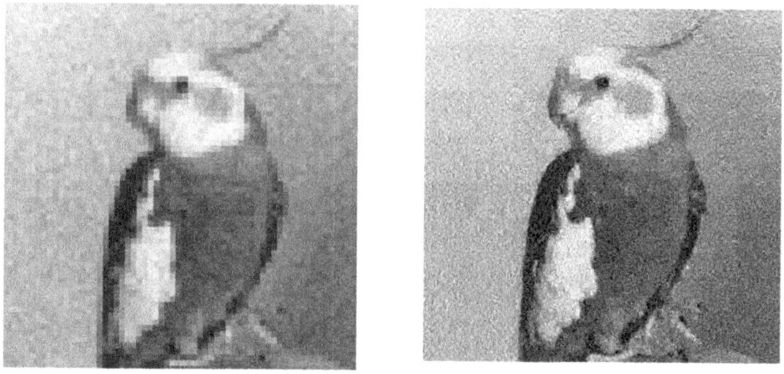

Fig. 4. Left:Restored image with removing two first level of wavelets coefficients, Right:Restored image with adaptive approach, where $m = 0.8; s = 10$

4 Conclusion

Using the approach of adaptive changeable regularization parameter in image restoration, it is more flexible to choose the model. We can choose the spaces with more smoothness order which have powerful ability in noise removal, at the same time, choosing changeable regularization function to containing more details and removing more noise.

References

1. A. N. Tikhonov and Vasiliy Y. Arsenin Solution of ill-posed problems, V. H. Winston & Sons Press, 1997;
2. R. A. Devore Fast wavelet techniques for near-optimal image processing, IEEE Military Communications Conference Record, 1992, P1129-1135; 418, 419
3. R. A. Devore, Image compression through wavelet transform coding, IEEE Transactions on Information Theory, vol. 38, 1992, P719-746; 418
4. Adaptive regularized constrained least squares image restoration, IEEE trans. on Image Processing, 1999, P1191-1203; 418
5. I. Daubechies, Ten lecture on wavelets CBMSNSF Series in Applied Math #61, SIAM, Pub1., Philudelphia, 1992;
6. Donoho D. L. De-noising by soft-thresholding, IEEE Trans. on Information Theory, 1993, 41(3); 419
7. Chui C. K. An Introuduction to wavelets, Xi'an Jiaotong Univ. Press, 1994. (in chinese) 421

zhaoly@cta.cq.cn

yytang@comp.hkbu.edu.hk

$$=$$

$$= \quad = \quad = \quad =$$

$$\sqrt{}$$

$$= \quad -$$

$$+ \quad = \quad -$$

$$=\phi \qquad =\psi$$

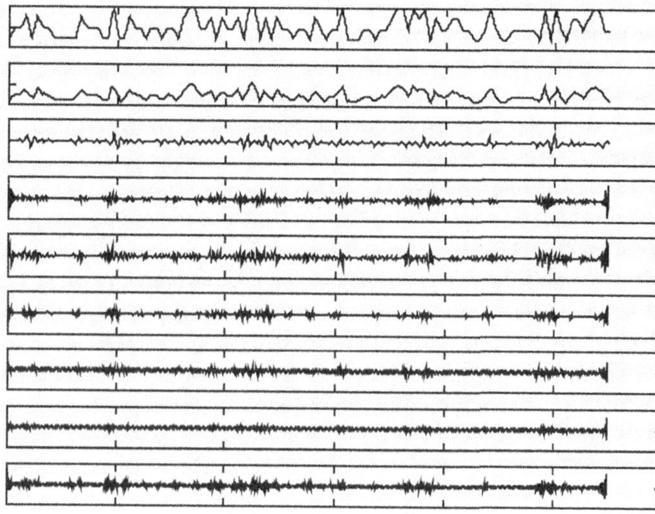

≤

≥

≤ ≤ ≤

≤ ≤ ≥ ≤

jpli2222@sina.com

ϕ ϕ

ϕ $<\phi$

ϕ ϕ

$$= \quad +\sigma \qquad = \quad \cdots \quad -$$

σ

$$\phi = \frac{\rho \quad -\rho}{\rho \quad -\rho}$$

ρ ρ

ρ

ϕ

ϕ

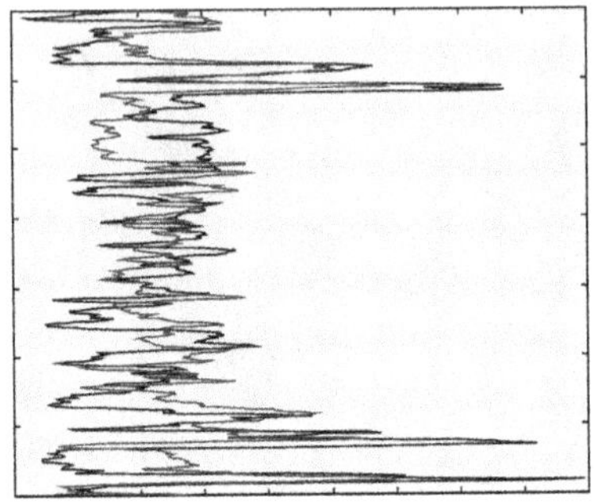

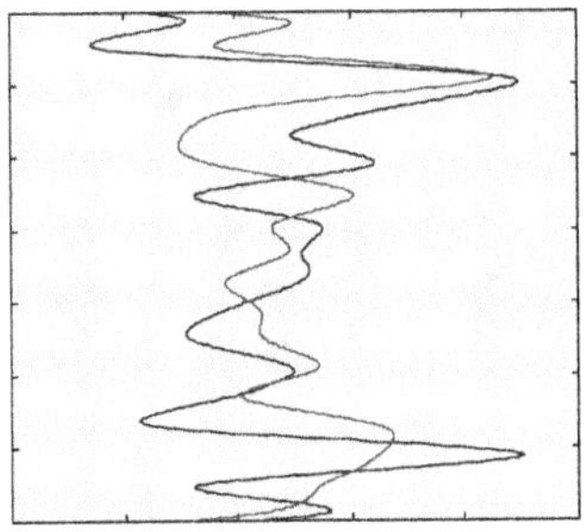

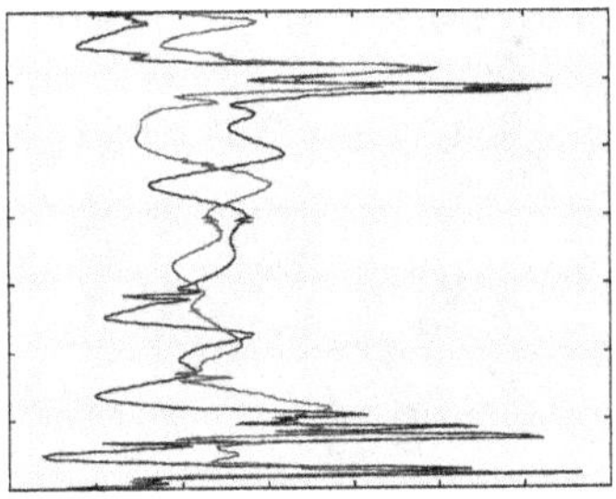

curve

PARAMETERIZATIONS OF M-BAND BIORTHOGONAL WAVELETS

Zeyin Zhang and Daren Huang

ABSTRACT. In this paper, we consider the structure of compactly supported wavelets. And we prove that any wavelet matrix (the polyphase matrix of the scaling filter and wavelet filters) can be factored as the product of fundamental biorthgonal matrices and a constant valued matrix.

1. INTRODUCTION

Fixed an integer $m \geq 2$. A compactly supported function $\varphi \in L^2(\mathbb{R})$ is an m-band scaling function if there exists a finite length sequence $\{h_k^0\}$ such that

$$\varphi(x) = \sum_k h_k^0 \varphi(m \cdot -k),$$

the z-transform

$$\sum_k h_k^0 z^{-k}$$

is a Laurent polynomial which is called scaling filter of φ. Let $\tilde{\varphi}(x) \in L^2(\mathbb{R})$ be another compactly scaling function with Laurent polynomial scaling filter

$$\sum_k g_k^0 z^{-k}.$$

The pair of φ and $\tilde{\varphi}$ is said to be a biorthogonal pair if

$$\int_{\mathbb{R}} \overline{\varphi(x)} \tilde{\varphi}(x-k) dx = \delta_{0,k}$$

for $k \in \mathbb{Z}$, where $\delta_{0,0} = 1$, and $\delta_{0,k} = 0$ if $k \in \mathbb{Z} \setminus \{0\}$.

Corresponding to the biorthogonal scaling functions, there exist compactly supported wavelets

$$\psi_i(x) = \sum_k h_k^i \varphi(m \cdot -k)$$

1991 *Mathematics Subject Classification.* Primary 42C15, 46A35, 46E15.

Key words and phrases. Wavelet, polyphase matrix, Parameterizations, Filter bank.

Y. Y. Tang et al. (Eds.): WAA 2001, LNCS 2251, pp. 435-447, 2001.

with finite length coefficient sequences $\{h_k^i\}$ for $1 \leq i \leq m-1$, and wavelets

$$\tilde{\psi}_i(x) = \sum_k g_k^i \tilde{\varphi}(m \cdot -k)$$

with finite length sequences $\{kg_k^i\}$ for $1 \leq i \leq m-1$, such that the family

$$\{m^{j/2}\psi_i(m^j \cdot -k),\ j,k \in \mathbb{Z}, 1 \leq i \leq m-1\}$$

and the family

$$\{m^{j/2}\tilde{\psi}_i(m^j \cdot -k),\ j,k \in \mathbb{Z}, 1 \leq i \leq m-1\}$$

are biorthogonal bases in $L^2(\mathbb{R})$. Now we introduce the polyphase Laurent polynomials

$$H_{i,j}(z) = \sum_k h^j_{mk+i} z^{-k} \tag{1.1}$$

$$G_{i,j}(z) = \sum_k g^j_{mk+i} z^{-k} \tag{1.2}$$

for $0 \leq i,j \leq m-1$. Let

$$H(z) = (H_{i,j})_{0 \leq i,j \leq m-1}, \quad G(z) = (G_{i,j})_{0 \leq i,j \leq m-1}. \tag{1.3}$$

By the biorthogonality we get

$$G^*(z^{-1})H(z) = mI_m \tag{1.4}$$

and the first column vectors of $H(1)$ and $G(1)$ is $(1, 1, \ldots, 1)^*$. $G^*(z^{-1})= G(\bar{z}^{-1})^*$, Here and hereafter, for a matrix or vector A, A^* denote the Hermite transpose of A, I_m is an m square identity matrix.

The theoretical work of orthogonal wavelets was done in the late eighties [1, 2, 4-6, 11, 15] and the framework of biorthogonal wavelets was established in the early nineties [3, 8, 10]. The invention of the polyphase decomposition is one of the reasons why multirate filter banks processing became practically attractive. It is valuable not only in practical design and actual implementation of filter banks, but also in theoretical study[14]. Actually with the polyphase decomposition, P. P. Vaidyanathan and his colleagues [9, 13] derive factorizations of paraunitary matrices and apply such factorizations to design quadrature mirror filter (QMF) banks for digital signal processing problems. P. N. Heller, H. L. Resnikoff, and R. O. Wells, Jr. [7, 12] use the polyphase decomposition to develop a parametrization theory of compactly supported orthonormal wavelets. The purposes of this paper is to factorize A pair of matrices $H(z)$ and $G(z)$ satisfying (1.4) into some simple building block. The building block used in this paper are of the form $I_m - P + Pz^{\pm 1}$, where P is an one order idempotent matrix. This paper is organized as follows. in section 2, we give a some definition and lemmas for the later use. then discuss parameterizations of dual Laurent polynomial pairs (section 3), and derive parametric decomposition of biorthogonal wavelet filter matrix(section 4), At last some final remarks are given (section 5)

2. SOME LEMMAS

For the convenience in the following we give some definitions.

Definition 1. The pair $(H(z), G(z))$ *of matrices consist of polyphase Laurent polynomials (1.1) and (1.2) of scaling filter and wavelet filters defined as in (1.3) is said to be a biorthogonal wavelet matrix pair.* $H(z)$, $G(z)$ *are said to be biorthogonal wavelet matrices.*

Now we consider a pair of Laurent polynomials vectors $\alpha(z)$ and $\beta(z)$ with vector valued coefficients,

$$\alpha(z) = \alpha_s z^{-s} + \alpha_{s+1} z^{-s-1} + \cdots + \alpha_k z^{-k}$$

$$\beta(z) = \beta_p z^{-p} + \beta_{p+1} z^{-p-1} + \cdots + \beta_q z^{-q}$$

with $\alpha_i, \beta_j \in \mathbb{R}^m$, for $s \le i \le k, p \le j \le q$.

Let V_α be a subspace of $\mathbb{R}^m$ spanned by $\{\alpha_i, s \le i \le k\}$, V_β be a subspace of $\mathbb{R}^m$ spanned by $\{\beta_j, p \le j \le q\}$.

Definition 2. We say $(\alpha(z), \beta(z))$ *of Laurent polynomial vectors is a dual pair if* $\alpha^*(z^{-1})\beta(z) = m$ *where* $\alpha^*(z) = \alpha(\bar{z})^*$.

Now if we rewrite (1.3) into

$$H(z) = (\alpha_0(z), \alpha_1(z), \ldots, \alpha_{m-1}(z)) \tag{2.1}$$

and

$$G(z) = (\beta_0(z), \beta_1(z), \ldots, \beta_{m-1}(z)) \tag{2.2}$$

then

$$\alpha_i^*(z^{-1})\beta_j(z) = m\delta_{i,j}, 0 \le i, j \le m-1 \tag{2.3}$$

by (1.4), so $(\alpha_i(z), \beta_i(z)), 0 \le i \le m-1$ are m dual pairs of Laurent polynomial vectors.

Lemma 1. *Let* $U = V_\alpha \cap V_\beta^\perp$, $W = V_\beta \cap V_\alpha^\perp$. *If* $(\alpha(z), \beta(z))$ *is a dual pair of Laurent polynomial vectors, then the difference spaces* $V_\alpha \ominus U$, $V_\beta \ominus W$ *are adjoint.* Here and hereafter, we say two subspaces $V_1, V_2 \subseteq \mathbb{R}^m$ are adjoint if there exist a basis $\{\alpha_i\}_1^k$ of V_1 and a basis $\{\beta_i\}_1^k$ of V_2 such that

$$\alpha_i^* \beta_j = \delta_{i,j}, \quad 1 \le i, j \le k.$$

Proof. Let k be the rank of matrix $(\alpha_i^*\beta_j)_{s\le i\le k,p\le j\le q}$. Then there exist a k order invertable block $(\alpha_i'^*\beta_j')_{1\le i,j\le k}$ of matrix $(\alpha_i^*\beta_j)$ and such that

$$\alpha_i^*\beta_j = \sum_{m=1}^{k} a_{m,j}\alpha_m'^*\beta_j, \quad p \le j \le q; \tag{2.4}$$

and

$$\alpha_i^*\beta_j = \sum_{m=1}^{k} b_{i,m}\alpha_i^*\beta_m', \quad s \le i \le k. \tag{2.5}$$

By (2.4) and (2.5), it is easy to verify that $\{\alpha_i - \sum_{l=1}^k b_{i,l}\alpha_l', s \le i \le k\} \subset U$, and $\{\beta_j - \sum a_{l,j}\beta_l', p \le j \le q\} \subset W$, therefore $V_\alpha \ominus U$ is the subspace spanned by $\{\alpha_i'\}_{i=1}^k$, $V_\beta \ominus W$ is the subspace spanned by $\{\beta_i'\}_{i=1}^k$. To prove $V_\alpha \ominus U$ and $V_\beta \ominus W$ are adjoint, it is need to prove that there are two bases of the two subspaces respectively which are biorthogonal. In fact, let $\{\bar\beta_j'\}_1^k$ be defined by

$$(\bar\beta_1', \bar\beta_2', \cdots, \bar\beta_k') = (\beta_1', \beta_2', \cdots, \beta_k')A^{-1}$$

where $A = (\alpha_i'^*\beta_j')_{1\le i,j\le k}$. Then $\{\alpha_i'\}_{i=1}^k$ and $\{\bar\beta_j'\}_1^k$ are bases of $V_\alpha \ominus U$ and $V_\beta \ominus W$ respectively, and they are biorthogonal, the proof is completed.

Definition 3. Under the condition in Lemma 1. The dual order of dual pair $(\alpha(z), \beta(z))$ of Laurent polynomial vectors is defined as the dimension of $V_\alpha \ominus U$.

For a subspace $V \subseteq \mathbb{R}^m$, define

$$V^\perp = \{\alpha \in \mathbb{R}^m;\ \alpha\beta^* = 0,\ \forall \beta \in V\}.$$

By the argument in Lemma 1, we see that the dual order of $(\alpha(z), \beta(z))$ is equal to the rank of matrix $(\alpha_i^*\beta_j)_{s\le i\le k,p\le j\le q}$. By the result of Lemma 1, we have

Lemma 2. *Let k be the order of dual pair $(\alpha(z), \beta(z))$ of Laurent polynomial vectors is k, then there exist birothogonal bases $\alpha_1', \ldots, \alpha_k' \in V_\alpha$ and $\beta_1', \ldots, \beta_k' \in V_\beta$ satisfying*

$$\alpha_i'^*\beta_j = \delta_{i,j}, 1 \le i, j \le k$$

such that

$$\begin{cases} \alpha(z) = \sum_{i=1}^k H_i(z)\alpha_i' + \sum_i \tilde\alpha_i z^i \\ \beta(z) = \sum_{i=1}^k G_i(z)\beta_i' + \sum_j \tilde\beta_j z^j \end{cases} \tag{2.6}$$

where $\tilde\alpha_i \in V_\alpha \cap V_\beta^\perp$ and $\tilde\beta_j \in V_\beta \cap V_\alpha^\perp$, and H_i, G_i are Laurent polynomials.

In Lemma 2, if $k = 1$, then $H_1(z) = cz^{-n}$, $G_1(z) = \frac{m}{c}z^{-n}$ for a nonzero constant c and an integer n.

Especially if $(\alpha(z), \alpha(z))$ is a dual pair of Laurent polynomial vectors, by the fact $V_\alpha \cap V_\alpha^\perp = \{0\}$, we have

Lemma 3. *If Let k be the dual order of $(\alpha(z), \alpha(z))$. There exist $\alpha_1, \ldots, \alpha_k \in V_\alpha$ such that*

$$\alpha_i^* \beta_j = \delta_{i,j}, 1 \leq i, j \leq k$$

and

$$\alpha(z) = \sum_{i=1}^{k} H_i(z)\alpha_i$$

where $H_i, 1 \leq i \leq k$ are Laurent polynomials.

A matrix $P \in \mathbb{R}^{m\times m}$ is said idempotent if $P^2 = P$. for a given matrix $Q \in \mathbb{R}^{m\times m}$ and a subspace $V \subseteq \mathbb{R}^m$, define

$$PV = \{P\alpha;\ \alpha \in V\}.$$

For a subspace $V \subseteq \mathbb{R}^m$, $Q \in \mathbb{R}^{m\times m}$ is said to be an annihilator on V, if $QV = \{0\}$. Denoted by $\mathcal{N}(V)$ the set of all annihilators on V.

3. Parameterizations of dual pair of Laurent polynomial vectors with one rank idempotent matrices

Theorem 1. *If $(\alpha(z), \beta(z))$ is a dual pair of Laurent polynomial vectors, then there exist one rank idempotent matrices $P_1, P_2, \ldots, P_d$ with that $P_i \in \mathcal{N}(V_\beta^\perp)$, $P_i^* \in \mathcal{N}(V_\alpha^\perp)$, $1 \leq i \leq d$ such that*

$$\alpha(z) = V_d(z)V_{d-1}(z)\cdots V_1(z)\delta(z)$$
$$\beta(z) = V_d^*(z)V_{d-1}^*(z)\cdots V_1^*(z)\gamma(z)$$

where $(\delta(z), \gamma(z))$ is a one order dual pair of Laurent polynomial vectors, $V_\delta \subseteq V_\alpha$, $V_\gamma \subseteq V_\beta$ and

$$V_i(z) = I_m - P_i + P_i z^{-1},\ 1 \leq i \leq d.$$

Let P be a one order idempotent matrix, that is, there exist $u, v \in R^m, u^*v = 1$ such that $P = uv^*$. Define

$$V(z) = I_m - P + Pz^{-\tau},\ \tau \in \{-1, +1\} \tag{3.1}$$

then $V^*(z) = I_m - P^* + P^*z^{-\tau}$, so

$$V(z)V(z^{-1}) = 1, \quad \det(V(z)) = z^{-\tau} \tag{3.2}$$

we will say that the matrix $V(z)$ of the form (3.1) as primitive biorthogonal matrix.

Proof of Theorem 1. Let k be the dual order of $(\alpha(z), \beta(z))$. By Lemma 2, we can represent $\alpha(z)$ and $\beta(z)$ in the form (2.6)with coefficient Laurent polynomials $H_i, G_i, 1 \leq i \leq k$. Now write

$$(H_1(z), H_2(z), \ldots, H_k(z))^T = z^{-n} \sum_0^r \eta_i z^{-i}, \tag{3.3}$$

here and hereafter, for a vector α, α^T denote the transpose of α, n is an integer, $\eta_i \in \mathbb{R}^k, i = 0, 1, \ldots, r$. The scheme of the proof is to decrease the length $r+1$ to 1 recursively.

If $r = 0$, the length of (3.3) is just one, there is need do nothing. Assume $r \geq 1$.

case 1. η_0, η_r is independent.

Let

$$(b_1, b_2, \ldots, b_k)^T = \eta_r$$

and

$$(a_1, a_2, \ldots, a_k)^T = \frac{u}{\eta_r^* u},$$

where

$$u = (\eta_0^* \eta_0)\eta_r - (\eta_0^* \eta_r)\eta_0.$$

Define

$$\beta = \sum_1^k a_i \beta_i', \quad \alpha = \sum_1^k b_i \alpha_i',$$

then $\beta \in V_\beta$ and $\alpha \in V_\alpha$.

Let $P = \alpha\beta^*$, then P is a one order idempotent matrix, and $P \in \mathcal{N}(V_\beta^\perp), P^* \in \mathcal{N}(V_\alpha^\perp)$. Define

$$V(z) = I_m - P + Pz^{-1},$$

then $V(z)$ is a primitive biorthogonal wavelet matrix. And define

$$\alpha'(z) = V(z^{-1})\alpha(z), \quad \beta'(z) = V^*(z^{-1})\beta(z),$$

it follows that $(\alpha'(z), \beta'(z))$ is a dual pair of Laurent polynomial vectors, $V_{\alpha'} \subseteq V_\alpha, V_{\beta'} \subseteq V_\beta$ and

$$\alpha(z) = V(z)\alpha'(z), \quad \beta(z) = V^*(z)\beta'(z).$$

Note that

$$\alpha'(z) = \sum_1^k H_i'(z)\alpha_i' + \sum \tilde{\alpha}_i z^{-i}$$

where

$$(H_0'(z), H_1'(z), \ldots, H_k'(z))^T = z^{-n}(\eta_0 + \frac{u^*\eta_1}{u^*\eta_r}\eta_r + \cdots + (\eta_r + \eta_{r-1} - \frac{u^*\eta_{r-1}}{u^*\eta_r}\eta_r)z^{-r+1}).$$

Thus the length of $(H_0'(z), H_1'(z), \ldots, H_k'(z))^T$ is decreased by 1.

case 2. η_0, η_r are dependent.

Write Laurent polynomials $G_i, 1 \le i \le k$ in (2.6) as

$$(G_1(z), \dots, G_k(z))^T = z^{-n_1} \sum_0^s \gamma_s z^{-i},$$

then $\eta_0^* \gamma_s = \eta_r^* \gamma_s = 0$ or $\eta_r^* \gamma_0 = \eta_0^* \gamma_0 = 0$. we only consider $\eta_0^* \gamma_s = \eta_r^* \gamma_s = 0$, for another is similar.

There exist an l such that $\eta_i^* \gamma_s = 0, \quad 0 \le i \le l-1$ and $\eta_l^* \gamma_s \ne 0$.

Let

$$\eta_l = (c_1, c_2, \dots, c_k)^T$$

and

$$\gamma_s = (d_1, d_2, \dots, d_k)^T$$

Define

$$\alpha = \sum_1^k c_i \alpha_i', \quad \beta = \sum_1^k d_i \beta_i',$$

then $\alpha \in V_\alpha$ and $\beta \in V_\beta$. Now if we set $P = \frac{\alpha\beta^*}{\gamma_s^* \eta_l}$, then P is a one rank idempotent matrix, $P \in \mathcal{N}(V_\beta^\perp), P^* \in \mathcal{N}(V_\alpha^\perp)$. Define

$$V(z) = I_m - P + Pz^{-n} = (I_m - P + Pz^{-1})^n$$

then it is a power of primitive biorthogonal wavelet matrix, and define

$$\alpha'(z) = V(z^{-1})\alpha(z), \quad \beta'(z) = V^*(z^{-1})\alpha(z).$$

It follows that $(\alpha'(z), \beta'(z))$ is a dual pair of Laurent polynomial vectors, $V_{\alpha'} \subseteq V_\alpha, V_{\beta'} \subseteq V_\beta$ and

$$\alpha(z) = V(z)\alpha'(z), \quad \beta(z) = V^*(z)\alpha'(z).$$

Note that

$$\alpha'(z) = \sum_1^k H_i'(z)\alpha' + \sum \tilde{\alpha}_i z^{-i}$$

where

$$(H_1'(z), H_2'(z), \dots, H_k'(z))^T = z^{-n}((\eta_0 + \eta_n) + \eta_1 z^{-1} + \dots + \eta_r z^{-r})$$

thus the length of $(H_1'(z), H_2'(z), \dots, H_k'(z))^T$ is the same as the length of (3.3), but $\eta_0 + \eta_n$ and η_r are independent,which transform to condition in the case 1.

Recursively proceeding in this fashion, we decrease the length of (3.3) to 1, that is

$$\alpha(z) = V_d(z)V_{d-1}(z) \cdots V_1(z)\delta(z),$$
$$\beta(z) = V_d^*(z)V_{d-1}^*(z) \cdots V_1^*(z)\gamma(z),$$

where

$$\delta(z) = z^{-n}\sum_1^k c_i\alpha_i + \sum \tilde{\alpha}_i z^{-i}$$

and

$$\gamma(z) = \sum_1^k G_i'(z)\beta_i + \sum \tilde{\beta}_i z^{-i},$$

therefore $(\delta(z), \gamma(z))$ is a dual pair of Laurent polynomial vectors with one order, and $V_\delta \subseteq V_\alpha, V_\gamma \subseteq V_\beta$. The proof is completed.

In the following, we consider the parameterizations of one order dual pair of Laurent polynomial vectors.

Theorem 2. *Let $(\alpha(z), \beta(z)$ be a dual pair of Laurent polynomial vectors. If the dual order is one, then there exist idempotent matrices $P_i, 1 \le i \le d$ with rank one, and $P_i \in \mathcal{N}(V_\beta^\perp), P_i^* \in \mathcal{N}(V_\alpha^\perp)$, $1 \le i \le d$ such that*

$$\alpha(z) = z^{-k}V_d(z)V_{d-1}(z)\cdots V_1(z)\alpha(1)$$
$$\beta(z) = z^{-k}V_d^*(z)V_{d-1}^*(z)\cdots V_1^*(z)\beta(1)$$

where k is an integer, and

$$V_i = I_m - P_i + P_i z^{-\tau_i},\ \ \tau_i \in \{1, -1\},\ \ i = 1, 2, \cdots, d.$$

Proof. Since $(\alpha(z), \beta(z))$ is a one order dual pair of Laurent polynomial vectors with order one, then by Lemma 2, we have

$$\begin{cases} \alpha(z) = \alpha_{k-r}z^{-k+r} + \cdots + \alpha_k z^{-k} + \cdots + \alpha_{k+s}z^{-k-s} \\ \beta(z) = \beta_{k-p}z^{-k+p} + \cdots + \beta_k z^{-k} + \cdots + \beta_{k+q}z{-k-q} \end{cases}$$

where $\alpha_i^*\beta_j = m\delta_{i,k}\delta_{j,k}$ for $k-r \le i \le k+s, k-p \le j \le k+q$.

Define

$$V_i(z) = I_m - P_i + P_i z^{-1},\quad U_j(z) = I_m - Q_j + Q_j z^{-1}$$

as primitive biorthogonal wavelet matrix, where

$$P_i = (\sum_{l=k-i}^{k} \alpha_l)(\sum_{l=k-i}^{k} \beta_l)^*,\ 0 \le i \le r-1$$

and

$$Q_j = (\sum_{l=k-r}^{k+j} \alpha_l)(\sum_{l=k-r}^{k+j} \beta_l)^*,\quad 1 \le j \le s-1.$$

Then $P_i, Q_i \in \mathcal{N}(V_\beta^\perp),\ P_i^*, Q_i^* \in \mathcal{N}(V_\alpha^\perp),\ 0 \le i \le r-1,\ 1 \le j \le s-1.$

Define

$$\tilde{\alpha}(z) = \prod_{1}^{s-1} U_j(z) V_r(z)^r \prod_{r-1}^{0} V_i(z^{-1}) \alpha(z),$$

and define

$$\tilde{\beta}(z) = \prod_{0}^{r-1} V_i^*(z^{-1})(V_r^*(z)^r) \prod_{s-1}^{1} U_j^*(z) \beta(z).$$

It follows that $(\tilde{\alpha}(z), \tilde{\beta}(z))$ is a dual pair, $V_{\tilde{\alpha}} \subseteq V_\alpha$, $V_{\tilde{\beta}} \subseteq V_\beta$, and

$$\alpha(z) = \prod_{0}^{r-1} V_i(z)(V_r(z^{-1})^r) \prod_{s-1}^{1} U_j(z^{-1}) \alpha(z),$$

$$\beta(z) = \prod_{1}^{s-1} U_j^*(z^{-1}) V_r^*(z^{-1})^r \prod_{r-1}^{0} V_i^*(z) \alpha(z).$$

Note that $\tilde{\alpha}(z) = \alpha'_n z^{-n}$, where $n = k + s$, and $\alpha'_n = \sum_{k-r}^{k+s} \alpha_i$ and

$$\tilde{\beta}(z) = \beta'_{n-r} z^{r-n} + \cdots + \beta'_n z^{-n} + \cdots + \beta'_{n+s} z^{-n-s}$$

with that $\alpha'^*_n \beta'_i = m\delta_{n,i}$.

The next step is to factor $\tilde{\beta}(z)$. Define

$$V'_i(z) = I_m - P'_i + P'_i z^{-1}, \quad U'_j(z) = I_m - Q'_j + Q'_j z^{-1}$$

as primitive biorthogonal wavelet matrices, where

$$P'_i = \alpha'_n (\sum_{n-i}^{n} \beta'_j)^*, \quad 0 \le i \le r$$

and

$$Q'_j = \alpha'_n (\sum_{n-r}^{n+j} \beta'_l)^*, \quad 1 \le j \le s-1.$$

It follows that $P'_i, Q'_j \in \mathcal{N}(V_{\tilde{\beta}}^\perp), P'^*_i, Q'^*_j \in \mathcal{N}(V_{\tilde{\alpha}}^\perp)$, $0 \le i \le r, 1 \le j \le s-1$. And P'_i, Q'_j are idempotent matrices with rank one. Note that

$$\prod_{s-1}^{1} U'^*_i(z)(V'^*_r(z)^r) \prod_{0}^{r-1} V'^*_i(z^{-1}) \tilde{\beta}(z) = \tilde{\beta}_{n+s} z^{-n-s},$$

where $\tilde{\beta}_{n+s} = \sum_j \beta'_j$ and

$$\prod_{r-1}^{0} V'_i(z^{-1}) V'_r(z)^r \prod_{1}^{s-1} U'_i(z) \tilde{\alpha}(z) = \tilde{\alpha}_n z^{-n-s}.$$

By the fact of biorthogonal matrix (3.2), we obtain

$$\alpha(z) = z^{-n-s}V_d(z)V_{d-1}(z)\cdots V_1(z)\gamma,$$
$$\beta(z) = z^{-n-s}V_d^*(z)V_{d-1}^*(z)\cdots V_1^*(z)\delta.$$

At last letting $z = 1$ above to get $\gamma = \alpha(1), \delta = \beta(1)$.

Together with Theorem 1 and Theorem 2, we get

Theorem 3. *For any dual pair of Laurent polynomial vectors $\alpha(z)$ and $\beta(z)$, we have*

$$\alpha(z) = z^{-n}V_d(z)V_{d-1}(z)\cdots V_1(z)\alpha(1),$$
$$\beta(z) = z^{-n}V_d^*(z)V_{d-1}^*(z)\cdots V_1^*(z)\beta(1).$$

Where n is an integer, and there exist one rank idempotent matrices P_i satisfying $P_i \in \mathcal{N}(V_\beta^\perp), P_i^* \in \mathcal{N}(V_\alpha^\perp)$ such that $V_i(z) = I_m - P_i + P_i z^{-\tau_i}$, with that $\tau_i \in \{1, -1\}, 1 \le i \le d$.

4. Parameterizations of biorthogonal wavelet matrix

Theroem 4. *If $H(z)$ and $G(z)$ is a pair of biorthogonal wavelet matrices, then there exist one rank idempotent matrices P_i and integers $k_i, i = 1, 2, \cdots, m$ such that*

$$H(z) = V_d(z)V_{d-1}(z)\cdots V_2(z)V_1(z)diag(z^{-k_1}, z^{-k_2}, \cdots, z^{-k_m})H(1),$$

and

$$G(z) = V_d^*(z)V_{d-1}^*(z)\cdots V_2^*(z)V_1^*(z)diag(z^{-k_1}, z^{-k_2}, \cdots, z^{-k_m})G(1),$$

where $V_i(z) = I_m - P_i + P_i z^{\tau_i}, \tau_i \in \{1, -1\}$, $1 \le i \le d$.

Proof: Writing $H(z), G(z)$ in the form as (2.1)and (2.2) respectively, then (2.3)holds, $(\alpha_i, \beta_i), 0 \le i \le m-1$ are dual pairs of Laurent polynomial vectors. By theorem 3, for the dual pair $(\alpha_0(z), \beta_0(z))$ of Laurent polynomial vectors, there exist primitive biorthogonal matrices

$$V_{0,1}(z), V_{0,2}(z), \cdots, V_{0,d_1}(z)$$

and an integer k such that

$$\alpha_0(z) = z^{-k_1}V_{0,d_1}(z)V_{0,d_1-1}(z)\cdots V_{0,1}(z)\alpha_0(1),$$
$$\beta_0(z) = z^{-k_1}V_{0,d_1}^*(z)V_{0,d_1-1}^*(z)\cdots V_{0,1}^*(z)\beta_0(1)$$

where k_1 and d_1 are non-negative integers. Define

$$H_1(z) = V_1(z^{-1})V_2(z^{-1})\cdots V_{d_1}(z^{-1})H(z)$$

and

$$G_1(z) = V_1^*(z^{-1})V_2^*(z^{-1})\cdots V_{d_1}^*(z^{-1})G(z),$$

then $(H_1(z), G_1(z))$ is a pair of biorthogonal wavelet matrices and

$$H(z) = V_{d_1}(z)V_{d_1-1}(z)\cdots V_1(z)H_1(z),$$

$$G(z) = V_{d_1}^*(z)V_{d_1-1}^*(z)\cdots V_1^*(z)G_1(z).$$

It follows that

$$H_1(z) = \left(z^{-k_1}\alpha_0(1), \alpha_{1,1}(z), \cdots, \alpha_{1,m-1}(z)\right)$$

and

$$G_1(z) = \left(z^{-k_1}\beta_0(1), \beta_{1,1}(z), \cdots, \beta_{1,m-1}(z)\right).$$

By the biorthogonality we get

$$\alpha_0(1) \in V_{\beta_{1,k}}^{\perp},\ \beta_0(1) \in V_{\alpha_{1,k}}^{\perp}, 1 \leq k \leq m \tag{4.1}$$

and

$$\beta_{1,i}^*(z^{-1})\beta_{1,j}(z) = \begin{cases} m, & if \quad i = j \\ 0, & if \quad i \neq j. \end{cases}$$

therefore, $(\beta_{1,1}(z), \beta_{1,1}(z))$ is a dual pair of Laurent polynomial vectors. By Theorem 3, there exist primitive biorthogonal matrices

$$V_{1,1}(z), V_{1,2}(z), \cdots, V_{1,d_2}(z)$$

such that

$$\alpha_{1,1}(z) = z^{-k_2}V_{1,d_2}(z)V_{1,d_2-1}(z)\cdots V_{1,1}(z)\alpha_{1,1}(1),$$

$$\beta_{1,1}(z) = z^{-k_2}V_{1,d_2}^*(z)V_{1,d_2-1}^*(z)\cdots V_{1,1}^*(z)\beta_{1,1}(1),$$

where k_2 and d_2 are non-negative integers. Define

$$H_2(z) = V_{1,1}(z^{-1})V_{1,2}(z^{-1})\cdots V_{1,d_2}(z^{-1})H_1(z)$$

and

$$G_2(z) = V_{1,1}^*(z^{-1})V_{1,2}^*(z^{-1})\cdots V_{1,d_2}^*(z^{-1})G_1(z)$$

then $(H_2(z), G_2(z))$ is a pair of biorthogonal wavelet matrices and

$$H_1(z) = V_{1,d_2}(z)V_{1,d_2-1}(z)\cdots V_{1,1}(z)H_2(z),$$

$$G_1(z) = V_{1,d_2}^*(z)V_{1,d_2-1}^*(z)\cdots V_{1,1}^*(z)G_2(z).$$

Note the fact (4.1) and Theorem 3, we get

$$H_1(z) = \left(z^{-k_1}\alpha_0(1), z^{-k_2}\alpha_{1,1}(1), \cdots, \alpha_{1,m-1}(z)\right)$$

and

$$G_1(z) = \left(z^{-k_1}\beta_0(1), z^{-k_2}\beta_{1,1}(1), \cdots, \beta_{1,m-1}(z)\right).$$

Proceeding in the same fashion, we get primitive biorthogonal matrices

$$V_{i,j}(z), \quad j = 1, 2, \cdots, d_i,\ 1 \leq i \leq r$$

with nonnegative integers $d_i, 1 \leq i \leq r$ and integers $k_i, 1 \leq i \leq m$ such that

$$\begin{aligned} H(z) &= V_{r-1,d_r}(z)V_{r-1,d_r-1}(z) \times \\ &\cdots V_{r-1,1}(z)\cdots V_{1,d_2}(z)V_{1,d_2-1}(z) \times \\ &\cdots V_{1,1}(z)V_{d_1}(z)V_{d_1-1}(z)\cdots V_1(z) \times \\ &\text{diag}(z^{-k_1}, z^{-k_2}, \cdots, z^{-k_m})U \end{aligned}$$

and

$$\begin{aligned} G(z) &= V^*_{r-1,d_r}(z)V^*_{r-1,d_r-1}(z) \times \\ &\cdots V^*_{r-1,1}(z)\cdots V^*_{1,d_2}(z)V^*_{1,d_2-1}(z) \times \\ &\cdots V^*_{1,1}(z)V^*_{d_1}(z)V^*_{d_1-1}(z)\cdots V^*_1(z) \times \\ &\text{diag}(z^{-k_1}, z^{-k_2}, \cdots, z^{-k_m})W. \end{aligned}$$

By taking $z = 1$ above we get $U = A(1)$ and $W = B(1)$, therefore U, W is a pair of constant-valued biorthogonal wavelet matrices.

Especially, for orthogonal wavelet matrices, by using Lemma 2 and the similarly procedure as above, we have

Theroem 5. *If $H(z)$ is an orthogonal wavelet matrix, then there exist symmetric idempotent matrices P_i with rank one and integers $k_i, i = 1, 2, \cdots, m$ such that*

$$H(z) = V_d(z)V_{d-1}(z)\cdots V_2(z)V_1(z)\,diag(z^{-k_1}, z^{-k_2}, \cdots, z^{-k_m})U,$$

where U is a constant-valued orthogonal wavelet matrix,

$$V_i = I_m - P_i + P_i z^{-1}, 1 \leq i \leq d.$$

5. Final remark

1. In [12], H. L. Resnikoff, J.Tian and R. O. Wells. Jr discussed the parameterizations and parameterizations in biorthogonal wavelet space, they proved that any biorthogonal wavelet matrix pair can be decomposed into four components: an orthogonal component, a pseudo identity matrix pair, an invertible matrix and a constant matrix. The result is modified into theorem 3 in this paper: Any biorthogonal wavelet matrix pair can be decomposed into two parts: an biorthogonal components $V(z)$ and an constant matrix H.

2. It was proved in [12] that any constant matrix in Theorem 4 can be decomposed into

$$H = \begin{pmatrix} 1 & 0 \\ 0 & U \end{pmatrix} \tilde{H}$$

where $\tilde{H} = (\gamma_0, \gamma_1, \ldots, \gamma_{m-1})$ with that $\gamma_0 = (1, 1, \ldots, 1)$,

$$\gamma_i = \sqrt{\frac{m}{(m-i)(m-i+1)}}(\underbrace{0, \cdots, 0}_{i-1\, terms}, -m+i, \underbrace{1, \cdots, 1}_{m-i\, terms})^T.$$

for $1 \leq i \leq m-1$, and U is an $(m-1) \times (m-1)$ nonsingular constant-valued matrix.

References

[1] Bi N., Dai X. and Sun Q., Construction of compactly supported M-band wavelets, *Appl. Comp. Harmonic Anal.* 6(1999), pp.113-131.

[2] Chui C. K. and Lian J., Construction of compactly supported symmetric and antisymmetric orthogonal wavelets with scale=3, *Appl. Comput. Harmonic Anal.*, 2(1995), pp.21-51.

[3] Cohen A., Daubechies I. and Feauveau J. C., Biorthogonal basis of compactly supported Wavelets, *Commun. Pure Appl. Math.*, 45(5)(1992), pp.485-560.

[4] Daubechies I., Ten lectures on wavelets, SIAM, Philadelphia, PA, 1992.

[5] Han B., Symmetric orthogonal scaling functions and wavelets with dilation factor 4, *Adv. Compt. Math.*, 8(1998), pp.221-247.

[6] Heller D. N., Rank m wavelets with n vanish moments, *SIAM J. Matrix Anal.* 16(2)(1994), pp.502-519.

[7] Heller P. N., Resnikoff H. L. and Wells R. O. Jr., Wavelet matrices and the representation of discrete functions, in *Wavelet- A Tutorial in theory and applications*, C. K. Chui (ed.), Academic Press, Inc.(1992), 15-50.

[8] Ji H. and Shen Z., Compactly supported (bi)orthogonal wavelets generated by interplatory refinable functions, *Adv. Comput. Math.*, 111999, pp.81-104.

[9] Soman A. K., Vaidyanathan P.P. and Nguyen T.Q., Linear phase paraunitary filter banks: theory, factorization and designs, *IEEE Trans. Signal Processing* 41(1993), pp.3480-3496.

[10] Soardi P., Biorthogonal M-channel compactly supported wavelets, *Constr. Approx.*, 16(2000), pp.283-311.

[11] Sun Q. and Zhang Z., M-Band scaling function with filter having vanishing moments two and minimal length, *J. Math. Anal.* 222(1998), pp.225-243.

[12] Resnikoff H. L., Tian J. and Wells R. O. Jr, An algebraic structure of orthogonal wavelet space, *Appl. Comput. Harmon. Anal.*, 8(2000), pp. 223–248.

[13] Vaidyanathan P. P., Multi-rate systems and filter banks, Prentice-Hall, Englewood Cliffs, NJ, 1993.

[14] Vetterli M. and Herley C., Wavelet and filter banks: Theory and design, *IEEE Trans. Acounst. Speech SignaL Processing*, 40(1992), pp. 2207-2232.

[15] Welland G. V. and Lundberg M., Construction of compact p-wavelets, *Constr. Approx.* 9(1993), pp.347-370.

Author Index

Lecture Notes in Computer Science

For information about Vols. 1–2179
please contact your bookseller or Springer-Verlag

Vol. 2215: N. Kobayashi, B.C. Pierce (Eds.), Theoretical Aspects of Computer Software. Proceedings, 2001. XV, 561 pages. 2001.

Vol. 2216: E.S. Al-Shaer, G. Pacifici (Eds.), Management of Multimedia on the Internet. Proceedings, 2001. XIV, 373 pages. 2001.

Vol. 2217: T. Gomi (Ed.), Evolutionary Robotics. Proceedings, 2001. XI, 139 pages. 2001.

Vol. 2218: R. Guerraoui (Ed.), Middleware 2001. Proceedings, 2001. XIII, 395 pages. 2001.

Vol. 2219: S.T. Taft, R.A. Duff, R.L. Brukardt, E. Ploedereder (Eds.), Consolidated Ada Reference Manual. XXV, 560 pages. 2001.

Vol. 2220: C. Johnson (Ed.), Interactive Systems. Proceedings, 2001. XII, 219 pages. 2001.

Vol. 2221: D.G. Feitelson, L. Rudolph (Eds.), Job Scheduling Strategies for Parallel Processing. Proceedings, 2001. VII, 207 pages. 2001.

Vol. 2223: P. Eades, T. Takaoka (Eds.), Algorithms and Computation. Proceedings, 2001. XIV, 780 pages. 2001.

Vol. 2224: H.S. Kunii, S. Jajodia, A. Sølvberg (Eds.), Conceptual Modeling – ER 2001. Proceedings, 2001. XIX, 614 pages. 2001.

Vol. 2225: N. Abe, R. Khardon, T. Zeugmann (Eds.), Algorithmic Learning Theory. Proceedings, 2001. XI, 379 pages. 2001. (Subseries LNAI).

Vol. 2226: K.P. Jantke, A. Shinohara (Eds.), Discovery Science. Proceedings, 2001. XII, 494 pages. 2001. (Subseries LNAI).

Vol. 2227: S. Boztaş, I.E. Shparlinski (Eds.), Applied Algebra, Algebraic Algorithms and Error-Correcting Codes. Proceedings, 2001. XII, 398 pages. 2001.

Vol. 2228: B. Monien, V.K. Prasanna, S. Vajapeyam (Eds.), High Performance Computing – HiPC 2001. Proceedings, 2001. XVIII, 438 pages. 2001.

Vol. 2229: S. Qing, T. Okamoto, J. Zhou (Eds.), Information and Communications Security. Proceedings, 2001. XIV, 504 pages. 2001.

Vol. 2230: T. Katila, I.E. Magnin, P. Clarysse, J. Montagnat, J. Nenonen (Eds.), Functional Imaging and Modeling of the Heart. Proceedings, 2001. XI, 158 pages. 2001.

Vol. 2232: L. Fiege, G. Mühl, U. Wilhelm (Eds.), Electronic Commerce. Proceedings, 2001. X, 233 pages. 2001.

Vol. 2233: J. Crowcroft, M. Hofmann (Eds.), Networked Group Communication. Proceedings, 2001. X, 205 pages. 2001.

Vol. 2234: L. Pacholski, P. Ružička (Eds.), SOFSEM 2001: Theory and Practice of Informatics. Proceedings, 2001. XI, 347 pages. 2001.

Vol. 2235: C.S. Calude, G. Păun, G. Rozenberg, A. Salomaa (Eds.), Multiset Processing. VIII, 359 pages. 2001.

Vol. 2236: K. Drira, A. Martelli, T. Villemur (Eds.), Cooperative Environments for Distributed Systems Engineering. IX, 281 pages. 2001.

Vol. 2237: P. Codognet (Ed.), Logic Programming. Proceedings, 2001. XI, 365 pages. 2001.

Vol. 2239: T. Walsh (Ed.), Principles and Practice of Constraint Programming – CP 2001. Proceedings, 2001. XIV, 788 pages. 2001.

Vol. 2240: G.P. Picco (Ed.), Mobile Agents. Proceedings, 2001. XIII, 277 pages. 2001.

Vol. 2241: M. Jünger, D. Naddef (Eds.), Computational Combinatorial Optimization. IX, 305 pages. 2001.

Vol. 2242: C.A. Lee (Ed.), Grid Computing – GRID 2001. Proceedings, 2001. XII, 185 pages. 2001.

Vol. 2243: G. Bertrand, A. Imiya, R. Klette (Eds.), Digital and Image Geometry. VII, 455 pages. 2001.

Vol. 2244: D. Bjørner, M. Broy, A.V. Zamulin (Eds.), Perspectives of System Informatics. Proceedings, 2001. XIII, 548 pages. 2001.

Vol. 2245: R. Hariharan, M. Mukund, V. Vinay (Eds.), FST TCS 2001: Foundations of Software Technology and Theoretical Computer Science. Proceedings, 2001. XI, 347 pages. 2001.

Vol. 2246: R. Falcone, M. Singh, Y.-H. Tan (Eds.), Trust in Cyber-societies. VIII, 195 pages. 2001. (Subseries LNAI).

Vol. 2247: C. P. Rangan, C. Ding (Eds.), Progress in Cryptology – INDOCRYPT 2001. Proceedings, 2001. XIII, 351 pages. 2001.

Vol. 2248: C. Boyd (Ed.), Advances in Cryptology – ASIACRYPT 2001. Proceedings, 2001. XI, 603 pages. 2001.

Vol. 2249: K. Nagi, Transactional Agents. XVI, 205 pages. 2001.

Vol. 2250: R. Nieuwenhuis, A. Voronkov (Eds.), Logic for Programming, Artificial Intelligence, and Reasoning. Proceedings, 2001. XV, 738 pages. 2001. (Subseries LNAI).

Vol. 2251: Y.Y. Tang, V. Wickerhauser, P.C. Yuen, C.Li (Eds.), Wavelet Analysis and Its Applications. Proceedings, 2001. XIII, 450 pages. 2001.

Vol. 2252: J. Liu, P.C. Yuen, C. Li, J. Ng, T. Ishida (Eds.), Active Media Technology. Proceedings, 2001. XII, 402 pages. 2001.

Vol. 2253: T. Terano, T. Nishida, A. Namatame, S. Tsumoto, Y. Ohsawa, T. Washio (Eds.), New Frontiers in Artificial Intelligence. Proceedings, 2001. XXVII, 553 pages. 2001. (Subseries LNAI).

Vol. 2254: M.R. Little, L. Nigay (Eds.), Engineering for Human-Computer Interaction. Proceedings, 2001. XI, 359 pages. 2001.

Vol. 2256: M. Stumptner, D. Corbett, M. Brooks (Eds.), AI 2001: Advances in Artificial Intelligence. Proceedings, 2001. XII, 666 pages. 2001. (Subseries LNAI).

Vol. 2258: P. Brazdil, A. Jorge (Eds.), Progress in Artificial Intelligence. Proceedings, 2001. XII, 418 pages. 2001. (Subseries LNAI).

Vol. 2259: S. Vaudenay, A.M. Youssef (Eds.), Selected Areas in Cryptography. Proceedings, 2001. XI, 359 pages. 2001.

Vol. 2260: B. Honary (Ed.), Cryptography and Coding. Proceedings, 2001. IX, 416 pages. 2001.

Vol. 2264: K. Steinhöfel (Ed.), Stochastic Algorithms: Foundations and Applications. Proceedings, 2001. VIII, 203 pages. 2001.